Chemical Fundamentals of Geology

and Environmental Geoscience

Robin Gill

Chemical Fundamentals of Geology

and Environmental Geoscience

Third Edition

WILEY Blackwell

This edition first published 2015 © 2015 by John Wiley & Sons, Ltd
Second edition, Chemical Fundamentals of Geology, published 1996 by Chapman & Hall
First edition, Chemical Fundamentals of Geology, published 1989 by Unwin Hyman

Registered Office
John Wiley & Sons, Ltd, The Atrium, Southern Gate, Chichester, West Sussex, PO19 8SQ, UK

Editorial Offices
9600 Garsington Road, Oxford, OX4 2DQ, UK
The Atrium, Southern Gate, Chichester, West Sussex, PO19 8SQ, UK
111 River Street, Hoboken, NJ 07030-5774, USA

For details of our global editorial offices, for customer services and for information about how to
apply for permission to reuse the copyright material in this book please see our website at
www.wiley.com/wiley-blackwell.

The right of the author to be identified as the author of this work has been asserted in accordance with
the UK Copyright, Designs and Patents Act 1988.

Library of Congress Cataloging-in-Publication Data

Gill, Robin, 1944–
    Chemical fundamentals of geology and environmental geoscience/Robin Gill. – Third edition.
        pages cm
    Previous edition: Chemical fundamentals of geology (London ; Glascow : Weinheim : Chapman and
Hall, 1996).
    Includes bibliographical references and index.
    ISBN 978-0-470-65665-5 (cloth)
1. Chemistry. 2. Geology. 3. Geochemistry. 4. Environmental geochemistry. I. Title.
    QD39.2.G55 2015
    540.24′553–dc23
                                    2014029366

A catalogue record for this book is available from the British Library.

Wiley also publishes its books in a variety of electronic formats. Some content that appears in print may
not be available in electronic books.

Cover images: Front cover image of Malachite and Azurite Deposits in a Boulder, ©iStock.com/chapin31;
                Images on the back cover from top to bottom: Pyroxene chain: photo by K. d'Souza;
                Black smoker: by permission of GEOMAR Helmholtz Centre for Ocean Research, Kiel;
                Andromeda galaxy: public-domain image courtesy of Adam Evans; X-ray map of zoned
                olivine: from Qian and Herma (2010) by permission of *Journal of Petrology*;
                Sulphur crystals: by permission of Vincenzo Ragone; Reaction rim between olivine
                and plagioclase: by permission of H.M. Helmy

Set in 10/13pt Palatino by SPi Publisher Services, Pondicherry, India

Printed and bound by CPI Group (UK) Ltd, Croydon, CR0 4YY

C9780470656655_081223

CONTENTS

PREFACE TO THE THIRD EDITION

Many Earth science departments today recruit significant numbers of environmental geoscience students alongside mainstream geology or Earth science students. This student-led change in academic emphasis suggests that a successor to *Chemical Fundamentals of Geology* will meet the needs of students better if it covers the chemical foundations of environmental Earth science as well as geology. The shift in the emphasis of this new edition is reflected in its expanded title.

This change in emphasis has mostly been accommodated by expanding the existing chapters of the book. Reviewers have, however, highlighted the thin coverage of isotopes in earlier editions and – since isotopes contribute so much to our understanding of processes within the Earth and on its surface – I have devoted a new chapter (Ch. 10) to simple concepts of isotope geochemistry. Although to many students this chapter may seem more challenging – conceptually and mathematically – than earlier chapters in the book, I hope it will nonetheless help them to appreciate the unique insights that isotopes contribute to our science.

This book was originally conceived to support Earth science students having no background at all in chemistry. Despite this intention, teachers using the book continue to mention students who "feel that it is too high-level for their needs". With the requirements of those students in mind, I have tried to lower still further the entry hurdles that the book places in their path, for example, by introducing boxes like Box 1.1 'What is energy?'

As in previous editions (despite criticism in one published review), where variables and units appear on graph axes they are separated by a forward slash (i.e. variable / unit) according to Royal Society convention.[1] The logic of this implied division (variable ÷ unit) is that, when distance x is divided by a metre, say, or time t is divided by a second, the result is a **dimensionless** pure number – numbers (not quantities) are after all what we actually plot on graph axes.

---

[1] See also www.animations.physics.unsw.edu.au/jw/graphs. htm#Units.

PREFACE TO THE SECOND EDITION

Since its first publication in 1989, this book has been used by a much wider range of students than I originally envisaged. My initial objective was simply to equip elementary Earth science undergraduates with the basic chemical understanding that their geological degree programme would require, but in the course of writing the first edition I fell into the trap of trying to entice them a little further so that they would embrace some of the underlying scientific principles. Accordingly, though somewhat to my surprise, the book has been adopted at the advanced undergraduate level and even as remedial reading for postgraduate studies as well as for its original purpose. It has also been adopted in some chemistry departments as a geochemistry primer.

In preparing this edition I have tried to increase the book's usefulness in two directions. Firstly, I have sought to make it more reader-friendly to those, such as mature students, who may remember very little chemistry from their school days or whose experiences of school chemistry were discouraging. Secondly, with more advanced students in mind, I have introduced a little more chemical rigour and more pointers to geological and geochemical applications, and have, I hope, also reflected some recent advances in cosmological, geological and environmental understanding. The change in format has allowed me to make more effective use of the boxes and to enhance some figures.

I am grateful to many people who have fed back to me their reactions to the first edition, particularly Paul Browning, Hilary Downes, Mike Henderson, Bob Major, Steven Richardson and Andy Saunders. Not every suggestion, of course, has seemed to me consistent with the original spirit of the book, but all comments have been valuable in giving me a reader's or a teacher's eye view. The encouragement I have received from colleagues such as Derek Blundell and Euan Nisbet has also been greatly appreciated. My family has been as understanding (or as resigned) as ever!

I thank Lynne Blything for her kind help with figure revision, and Ruth Cripwell and Ian Francis for their editorial advice and encouragement.

PREFACE TO THE FIRST EDITION

Chemical principles are fundamental to a large area of geological science, and the student who reads for a degree in geology without knowing any basic chemistry is handicapped from the start. This book has been written with such students in mind, but I hope it will also provide a helpful refresher course and useful background reading in 'geo-relevant' chemistry for all earth science students who believe in understanding rather than merely memorizing.

The book has been conceived in three broad sections. The first, Chapters 1–4, deals with the basic physical chemistry of geological processes, emphasizing how consideration of energy can aid our understanding. The second section, Chapters 5–8, introduces the wave-mechanical view of the atom and describes from that perspective the various types of chemical bonding that give minerals their distinctive properties. The final section, Chapters 9 and 10, surveys the geologically relevant elements, concluding with a chapter on why some are more abundant than others, in the universe as a whole and in the Earth in particular. The emphasis throughout is on geological and environmental relevance; laboratory reactions are hardly mentioned.

The book is designed to be accessible and stimulating, even to students who remember none of their school chemistry; for these readers, introductory background material and a glossary are provided in the Appendices. The more advanced material, which I hope will sustain the interest of the chemically literate reader, has been segregated into boxes that can be ignored on a first reading. The text is thoroughly cross-referenced to help the browser or the searcher to find what each wants. The book might also assist teachers of more advanced courses in geochemistry, mineralogy and petrology by relieving them of the need to cover elementary material in class.

Many colleagues have given me advice and encouragement during the writing of the book, and I am particularly grateful to the following colleagues who have read and commented on individual chapters and suggested innumerable improvements: David Alderton, Peter Barnard, Keith Cox, Giles Droop, Paul Henderson, Steve Killops, Robert Hutchison, Philip Lee and Eric Whittaker. Professor W.D. Carlson, Dr T.K. Halstead and Dr J.B. Wright read the whole manuscript and provided a wealth of helpful comment. The errors and eccentricities that remain are of course my responsibility alone. The book has grown from a lecture course I gave at the former Geology Department of Chelsea College, and I warmly acknowledge the opportunities which that department provided.

I would like to thank Joan Hirons, Sue Clay and Jennifer Callard for typing some of the chapters, and Neil Holloway and Christine Flood who prepared a number of the figures. I owe a great deal to Roger Jones of Unwin Hyman, whose faith in the project sustained me in the face of initial difficulties.

The biggest debt I owe is to Mary, Joanna and Tim, who have tolerated with remarkably good humour my neglect of family activities, the perpetual retreat up to the study, and the curtailment of holidays. I promise not to write another book for a long time.

ACKNOWLEDGEMENTS

I am most grateful to Dave Alderton for guidance on fluid inclusions and providing Figure 4.6.1, Hilary Downes for reviewing the new Chapter 10, Kevin d'Souza for photography, Chris Emlyn-Jones for advice on Latin and Greek etymology of scientific terms, Kelvin Matthews of Wiley-Blackwell for editorial support, and Derek Vance for sharing insights on carbonate equilibria. I have much appreciated the input of users of the book who have kindly offered helpful suggestions.

I am indebted to the following for providing, or giving me permission to reproduce, the images or figures specified: American Mineralogist (Figure 3.1.1), H.M. Helmy (Plate 1), Journal of Petrology (Plate 2), Princeton University Press (Figure 3.5a), M. Schoonen (Figure 3.5b), Springer Science and Business Media (Figure 3.6b), Elsevier (Figures 4.2, 10.15, 10.2.2), GEOMAR Helmholtz Centre for Ocean Research Kiel (Figure 4.2.1), D. Alderton (Figure 4.6.1), John Wiley & Sons, Ltd. (Figure 7.2a), McGraw Hill (Figure 8.6b), Holt Rinehart and Winston (Figure 8.7a), US National Oceanographic and Atmospheric Administration (Figure 9.2), Geological Society of America (Figures 10.5 and 10.11), the American Association for the Advancement of Science (Figure 10.7), Nature Publishing Group (Figures 10.8 and 10.12a), D. Johnson (Figure 10.14), and the Natural History Museum, London (Figure 11.1.1 and Plates 5 and 6).

ABOUT THE COMPANION WEBSITE

This book is accompanied by a companion website:

www.wiley.com/go/gill/chemicalfundamentals

The website includes:

- Powerpoints® of all figures from the book for downloading.
- PDFs of tables from the book.

1 ENERGY IN GEOCHEMICAL PROCESSES

Introduction

The purpose of this book is to introduce the average Earth science student to chemical principles that are fundamental to the sciences of geology and environmental geoscience. There can be no more fundamental place to begin than with the topic of energy (Box 1.1), which lies at the heart of both geology and chemistry. Energy plays a role in every geological process, from the atom-by-atom growth of a mineral crystal to the elevation and subsequent erosion of entire mountain chains. Consideration of energy provides an incisive intellectual tool for analysing the workings of the complex geological world, making it possible to extract from this complexity a few simple underlying principles upon which an orderly understanding of Earth processes can be based.

Many natural processes involve a flow of energy. The spontaneous melting of an ice crystal, for example, requires energy to be transferred by means of heat into the crystal from the 'surroundings' (the air or water surrounding the crystal). The crystal experiences an increase in its internal energy, which transforms it into liquid water. The process can be symbolized by writing down a formal reaction:

$$\underset{ice}{H_2O} \rightarrow \underset{water}{H_2O} \tag{1.1}$$

in which molecules of water (H_2O) are represented as migrating from the solid state (left-hand side) into the liquid state (right-hand side). Even at $0\,°C$, ice and water both possess internal energy associated with the individual motions of their constituent atoms and molecules. This energy content, which we loosely visualize as heat 'stored' in each of the substances, is more correctly called the **enthalpy**[1] (symbol H). Because the

---

[1] Words in **bold** type indicate terms that are defined in the Glossary.

Chemical Fundamentals of Geology and Environmental Geoscience, Third Edition. Robin Gill.
© 2015 John Wiley & Sons, Ltd. Published 2015 by John Wiley & Sons, Ltd.
Companion Website: www.wiley.com/go/gill/chemicalfundamentals

Box 1.1 What is energy?

The concept of energy is fundamental to all branches of science, yet to many people the meaning of the term remains elusive. In everyday usage it has many shades of meaning, from the personal to the physical to the mystical. Its scientific meaning, on the other hand, is very precise.

To understand what a scientist means by energy, the best place to begin is with a related – but more tangible – scientific concept that we call *work*. Work is defined most simply as *motion against an opposing force* (Atkins, 2010, p. 23). Work is done, for example, when a heavy object is lifted a certain distance above the ground against the force of gravity (Figure 1.1.1). The amount of work this involves will clearly depend upon how heavy the object is, the vertical distance through which its centre of gravity is lifted (Figure 1.1.1b), and the strength of the gravitational field acting on the object. The work done in this operation can be calculated using a simple formula:

$$\text{Work} = m \times h \times g \qquad (1.1.1)$$
$$\phantom{\text{Work} = } \text{J} \quad \text{kg} \quad \text{m} \quad \text{m s}^{-2}$$

where m represents the mass of the object (in kg), h is the distance through which its centre of gravity is raised (in m – see footnote)[2], and g, known as the *acceleration due to gravity* (metres per second per second $= \text{m s}^{-2}$), is a measure of the strength of the gravitational field where the experiment is being carried out; at the Earth's surface, the value of g is $9.81\,\text{m s}^{-2}$. The scientific unit that we use to measure work is called the joule (J), which as

Equation 1.1.1 shows is equivalent to $\text{kg} \times \text{m} \times \text{m s}^{-2} = \text{kg m}^2\,\text{s}^{-2}$ (see Table A2, Appendix A). Alternative forms of work, such as cycling along a road against a strong opposing wind, or passing an electric current through a resistor, can be quantified using comparably simple equations, but whichever equation we use, work is always expressed in joules.

The weight suspended in its elevated position (Figure 1.1.1b) can itself do work. When connected to suitable apparatus and allowed to fall, it could drive a pile into the ground (this is how a pile-driver works), hammer a nail into a piece of wood, or generate electricity (by driving a dynamo) to illuminate a light bulb. The work ideally recoverable from the elevated weight in these circumstances is given by Equation 1.1.1. If we were to raise the object twice as far above the ground (Figure 1.1.1c), we double its capacity for doing work:

$$\text{Work} = m \times 2h \times g \qquad (1.1.2)$$

Alternatively if we raise an object three times as heavy to a distance h above the ground (Figure 1.1.1d), the amount of work that this new object could perform would be three times that of the original object in Figure 1.1.1b:

$$\text{Work} = 3m \times h \times g \qquad (1.1.3)$$

The simple mechanical example in Figure 1.1.1 shows only one, simply understood way of doing work. Mechanical work can also be done by an object's motion, as a demolition crew's 'wrecking ball' illustrates. Electric current heating the element of an electric fire represents another form of work, as does an explosive charge used to blast a rock face in a mine.

---

[2] m in italics represents mass (a variable in this equation); m in regular type is the abbreviation for metres (units).

---

molecules in liquid water are more mobile than those in ice – that is, they have higher **kinetic energy** – the enthalpy of water (H_{water}) is greater than that of an equivalent amount of ice (H_{ice}) at the same temperature. The difference can be written:

$$\Delta H = H_{\text{water}} - H_{\text{ice}} \qquad (1.2)$$

The Δ symbol (the Greek capital letter 'delta'), when written in front of H, signifies the *difference* in enthalpy between the initial (solid) and final (liquid) states of the compound the H_2O. It represents the work (Box 1.1) that must be done in disrupting the chemical bonds that hold the crystal together. ΔH symbolizes the amount of heat that must be supplied from the surroundings for the crystal to melt completely; this is called the **latent heat of fusion**, or more correctly the *enthalpy of fusion*, a quantity that can be measured experimentally or looked up in tables.

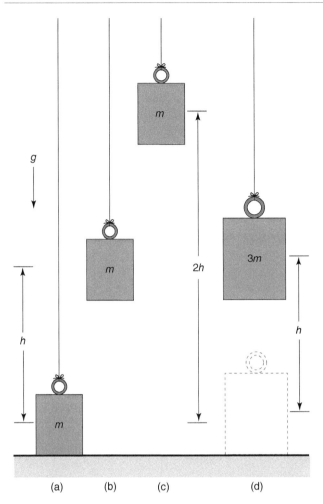

Figure 1.1.1 Work done in raising an object: (a) an object of mass m resting on the ground; (b) the same object elevated to height h; (c) the object elevated to height $2h$; (d) another object of mass $3m$ elevated to height h. *Note*: elevation is measured between each object's *centre of gravity* in its initial and final positions (note the centre of gravity of the larger weight is slightly higher than the smaller one).

Energy is simply the term that we use to describe a system's *capacity for doing work*. Just as we recognize different forms of work (mechanical, electrical, chemical …), so energy exists in a number of alternative forms, as will be illustrated in the following pages. The energy stored in an electrical battery, for example, represents the amount of work that it can generate before becoming exhausted. A system's capacity for doing work is necessarily expressed in the units of work (just as the capacity of a bucket is expressed as the number of litres of water it can contain), so it follows that energy is also expressed in joules $= \mathrm{kg\,m^2\,s^{-1}}$. When discussing large amounts of energy, we use larger units such as kilojoules ($\mathrm{kJ} = 10^3\,\mathrm{J}$) or megajoules ($\mathrm{MJ} = 10^6\,\mathrm{J}$).

This simple example illustrates how one can go about documenting the energy changes that accompany geological reactions and processes, as a means of understanding why and when those reactions occur. This is the purpose of *thermodynamics*, a science that documents and explains quantitatively the energy changes in natural processes, just as economics analyses the exchange of money in international trade. Thermodynamics provides a fundamental theoretical framework for documenting and interpreting energy changes in processes of all kinds, not only in geology but in a host of other scientific disciplines ranging from chemical engineering to cosmology.

Thermodynamics, because it deals with very abstract concepts, has acquired an aura of impenetrability in the eyes of many Earth science students, particularly those less at home in the realm of mathematics. With this in mind, one objective of these opening chapters will be to show that thermodynamics, even at a simple and approachable level, can contribute a lot to our understanding of *chemical reactions* and *equilibrium* in the geological world.

Energy changes in chemical systems are most easily introduced (as in Box 1.1) through analogy with mechanical forms of energy, which should be familiar from school physics.

Energy in mechanical systems

The energy of a body is defined as its capacity for doing work (Box 1.1). As we have discovered, 'work' can take various forms, but in simple mechanical systems it usually implies the movement of a body from one position to another against some form of physical resistance (friction, gravity, electrostatic forces, etc.). Then:

$$\underset{\text{J (joules)}}{\text{work done}} = \underset{\text{N (newtons)}=\text{kg m s}^{-2}}{\text{force required to move body}}$$
$$\times \underset{\text{m (metres)}}{\text{distance body is moved}} \tag{1.3}$$

So, for example, the work done in transporting a train-load of iron ore from A to B is the mechanical force required to keep the train rolling multiplied by the distance by rail from A to B. The energy required to do this is generated by the combustion of fuel in the engine.

One can distinguish two kinds of mechanical energy. Firstly, an object can do work by means of its motion. A simple example is the use of a hammer to drive a nail into wood. Work is involved because the wood resists the penetration of the nail. The energy is provided by the downward-moving hammer-head which, because of its motion, possesses **kinetic energy** (a name derived from the Greek *kinetikos*, meaning 'setting in motion'). The kinetic energy E_k possessed by a body of mass m travelling with velocity v is given by:

$$\underset{\text{(J)}}{E_k} = \frac{1}{2} \underset{\text{(kg)}}{m} \underset{\left(\text{m s}^{-1}\right)^2}{v^2} \tag{1.4}$$

Thus the heavier the hammer (m) and/or the faster it moves (v), the more kinetic energy it possesses and the further it will drive in the nail. For similar reasons, a fast-moving stream can carry a greater load of sediment than a slow-moving one.

Secondly, an object in a gravitational field possesses energy (i.e. can do work) by virtue of its position in that field, a property known as **potential energy**. The water held behind a hydroelectric dam has a high potential energy: under the influence of the Earth's gravitational field it would normally flow downhill until it reached sea-level, but the dam prevents this from happening. The fact that the *controlled* downward flow of this water under gravity can be made to drive turbines and generate electricity indicates that the water held behind the dam has the potential for doing work, and therefore possesses energy.

The potential energy E_p of an object of mass m at a height h above the ground is given by:

$$\underset{\text{(J)}}{E_p} = \underset{\text{(kg)}}{m} \times \underset{\left(\text{m s}^{-2}\right)}{g} \times \underset{\text{(m)}}{h} \tag{1.5}$$

where g is the acceleration due to gravity ($9.81\,\text{m s}^{-2}$). Similar equations can be written representing the potential energies of bodies in other types of force field, such as those in electric and nuclear fields.

An important aspect of potential energy is that its value as calculated from Equation 1.5 depends upon the baseline chosen for the measurement of height h. The potential energy calculated for an object in a second-floor laboratory, for example, will differ according to whether its height is measured from the laboratory floor, from the ground level outside, or from sea-level. The last of these alternatives seems at first sight to be the most widely applicable standard to adopt, but even that reference point fails to provide a baseline that can be used for the measurement of height and potential energy down a deep mine (where both quantities might have negative values measured relative to sea-level). This ambiguity forces us to recognize that potential energy is not something we can express on an absolute scale having a universal zero-point, as we do in the case of temperature or electric charge. The value depends upon the 'frame of reference' we choose to adopt. We shall discover that this characteristic applies to chemical energy as well. It seldom presents practical difficulties because in thermodynamics one is concerned with energy *changes*, from which the baseline-dependent element cancels out (provided that the energy values used have been chosen to relate to the same frame of reference).

In general, a body possesses kinetic *and* potential energy by virtue of its overall motion and position.

There is also an internal contribution to its total energy from the individual motions of its constituent atoms and molecules, which are continually vibrating, rotating and – in liquids and gases – migrating about. This internal component, the aggregate of the kinetic energies of all the atoms and molecules present, is what we mean by the **enthalpy** of the body. Enthalpy is closely related to the concept of heat (and was at one time referred to, rather misleadingly, as 'heat content'). Heat is one of the mechanisms by which enthalpy can be transferred from one body to another. The effect of heating a body is simply to increase the kinetic energy of the constituent atoms and molecules, and therefore to increase the enthalpy of the body as a whole.

Natural processes continually convert energy from one form into another. One of the fundamental axioms of thermodynamics, known as the First Law, is that *energy can never be created, destroyed or 'lost'* in such processes, but merely changes its form (Box 1.2). Thus the energy given out by a reaction is matched exactly by the amount of energy gained by the surroundings.

Energy in chemical and mineral systems: free energy

Experience tells us that mechanical systems in the everyday world tend to evolve in the direction that leads to a *net reduction in total potential energy*. Water runs downhill, electrons are drawn toward atomic nuclei, electric current flows from 'live' to 'neutral', and so on. The potential energy released by such changes reappears as other forms of energy or work: for example, the kinetic energy of running water, the light energy radiated by electronic transitions in atoms (Chapter 6), or the heat generated by an electric fire.

Thermodynamics visualizes chemical processes in a similar way. Reactions in chemical or geological systems arise from differences in what is called **free energy**, G, between products and reactants. The significance of free energy in chemical systems can be compared to that of potential energy in mechanical systems. A chemical reaction proceeds in the direction which leads to a net *reduction in free energy*, and the chemical energy so released reappears as energy in some other form – the

---

Box 1.2 The First Law of Thermodynamics

The most fundamental principle of thermodynamics is that energy is never created, lost or destroyed. It can be transmitted from one body to another, or one place to another, and it can change its identity between a number of forms (as for example when the potential energy of a falling body is transformed into kinetic energy, or when a wind turbine converts the kinetic energy of moving air into electrical energy). But we never observe new energy being created from scratch, nor does it ever just disappear. Accurate energy bookkeeping will always show that *in all known processes total energy is always conserved*. This cardinal principle is called the *First Law of Thermodynamics*. The energy given out by a reaction or process is matched exactly by the amount of energy gained by the surroundings.

Implicit in the First Law is the recognition that work is equivalent to energy, and must be accounted for in energy calculations. When a compressed gas at room temperature escapes from a cylinder, it undergoes a pronounced cooling, often to the extent that frost forms around the valve. (A smaller cooling effect occurs when you blow on your hand through pursed lips.) The cause of the cooling is that the gas has had to do work during escaping: it occupies more space outside the cylinder than when compressed inside it, and it must make room for itself by displacing the surrounding atmosphere. Displacing something against a resisting force (in this case atmospheric pressure) constitutes work, which the gas can only accomplish at the expense of its enthalpy. This is related directly to temperature, so that a drain on the gas's internal energy reserves becomes apparent as a fall in temperature.

A similar cooling effect may operate when certain gas-rich magmas reach high crustal levels or erupt at the surface. An example is kimberlite, a type of magma that commonly carries diamonds up from the mantle. Kimberlite penetrates to the surface from depths where the associated gases are greatly compressed, and the work that they do in expanding as the magma–gas system bursts through to the surface reduces its temperature; kimberlites found in subvolcanic pipes (diatremes) appear to have been emplaced in a relatively cool state.

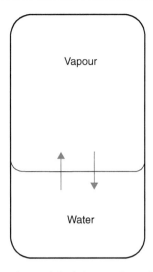

Figure 1.1 A simple model of chemical equilibrium between two coexisting phases, water and its vapour. The equilibrium can be symbolized by a simple equation.

$$\underset{liquid}{H_2O} \rightleftharpoons \underset{vapour}{H_2O}$$

At equilibrium, the migration of water molecules from the liquid to the vapour (evaporation, upward arrow) is balanced exactly by the condensation of molecules from vapour to liquid (downward arrow).

electrical energy obtained from a battery, the light and heat emitted by burning wood, and so on.

What form does free energy take? How can it be calculated and used? These questions are best tackled through a simple example. Imagine a sealed container partly filled with water (Figure 1.1). The space not filled by the liquid takes up water vapour (a gas) until a certain pressure of vapour is achieved, called the *equilibrium vapour pressure* of water, which is dependent only upon the temperature (assumed to be constant). H_2O is now present in two stable forms, each one having physical properties and structure distinct from the other: these two states of matter are called **phases**. From this moment on, unless circumstances change, the system will maintain a constant state, called *equilibrium*, in which the rate of evaporation from the liquid phase is matched exactly by the rate of condensation from the vapour phase: the relative volumes of the two phases will therefore remain constant.

In this state of equilibrium, the free energies associated with a given amount of water in each of these two phases must be equal. If that were not the case, a net

flow of water molecules would begin from the high-G phase to the low-G phase, allowing the total free energy of the system to fall in keeping with the general tendency of chemical systems to minimize free energy. Any such flow, which would alter the relative proportions of the two phases, is inconsistent with the steady state observed. Clearly, at equilibrium, equivalent amounts of the two phases must have identical free energies:

$$G_{\text{vapour}} = G_{\text{liquid}} \tag{1.6}$$

This statement is in fact the thermodynamic definition of 'equilibrium' in such a system.

But here we seem to have stumbled upon a paradox. Common sense tells us that to turn liquid water into vapour we have to supply energy, in the form of heat. The amount required is called the **latent heat of evaporation** (more correctly, the *enthalpy* of evaporation). This indicates that the vapour has a greater enthalpy (H_{vapour}) than an equivalent amount of the liquid (H_{liquid}):

$$H_{\text{vapour}} > H_{\text{liquid}} \tag{1.7}$$

The difference reflects the fact that water molecules in the vapour state have (a) greater potential energy, having escaped from the intermolecular forces that hold liquid water together, and (b) greater kinetic energy (owing to the much greater mobility of molecules in the gaseous state).

How can we reconcile Equations 1.6 and 1.7? Is it not common sense to expect the liquid state, in which the water molecules have much lower energies, to be intrinsically more stable than the vapour? What is it that sustains the vapour, in spite of its higher enthalpy, as a stable phase *in equilibrium with* the liquid?

The answer lies in the highly disordered state characteristic of the vapour. Molecules in the gas phase fly around in random directions, occasionally colliding but, unlike molecules in a liquid, free to disperse throughout the available volume. The vapour is said to possess a high **entropy** (S). Entropy is a parameter that quantifies the degree of internal disorder of a substance (Box 1.3). Entropy has immense significance in thermodynamics, owing to Nature's adherence to the *Second Law of Thermodynamics*. This states that *all spontaneous processes result in an increase of entropy*. The everyday consequences of the Second Law – so familiar that we often take them for granted – are discussed further in Box 1.3. In the present context, Nature's preference for disordered, high-entropy states of matter is what makes it possible for vapour to coexist with liquid. In a sense, the higher entropy of the vapour 'stabilizes' it in relation to the liquid state, compensating for the higher enthalpy required to sustain it.

Clearly, any analysis in energy terms, even of this simple example, will succeed only if the entropy difference (ΔS) between liquid and vapour is taken into account. This is why the definition of the free energy (alternatively called the 'Gibbs energy') of each phase therefore incorporates an entropy term:

$$G_{\text{liquid}} = H_{\text{liquid}} - T.S_{\text{liquid}} \qquad (1.8)$$

$$G_{\text{vapour}} = H_{\text{vapour}} - T.S_{\text{vapour}} \qquad (1.9)$$

H_{liquid} and H_{vapour} are the enthalpies of the liquid and vapour respectively. S_{liquid} and S_{vapour} are the corresponding entropies. (Take care not to confuse the similar-sounding terms 'enthalpy' and 'entropy'.) The absolute temperature T (measured in kelvins) is assumed to be uniform in a system in equilibrium (Chapter 2), and therefore requires no subscript.

The important feature of these equations is the negative sign. It means that the vapour can have higher enthalpy (H) *and* higher entropy (S) than the liquid, and yet have the same free energy value (G), which must be true if the two phases are to be in equilibrium. Perhaps a more fundamental understanding of the minus sign can be gained by rearranging Equations 1.8 and 1.9 into this form:

$$H = G + T.S \qquad (1.10)$$

The enthalpy of a phase can thus be seen as consisting of two contributions:

G The part that potentially can be released by the operation of a chemical reaction, which is logically called 'free' energy. This therefore provides a measure of the instability of a system (just as the potential energy of the water in a reservoir reflects its gravitational instability).

$T.S$ The part that is irretrievably bound up in the internal disorder of a phase at temperature T, and that is therefore not recoverable through chemical reactions.

Equations 1.8 and 1.9 express the fundamental contribution that disorder makes to the energetics of chemical and geological reactions, a question we shall take up again in the following sections.

Units

Enthalpy, entropy and free energy, like mass and volume, are classified as **extensive** properties. This means that their values depend on the amount of material present. On the other hand, temperature, density, viscosity, pressure and similar properties are said to be **intensive** properties, because their values are unrelated to the size of the system being considered.

In published tables of enthalpy and entropy (Chapter 2), the values given are those for one **mole** – abbreviated in the SI system to 'mol' – of the substance concerned (18 g in the case of water). One therefore speaks of **molar** enthalpy and entropy, and of molar free energy and molar volume as well. The units of molar enthalpy and molar free energy are joules per mole (J mol^{-1}); those of molar entropy are joules per kelvin per mole ($\text{J K}^{-1}\text{mol}^{-1}$). The most convenient units for expressing molar volume are $10^{-6}\,\text{m}^3\,\text{mol}^{-1}$ (which are the same as $\text{cm}^3\,\text{mol}^{-1}$, the units used in older literature).

In thermodynamic equations like 1.8, temperature is always expressed in *kelvins* (K). One kelvin is equal in magnitude to one °C but the scale begins at the absolute zero of temperature ($-273.15\,°\text{C}$), not at the freezing point of water ($0\,°\text{C}$). Therefore:

$$T \ln \text{K} = T \ln °\text{C} + 273.15 \qquad (1.11)$$

The SI units for pressure are pascals (Pa; see Appendix A).

Free-energy changes

For the reasons discussed above in relation to potential energy, the numerical values of G_{liquid} and G_{vapour} have no *absolute* significance. In considering whether water will evaporate or vapour will condense in specific circumstances, what concerns us is the *change* in free energy ΔG arising from the liquid-to-vapour 'reaction'. The first step in calculating free-energy changes is to write down the process concerned in the form of a

Box 1.3 Some properties of entropy

The concept of disorder is of fundamental importance in thermodynamics, because it allows us to distinguish those processes and changes that occur naturally – 'spontaneous' processes – from those that do not. We are accustomed to seeing a cup shattering when it falls to the floor, but we never see the fragments reassemble themselves spontaneously to form a cup hanging on the dresser hook. Nor is it a natural experience for the air in a cold room to heat up a warm radiator. The direction of change that we accept as natural always leads to a more disordered state than we began with.

   To apply such reasoning to the direction of chemical change, we need a variable that quantifies the degree of disorder in a chemical system. In thermodynamics this is defined by the **entropy** of the system. To define entropy rigorously lies beyond the scope of this book, but it is worth identifying the processes that lead to an increase of entropy. The entropy of a system depends upon:

  (i) the distribution of *matter* or of individual chemical species in the system; and
 (ii) the distribution of *energy*.

Entropy and the distribution of matter
(a) Entropy increases as a substance passes from the solid state to the liquid state to the gaseous state.*

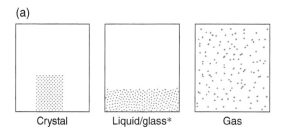

(a)

Crystal   Liquid/glass*   Gas

(b) Entropy increases when a gas expands.

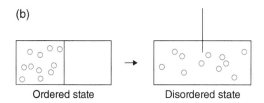

(b)

Ordered state   Disordered state

(c) Entropy increases when pure substance are mixed together.

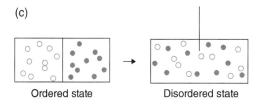

(c)

Ordered state   Disordered state

---

*   A glass is a solid having the disordered structure of a liquid, but deprived of atomic mobility (no flow). Its entropy is intermediate between liquid and crystalline solid.

Entropy and the distribution of energy

The entropy of a system increases

(d)

when a substance is heated:

because raising the temperature makes the random thermal motions of atoms and molecules more vigorous.

(e)
when heat flows from a hot body (*e.g.* a radiator) to a cold body (the surrounding air):

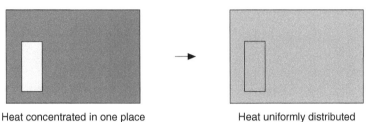

Heat concentrated in one place
Ordered state

Heat uniformly distributed
Disordered state

(f)
when chemical energy (fuel + oxidant) is transformed into heat:

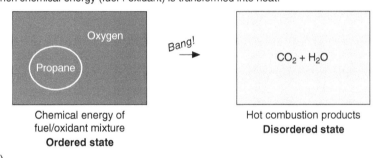

Chemical energy of
fuel/oxidant mixture
Ordered state

Hot combustion products
Disordered state

(g)
when mechanical energy is transformed into heat (*e.g.* friction).

The Second Law of Thermodynamics

The Second Law states that *the operation of any spontaneous process leads to an (overall) increase of entropy*. Our experience of this law is so intricately woven into the fabric of everyday life that we are scarcely aware of its existence, but its impact on science is nonetheless profound. The expansion of a gas is a spontaneous process involving an increase of entropy: a gas never spontaneously contracts into a smaller volume. Water never runs uphill. Applying heat to an electric fire will never generate electricity. All of these impossible events, were they to occur, would bring about a reduction of entropy and therefore violate the Second Law.

Entropy is lowest when energy is concentrated in one part of a system. This is a characteristic of all of the energy resources that we exploit: water retained behind a hydroelectric dam, chemical energy stored in a tank of gasoline or in a charged battery, nuclear energy in a uranium fuel rod, etc. Entropy is highest when energy is evenly distributed throughout the system being considered, and in such circumstances it cannot be put to good use. Spontaneous (entropy-increasing) changes are always accompanied by a degradation in the 'quality' of energy, in the sense that it becomes dispersed more widely and uniformly.

chemical reaction, as in Equation 1.1. For the water/vapour equilibrium:

$$\underset{liquid}{H_2O} \rightleftharpoons \underset{vapour}{H_2O} \qquad (1.12)$$

The equilibrium symbol ($\rightleftharpoons$) represents a balance between two competing opposed 'reactions' taking place at the same time:

'Forward' reaction: liquid $\rightarrow$ vapour   (evaporation)
                        reactant   product

'Reverse' reaction: liquid $\leftarrow$ vapour   (condensation)
                        product   reactant

By convention, the free-energy *change* for the forward reaction (ΔG) is written:

$$\Delta G = G_{products} - G_{reactants}$$
$$= G_{vapour} - G_{liquid} \qquad (1.13)$$

Each G can be expressed in terms of molar enthalpy and entropy values obtained from published tables (Equations 1.8 and 1.9). Thus

$$\Delta G = \left(H_{vapour} - T.S_{vapour}\right) - \left(H_{liquid} - T.S_{liquid}\right)$$
$$= \left(H_{vapour} - H_{liquid}\right) - T\left(S_{vapour} - S_{liquid}\right)$$
$$= \Delta H - T.\Delta S \qquad (1.14)$$

In this equation ΔH is the heat input per mole required to generate vapour from liquid (the latent heat of evaporation). In the context of a true chemical reaction, it would represent the *heat of reaction* (strictly the *enthalpy* of reaction). If ΔH for the forward reactions is negative, heat must be given out by the reaction, which is then said to be **exothermic** ('giving out heat'). A positive value implies that the reaction will proceed only if heat is drawn in from the surroundings. Reactions that absorb heat in this way are said to be **endothermic** ('taking in heat'). ΔS represents the corresponding entropy change between liquid and vapour states.

The values of H_{vapour}, H_{liquid}, S_{vapour} and S_{liquid} can be looked up as molar quantities for the temperature of interest (e.g. room temperature $\simeq 298\,K$) in published tables. In this case, ΔH and ΔS can be calculated by simple difference, leading to a value for ΔG (taking care to enter the value of T in kelvins, not °C). From the sign obtained for ΔG, it is possible to predict in which direction the reaction will proceed under the conditions being considered. A negative value of ΔG indicates that the products are more stable – have a lower free energy – than the reactants, so that the reaction can be expected to proceed in the forward direction. If ΔG is positive, on the other hand, the 'reactants' will be more stable than the 'products', and the reverse reaction will predominate. In either case, reaction will lead eventually to a condition where $\Delta G = 0$, signifying that equilibrium has been reached.

Now let us see how these principles apply to minerals and rocks.

Stable, unstable and metastable minerals

The terms 'stable' and 'unstable' have a more precise connotation in thermodynamics than in everyday usage. In order to grasp their meaning in the context of minerals and rocks, it will be helpful to begin by considering a simple physical analogue. Figure 1.2a shows a rectangular block of wood in a series of different positions relative to some reference surface, such as a table top upon which the block stands. These configurations differ in their potential energy, represented by the vertical height of the *centre of gravity* of the block – shown as a dot – above the table top. Several general principles can be drawn from this physical system which will later help to illuminate some essentials of mineral equilibrium:

(a) Within this frame of reference, configuration D has the lowest potential energy possible, and one calls this the *stable* position. At the other extreme, configurations A and C are evidently *unstable*, because in these positions the block will immediately fall over, ending up in a position like D. Both clearly have higher potential energy than D.

(b) In discussing stable and unstable configurations, one need not consider all forms of energy possessed by the wooden block, some of which (for example, the total electronic energy) would be difficult to quantify. Mechanical stability depends solely upon the relative potential energies of – or energy differences between – the several configurations, and not on their absolute energy values.

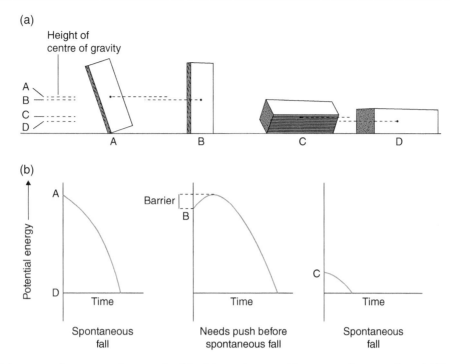

Figure 1.2 Potential energy of a rectangular wooden block in various positions on a planar surface. (a) Four positions of the block, showing the height of its centre of gravity (dot) in each case. (b) The pattern of potential-energy change as the block topples, for the unstable (A, C) and **metastable** (B) configurations.

(c) Configuration B presents something of a paradox. It has a potential energy greater than the unstable state C, yet, if left undisturbed, it will persist indefinitely, maintaining the appearance of being stable. The introduction of a small amount of energy, such as a person bumping into the table, may however be sufficient to knock it over. The character of configuration B can be clarified by sketching a graph of potential energy against time as the block topples over (Figure 1.2b). For both unstable positions A and C, the potential energy falls continuously to the value of position D; but in the case of position B the potential energy must first *rise* slightly, before falling to the minimum value. The reason is that the block has to be raised on to its corner (position A) before it can fall over, and the work involved in so raising its centre of gravity constitutes a potential energy 'hurdle' which has to be surmounted before the block can topple. By inhibiting the spontaneous toppling of the block, this hurdle stabilizes configuration B. One uses the term **metastable** to describe any high-potential-energy state that is stabilized by such an energy hurdle.

The application of this reasoning to mineral stability can be illustrated by the minerals calcite and aragonite, whose ranges of stability in pressure–temperature space are shown in the form of a *phase diagram* in Figure 1.3a. These minerals are alternative crystallographic forms of calcium carbonate ($CaCO_3$), stable under different physical conditions. The phase diagram shown in Figure 1.3a is divided into two areas called *stability fields*, one representing the range of applied pressure and temperature under which calcite is the stable mineral; the other – at higher pressures – indicating the range of conditions favouring aragonite. The stability fields are separated by a line, called a **phase boundary**, which defines the restricted set of circumstances under which calcite *and* aragonite can coexist together *in equilibrium* with each other.

The energetics of the calcite–aragonite system are illustrated in Figure 1.3b, which shows how the molar free energies of the two minerals vary along the line X–Y in Figure 1.3a. At high pressure (Y), deep within the crust, the molar free energy of aragonite is less than that of calcite, and thus aragonite is the stable mineral under these conditions, analogous to configuration D

2 EQUILIBRIUM IN GEOLOGICAL SYSTEMS

The significance of mineral stability

Igneous and metamorphic rocks form in places that are not, on the whole, directly accessible to the investigating geologist. To discover how such rocks are produced within the Earth, one must resort to indirect lines of enquiry. The most important clues consist of the minerals that the rocks themselves contain. A given mineral crystallizes as a stable phase only within restricted ranges of pressure and temperature, as we saw in the case of aragonite which occurs *stably* only at high pressures (Figure 1.3a). Subjecting the mineral to conditions that fall outside its stability range will cause another mineral, which is stable under those conditions (like calcite at low pressure), to begin crystallizing in its place. The stability of other types of minerals may depend in a similar way on the pressure of water vapour or some other gaseous component present during crystallization, and such a mineral will occur in a rock only if the vapour pressure present during its formation falls within the appropriate range.

The sensitivity of such minerals to the physical circumstances of their formation offers the petrologist tremendous opportunities because, when found in an igneous or metamorphic rock now exposed at the surface, they provide a means of establishing quantitatively the characteristics of the physical environment in which that rock originally crystallized. The study of mineral stability therefore offers the key to a veritable library of information, sitting in the rocks waiting to be utilized, about conditions and processes deep within the Earth's crust and upper mantle.

The usual way to establish the physical limits within which a mineral is stable – and beyond which it is unstable – is to cook it up in a laboratory experiment (Box 2.1). Technology today is capable of reproducing in the laboratory the physical conditions (temperature T, pressure P, water vapour pressure P_{H_2O}, and so on)

Chemical Fundamentals of Geology and Environmental Geoscience, Third Edition. Robin Gill.
© 2015 John Wiley & Sons, Ltd. Published 2015 by John Wiley & Sons, Ltd.
Companion Website: www.wiley.com/go/gill/chemicalfundamentals

Box 2.1 Phase equilibrium experiments with minerals

The process of mapping out a phase diagram like that in Figure 2.1 is illustrated in Figure 2.1.1a. Each of the filled and open circles represents an individual experiment in which a sample of the relevant composition is heated in a pressure-vessel at the pressure and temperature indicated by the coordinates, for a sufficient time for the phases to react and equilibrate with each other. At the end of each experiment – which may last hours, days or even months, depending upon the time needed to reach equilibrium – the sample is 'quenched', meaning that it is cooled as quickly as possible to room temperature in order to preserve the phase assemblage formed under the conditions of the experiment (which on slower cooling might recrystallize to other phases – see Chapter 3). The sample is removed from its capsule, and the phase assemblage is identified under the microscope or by other methods. The symbol for each experiment is ornamented on the diagram in such a way that it indicates the nature of the phase assemblage observed, so that the results of a series of experiments allow the position of the phase boundary to be determined. Conditions can be chosen for later experiments which allow accurate bracketing of its position in *P–T* space.

Experiments in the laboratory must necessarily be concluded in much shorter times than nature can take to do the same job. Even at high temperatures, silicate reactions are notoriously sluggish, and the assemblage observed at the end of an experimental run might reflect an incomplete reaction or a metastable intermediate state rather than a true equilibrium assemblage. The proportions that this problem can sometimes assume are illustrated by the disagreement among the published determinations of the kyanite–sillimanite–andalusite **triple point** shown in Figure 2.1.1b. The present consensus places the triple point at about 4×10^8 Pa and 500 °C (Figure 2.1).

One precaution that the experimenter can take is to ensure that the position of every phase boundary is established by approaching it from both sides, a procedure known as 'reversing the reaction'. In locating the kyanite–sillimanite phase boundary, for example, it is insufficient just to measure the temperature at which kyanite changes into sillimanite; the careful experimenter will also measure the temperature at which sillimanite, on cooling, inverts to kyanite.

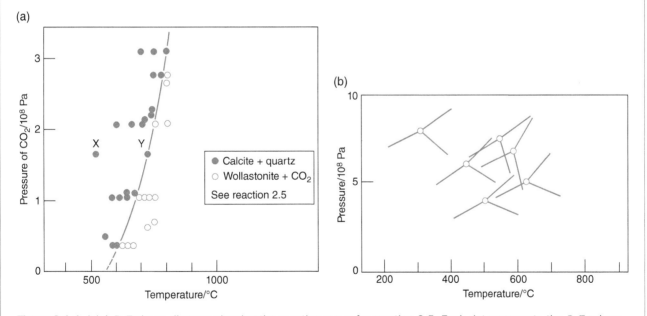

Figure 2.1.1 (a) A *P–T* phase diagram showing the reaction curve for reaction 2.5. Each dot represents the *P–T* values for an individual experiment. All experiments were conducted in a CO_2 atmosphere; the pressure of CO_2 gas present (symbolized as P_{CO_2}) is equal to the applied pressure. Solid dots identify runs that produced calcite and quartz; open dots represent runs in which wollastonite was formed. The curve is drawn to run between filled and open dots. (b) Differences between published *P–T* values of the kyanite–sillimanite–andalusite **triple point** shown in Figure 2.1.

encountered anywhere in the crust and in most of the upper mantle. Experiments can be conducted in which minerals are synthesized at a series of accurately known temperatures and pressures, from starting materials of known composition. In each case, the *phases* (minerals) formed are considered to crystallize in chemical *equilibrium* with each other, and/or with a molten silicate liquid, and for this reason the experiments are called *phase-equilibrium experiments*. Experimental information like this is vital to the petrological interpretation of naturally occurring rocks. The results are used to map out, in phase diagrams similar to Figure 1.3a, the areas within which particular mineral assemblages are stable. In crossing a boundary from one stability field into a neighbouring one as physical conditions change, one mineral assemblage recrystallizes into another, a chemical reaction between coexisting minerals which transforms an unstable – or metastable – assemblage into a new, stable one. These boundaries, like the diagonal line in Figure 1.3a, are called *phase boundaries* or, more generally, *reaction boundaries*.

Phase diagrams form a key part of the literature of petrology, a powerful means of portraying and interpreting the phase-equilibrium data relevant to igneous and metamorphic petrogenesis. Reading and interpreting such diagrams, in the light of underlying thermodynamic principles, is one of the basic skills essential to every geologist.

Just as atmospheric pressure represents the weight of the column of air above us under the influence the Earth's gravitational field, so the pressure experienced by a rock buried at some depth within the Earth reflects the total weight of the column of rock + ocean + atmosphere resting upon it. This *lithostatic* pressure P therefore increases with depth d in a simple manner conveniently approximated as:

$$P \approx 3 \times d \tag{2.1}$$

where P is expressed in units of $10^8\,Pa$ and d is in km. The high pressure applied in some phase-equilibrium experiments is simply the means by which we simulate in the laboratory the effect of depth in the Earth.

Systems, phases and components

To avoid confusion, one must be clear about the meaning of several terms which are used in a specific sense in the context of phase equilibrium.

System

This is a handy word for describing any part of the world to which we wish to confine attention. Depending on the context, **system** could mean the whole of the Earth's crust, or the oceans, or a cooling magma chamber, or an individual rock, or a sample undergoing a phase-equilibrium experiment. In most cases the term will refer to a collection of geological **phases** (see below) interacting with each other.

An *open system* is one that is free to exchange both material and energy with its surroundings. The sea is an open system; it may be useful to consider it as an isolated entity for the purposes of discussion, but one must recognize that it receives both sunlight and river water from outside, and loses heat, water vapour and sediment to the atmosphere and crust.

A *closed system* is one that is sealed with respect to the transfer of matter, but that can still exchange energy with the surroundings. A sealed magma chamber would be a good example, its only interaction with its environment (ideally) being the gradual loss of heat. An *isolated system* is one that exchanges neither mass nor energy with its surroundings, a notion of little relevance to the real world.

'System' may alternatively be used to denote a domain of chemical (rather than physical) space. Petrologists use the word to distinguish a particular region of '*compositional space*' to which attention is to be confined. Thus one speaks of 'the system $MgO-SiO_2$', referring to the series of compositions that can be generated by mixing these two chemical components in all possible proportions. The system so designated includes various minerals whose compositions lie within this range (the silica minerals, and olivine and pyroxene).

Phase

The meaning of **phase** in physical chemistry and petrology is easy to grasp although cumbersome to put into words. In formal terms, a phase can be defined as 'a part or parts of a system occupying a specific volume and having *uniform physical and chemical characteristics* which distinguish it from all other parts of the system'.

Each individual mineral in a rock thus constitutes a separate phase, but that is not the end of the story. A lump of basalt collected from a solidified lava flow might be found to consist of four minerals, say plagioclase, augite, olivine and magnetite. But the igneous texture tells us that these minerals crystallized from a melt, which itself had 'uniform physical and chemical properties' distinct from each of the crystalline minerals present. So in considering the phase relations that dictated the present character of the rock we must count the melt as a phase too, although it is now no longer present as a constituent. If the basalt is *vesicular* (that is, it contains empty bubbles or 'vesicles'), we have evidence that a sixth phase, water vapour, was also present as the rock crystallized.

(Note that water can be present as a dissolved species within a melt or a hydrous crystalline mineral. Being accommodated within the volume occupied by that phase, it does not then count as a phase in its own right. Only when separate bubbles of water vapour appear can one accord water the status of a separate phase. When this happens, as with the vesiculating lava, it is a sign that all other phases present contain as much water as they can accommodate, and the system is **saturated** with water.)

One must therefore be careful not to overlook additional phases which – though no longer present in a particular rock – may have influenced its formation or development. Here are some examples where this could happen:

(a) Some igneous rocks show textural evidence for the existence of two distinct (presumably immiscible) silicate liquids that once existed together in mutual equilibrium.

(b) A metamorphic rock may have developed in the presence of a vapour phase, permeating the grain boundaries between crystals, of which no visible trace now remains.

(c) Mineral veins are deposited from a fluid phase preserved only in occasional microscopic inclusions within crystals (Box 4.6).

(d) When a magma is produced by melting deep inside the Earth, it may be in pressure-dependent equilibrium with minerals quite different from those that crystallize from the same melt at the surface.

Thus the minerals we see in a rock today may represent only a part (or even no part at all) of the original phase equilibria to which the rock owes its present constitution.

It is usual to refer to solid phases by the appropriate mineral name – quartz, kyanite, olivine, and so on. The

essential distinction between them is crystallographic structure, not their chemical composition (which several minerals may have in common). Any molten phase present – regardless of composition – is referred to as 'melt'. By convention one refers to any gaseous phase involved as '**vapour**'.

Component

The basic chemical constituents of a system, of which the various phases are composed, are called its **components**. The concept of a component is defined in a precise but rather roundabout way: 'the components of a system comprise the *minimum number* of chemical (atomic and molecular) species required to specify completely the compositions of all the phases present'.

Consider a crystal of olivine, which at its simplest consists of the elements magnesium (chemical symbol Mg), iron (Fe), silicon (Si) and oxygen (O). One way to define the components of the olivine would be to regard each chemical element as a separate component because the composition of any olivine can be stated in terms of the concentrations[1] of four elements:

Mg Fe Si O

However, defining the components in this way fails to recognize an important property of all silicate minerals, including olivine: that the oxygen content is not an independent quantity, but is tied by valency (Chapter 6) to the amounts of Mg, Fe and Si present, being just sufficient to generate the oxides of each of these elements (this is explained in Box 8.4). So, in describing the composition of an olivine the same information can be conveyed more economically in terms of the concentrations of only three components:

MgO FeO SiO_2

By using a property specific to olivines, however, a still more economical statement of olivine composition can be devised. The crystal chemistry of olivine (see Chapter 8) requires an olivine composition to conform to a general formula that we can represent by X_2SiO_4. X represents a type of atomic site in the olivine crystal

---

[1] The ways in which concentration can be expressed are summarized at the beginning of Chapter 4.

structure that accommodates either Mg or Fe, but not Si. For every Si atom in the olivine structure, there have to be two divalent atoms present, each of which can be either Mg or Fe. Another way to symbolize this constraint is to write the formula as $(Mg,Fe)_2SiO_4$, in which '(Mg,Fe)' represents an atom of either Mg or Fe. One can now express the composition of an olivine as a combination of just two components:

$$Mg_2SiO_4 \quad Fe_2SiO_4$$

Mineralogists call these components the 'end-members' of the olivine series, and give them the names forsterite (Fo) and fayalite (Fa) respectively.

In analysing the arithmetic of chemical equilibrium between minerals, it is important to formulate the components of a system in such a way as to minimize their number, as the definition implies. What constitutes the minimum number depends upon the nature of the system. In an experiment involving the melting of an olivine crystal on its own, the composition of the melt, though different from the solid, still conforms to the olivine formula X_2SiO_4. The compositions of both phases present, olivine and melt, can therefore be expressed as proportions of only two components, Mg_2SiO_4 and Fe_2SiO_4 (Box 2.4). Systems consisting of only two components are called *binary* systems.

If, however, olivine coexists with, let us say, orthopyroxene, the formulation of components becomes less straightforward. Orthopyroxenes are composed of the same four elements as olivine, but they combine in different proportions. The general formula of orthopyroxene, $X_2Si_2O_6$, reveals an X:Si ratio (1:1) lower than for olivine (2:1). The composition of a pyroxene cannot therefore be expressed in terms of just the two olivine end-members. To represent the separate compositions of olivine and orthopyroxene in this system, three components will be needed:

Either : Mg_2SiO_4 Fe_2SiO_4 SiO_2
Or :   MgO   FeO SiO_2

The identity of the components is less important here than their number. A system like this requiring three components to express all possible compositions is said to be *ternary*.

There are circumstances in which four components would be necessary, such as when olivine coexists with metallic iron (for example in certain meteorites). The amount of oxygen present is no longer determined solely by the metals present, as it would be in a system consisting entirely of silicates. One cannot express the composition of metallic iron as a mixture of oxides, so one must resort to using four components, Mg, Fe, Si and O, in order to describe all possible compositions in this *quaternary* system.

In this book, the general practice will be to refer to components by means of their chemical formulae. This avoids confusion between phases and components, which can arise when a phase (for example the mineral quartz) happens to have the same chemical composition as one of the components (SiO_2) in the same system. However, in other books it is quite common for end-member names to be used in this way as well (for example, 'forsterite' for Mg_2SiO_4).

Equilibrium

It is useful to distinguish between two aspects of equilibrium: thermal equilibrium and chemical equilibrium.

Thermal equilibrium

All parts of a system in thermal equilibium have *the same temperature*: in these circumstances heat flowing from one part of the system, A, to another part, B, is exactly balanced by the heat passing from part B to part A, so there is no *net* transfer of heat. Net heat transfer only occurs when there is a difference in temperature between different parts of the system.

Chemical equilibrium

This describes a system in which the distribution of chemical components among the phases of a system has become constant, showing no net change with time. This steady state does not mean that the flow of components from one phase to another has ceased: equilibrium is a dynamic process. An olivine suspended in a magma is constantly exchanging components with the melt. At melt temperatures, atoms will diffuse across the crystal boundary, both into the crystal and out of it into the liquid. If the diffusion rates of element X in and out of the crystal are unequal, there will be a net change of the composition of each phase with time, a condition known as *disequilibrium*. Such changes usually lead eventually to a situation where, for every element present, the flux of atoms across the crystal boundary is the same in both directions, resulting in zero net flow, and no change of composition with time. This is what we mean by *equilibrium*.

The rate at which equilibrium is achieved varies widely and, as Chapter 3 will show, disequilibrium is found to be a common condition in geological systems, particularly at low temperatures.

The Gibbs Phase Rule

A natural question to ask is: how many phases can be in equilibrium with each other at any one time? In Figure 1.3a we looked at a simple system in which only two phases occurred. Most actual rocks, however, are not so simple. What factors determine the mineralogical complexity of a natural rock? Which aspect of a chemical equilibrium controls the number of phases that participate in it?

This question was addressed in the 1870s by the American engineer J. Willard Gibbs, the pioneer of modern thermodynamics. The outcome of his work was a simple but profoundly important formula called the *Phase Rule*, which expresses the number of phases that can coexist in mutual equilibrium (ϕ) in terms of the number of components (C) in the system and another property of the equilibrium called the *variance* (F). The Phase Rule can be stated symbolically as:

$$\phi + F = C + 2 \qquad (2.2)$$

The variance is alternatively known as the *number of degrees of freedom* (hence the symbol F used to represent it). The concept is most easily introduced through an example. Figure 2.1 illustrates the equilibrium phase relations between the minerals kyanite, sillimanite and andalusite. These minerals are all **aluminium silicate**[2] **polymorphs** of identical composition. A single component (Al_2SiO_5) is therefore sufficient to cover the compositional 'range' of the entire system.

Points A, B and C are three different points in the 'P–T space' covered by the diagram; they represent three classes of equilibrium that can develop in the system. The obvious difference between them is the nature

---

[2] Not to be confused with **aluminosilicate** minerals discussed in Chapter 9.

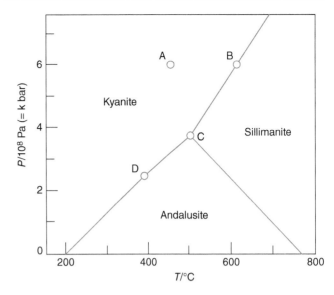

Figure 2.1 A P–T diagram showing phase relations between the **aluminium silicate** minerals (composition Al_2SiO_5). The pressure axis is graduated in units of 10^8 pascals, equal in magnitude to the traditional pressure units, kilobars (1 kbar = 10^3 bars = 10^8 Pa). Kyanite is a triclinic mineral that usually occurs as pale blue blades in hand specimen. Sillimanite is orthorhombic and is commonly fibrous or prismatic in habit. Andalusite is also orthorhombic and is characteristically pink in hand specimen.

of the equilibrium assemblage. Point A lies within a field where only one phase, kyanite, is stable. Point B lies on the phase boundary between two stability fields, where two minerals, kyanite and sillimanite, are stable together. Point C, at the **triple point** where the three stability fields (and the three phase boundaries) meet, represents the only combination of pressure and temperature in this system at which all three phases can exist stably together.

It is clear that the three-phase assemblage (kyanite + sillimanite + andalusite), when it occurs, indicates very precisely the state of the system (that is, the values of P and T) in which it is produced, because there is only one set of conditions under which this assemblage will crystallize in equilibrium. Using the Phase Rule (Equation 2.2), one can calculate that the variance F at point C is zero:

Point C	$\phi = 3$	(3 phases, ky + sill + andal)
	$C = 1$	(1 component, Al_2SiO_5)
	$3 + F = 1 + 2$	
Therefore	$F = 0$	an *invariant* equilibrium

A variance of zero means that the three-phase equilibrium assemblage completely constrains the state of the system to a particular combination of P and T. Such a situation is called **invariant**. There is no latitude (no degree of freedom) for P or T to change at all, if the assemblage is to remain stable. Any such variation would lead to the disappearance of one or more of the three phases, thus altering the character of the equilibrium. So the discovery of this mineral assemblage in a metamorphic rock (a rare event, since only about a dozen natural occurrences are known) ties down very precisely the conditions under which the rock must have been formed, assuming that:

(a) the kyanite–andalusite–sillimanite assemblage represents an *equilibrium state* obtained as the rock formed, and not simply an uncompleted reaction from one assemblage to another; and
(b) the *P–T* coordinates of the invariant point in Figure 2.1 are accurately known from experimental studies. (Whether this requirement is satisfied in the case of the Al_2SiO_5 polymorphs is arguable – see Box 2.1 – but in the present discussion we shall ignore this difficulty.)

The two-phase equilibrium between, say, kyanite and sillimanite is less informative. The coexistence of these two minerals indicates that the state of the system in which they crystallized must lie somewhere on the kyanite–sillimanite phase boundary, but exactly where along this line remains uncertain unless we can specify one of the coordinates of point B. We only need to specify one coordinate (for example, temperature) because the other is then fixed by intersection of the specified coordinate with the phase boundary. According to the Phase Rule, the variance at B is equal to 1:

Point B	$\phi = 2$	(2 phases, ky + sill)
	$C = 1$	(1 component, Al_2SiO_5)
	$2 + F = 1 + 2$	
Therefore	$F = 1$	a *univariant* equilibrium

The one degree of freedom indicates that the state of the system is only unconstrained in one direction in Figure 2.1, along the phase boundary. One additional piece of information is required (either T or P) to tie down the state of the system completely. The coexistence of kyanite and sillimanite in a rock will pinpoint the exact conditions of origin only in conjunction with other information about P or T.

At point A, where kyanite occurs alone, the variance is equal to 2:

Point A	$\phi = 1$	(1 phase, ky)
	$C = 1$	(1 component, Al_2SiO_5)
	$1 + F = 1 + 2$	
Therefore	$F = 2$	a *divariant* condition

Within the bounds of the *divariant* kyanite field, therefore, P and T can vary independently (two degrees of freedom) without upsetting the equilibrium phase assemblage (just kyanite). The one-phase assemblage is therefore little help in establishing the precise state of the system, because it leaves two variables (P and T) still to be specified.

The variance cannot be greater than 2 in a one-component system like Figure 2.1. In the more complex systems we shall meet in the following section, the phases present may consist of different proportions of several components. A complete definition of the state of such a system must then include the compositions of one or more phases, in addition to values of P and T. Such compositional terms (X_a, X_b, etc., representing the **mole fractions** of a, b, etc. in a phase) contribute to the total variance, which can therefore, in multicomponent systems, adopt values greater than 2.

Variance can be summarized in the following way. The 'state' of a system – whether we consider a simple experimental system or a real metamorphic rock in the making – is defined by the values of certain key intensive variables, including pressure (P), temperature (T) and, in multi-component systems, the compositions (X values) of one or more phases. For a given equilibrium between specific phases, some of these values are automatically constrained in the phase diagram by the equilibrium phase assemblage. The variance of this equilibrium is the number of the variables that remain free to adopt arbitrary values, which must be determined by some other means if the state of the system is to be defined completely.

Phase diagrams in *P–T* space

The need to represent phase equilibrium data in visual form on a two-dimensional page leads to the use of various forms of phase diagram, each having its own merits and limitations. We begin by looking at P–T diagrams.

The two phase diagrams so far considered (Figures 1.3a and 2.1) both show the effects of varying pressure and temperature on a system consisting of only one component ($CaCO_3$ or Al_2SiO_5). Other important examples of such *unary* systems are discussed in Box 2.2.

P–T diagrams can also be used to show the pressure–temperature characteristics of multicomponent reactions and equilibria. An example is shown in Figure 2.2. The univariant boundary in this diagram represents not a phase transition between different forms of the same compound, but a reaction or equilibrium between a number of different compounds:

$$NaAlSi_2O_6 + SiO_2 \rightleftharpoons NaAlSi_3O_8$$

jadeite (*a pyroxene*)   quartz   albite (*a feldspar*)

$$(2.3)$$

For this reason the term *reaction boundary* (or *equilibrium boundary*) is used. It marks the P–T threshold across which reaction occurs, or the conditions at which univariant equilibrium can be established.

Two components are sufficient to represent all possible phases in this system. We can choose them in a number of equivalent ways; selecting $NaAlSi_2O_6$ and SiO_2 is as good a choice as any. Applying the Phase Rule to point X:

Point X	$\phi = 2$	(2 phases, jadeite + quartz)
	$C = 2$	(2 components, $NaAlSi_2O_6$ and SiO_2)
	$2 + F = 2 + 2$	
Therefore	$F = 2$	signifying a *divariant* field.

The jadeite + quartz field is therefore a divariant field like that of kyanite in Figure 2.1. At point Y on the phase boundary, however, three phases are in equilibrium together:

Point Y	$\phi = 3$	(3 phases, jadeite + quartz + albite)
	$C = 2$	(2 components, $NaAlSi_2O_6$ and SiO_2)
	$3 + F = 2 + 2$	
Therefore	$F = 1$	a *univariant* equilibrium.

The three-phase assemblage represents a univariant equilibrium: only one variable, P or T, needs to be specified to determine completely the physical state of the system. The value of the other can be read off the reaction boundary. The existence of two components in

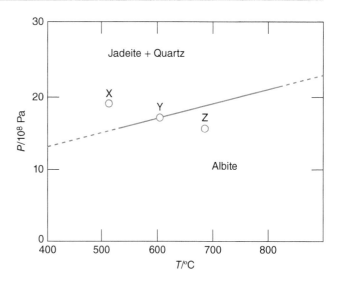

Figure 2.2 P–T diagram showing the experimentally determined reaction boundary (solid line) for the reaction

jadeite + quartz → albite

Jadeite ($NaAlSi_2O_6$) and albite ($NaAlSi_3O_8$) are both **aluminosilicates** of sodium (Na).

this system means that a three-phase assemblage is no longer invariant, as it was in Figure 2.1.

At first glance one might expect the albite field (point Z) to be divariant like the jadeite + quartz field, but here the Phase Rule springs a surprise:

Point Z	$\phi = 1$	(1 phase, albite)
	$C = 2$	(2 components, $NaAlSi_2O_6$ and SiO_2)
	$1 + F = 2 + 2$	
Therefore	$F = 3$	a *trivariant* equilibrium.

Analysing the albite field in this way, it appears necessary to specify the values of *three* variables to define the state of the system in this condition. P and T account for two of them, but what can the third variable be? The answer becomes clear if we ask what requirements must be met if, in passing from X to Z, we are to generate *albite alone*. If the mixture of jadeite and quartz contains more molecules of SiO_2 than $NaAlSi_2O_6$, a certain amount of quartz will be left over after all the jadeite has been used up. The resultant assemblage at Z will therefore be albite + quartz. The presence of two phases leads to a variance of 2 for this field, as originally expected. Conversely, if we react SiO_2 with an excess of $NaAlSi_2O_6$ molecules, the resultant assemblage at Z

Therefore $\Delta S = +32.4\ \mathrm{J\,K^{-1}\,mol^{-1}}$ and $\Delta V = +17.0$ $10^{-6}\,\mathrm{m^3\,mol^{-1}}$

Thus:

$$\frac{\mathrm{d}P}{\mathrm{d}T} = \frac{32.4}{17.0}\frac{\mathrm{J\,mol^{-1}K^{-1}}}{10^{-6}\,\mathrm{m^3\,mol^{-1}}}$$
$$= 1.91 \times 10^6\,\mathrm{J\,m^{-3}\,K^{-1}}$$
$$= 1.91 \times 10^6\,\mathrm{N\,m^{-2}\,K^{-1}}$$
$$= 19.1 \times 10^5\,\mathrm{Pa\,K^{-1}}$$

The units $(10^5\,\mathrm{Pa\,K^{-1}} = \mathrm{bar\,K^{-1}})$ relate to the gradient of a line in $P\text{–}T$ space (Figure 2.2). The sign of the gradient is positive, consistent with Figure 2.2 (P rises with T), and the magnitude $(19.1 \times 10^5\,\mathrm{Pa\,K^{-1}})$ agrees well with the value of about $20 \times 10^5\,\mathrm{Pa\,{}^\circ C^{-1}}$ as measured from Figure 2.2, noting that kelvins and degrees Celsius are units of equal size (Appendix A).

The positive slope of Figure 2.2 thus reflects the observation that both ΔS and ΔV for this reaction are positive (or both negative, if we write the reaction the other way round). A negative slope would signify that ΔS and ΔV had opposite signs, as is the case for the andalusite–sillimanite reaction (Figure 2.1).

The most striking difference between Figures 1.3a, 2.1 and 2.2, on the one hand, and Figure 2.3 on the other, is that the latter has a curved reaction boundary, whereas the others are straight. The reaction boundary is curved because the volume change for such a reaction (and therefore $\mathrm{d}P/\mathrm{d}T$) is very pressure-sensitive:

$$\mathrm{muscovite} \rightleftharpoons \mathrm{sanidine} + \mathrm{corundum} + \mathrm{vapour}$$
$$\Delta V = V_{\mathrm{vapour}} + V_{\mathrm{sanidine}} + V_{\mathrm{corundum}} - V_{\mathrm{muscovite}} \qquad (2.9)$$

At low pressures the volume of the 'vapour' phase (actually a supercritical fluid – see Box 2.2) is much greater than those of the solid phases, and therefore dominates the value of ΔV.

$$\Delta V \cong V_{\mathrm{vapour}}$$

Because this term is large, the reaction boundary at low pressure has only a moderate slope. But the vapour, like any gas, is much more *compressible* than the solid phases. At higher pressures, V_{vapour} and ΔV will get progressively less, and the slope of the dehydration boundary will get correspondingly steeper. This general shape is a feature of all reactions involving the generation of a 'vapour' phase (see also Figure 2.1.1). Curved boundaries in $P\text{–}T$ diagrams always signify the involvement of a highly compressible phase, usually a gas (e.g. Figure 2.2.2).

Phase diagrams in *T*–*X* space

Crystallization in systems with no solid solution

$P\text{–}T$ and $P_{\mathrm{V}}\text{–}T$ diagrams make no provision for changes in the compositions of individual phases during reactions. Such changes are an important feature of igneous processes, and make it necessary to introduce another type of diagram in which the temperature of equilibrium is plotted as a function of phase composition ('*X*'). An example is shown in Figure 2.4, which shows the phase relations at atmospheric pressure for the binary system $CaMgSi_2O_6\text{–}CaAl_2Si_2O_8$. Because this system is relevant to igneous rocks (it includes simple analogues of basalt), the temperature range extends up far enough to include melting.

If the temperature is sufficiently high, it is possible to make a homogeneous melt containing the two components $CaMgSi_2O_6$ and $CaAl_2Si_2O_8$ in any desired proportion. These compounds are said to be completely **miscible** in the melt phase. Consequently, in the field marked 'melt', only this single phase is stable. In the solid state, however, the two components exist as the separate phases diopside (ideal composition $CaMgSi_2O_6$) and anorthite (composition $CaAlSi_2O_8$): there is no stable homogeneous solid of intermediate composition. The area below $1274\,{}^\circ C$ is therefore a two-phase field.

The two areas ABE and ECD are also two-phase fields, each representing equilibrium between a melt and one of the crystalline phases. To see how, consider the line *xy*. This is an **isothermal** line at a temperature whose precise value is unimportant (in this case it is $1400\,{}^\circ C$). We call this a **tie-line**, because it links ('ties') together the compositions of two phases which can coexist stably at this temperature. *x* represents the only composition of the melt that can be in equilibrium with anorthite (composition *y*) at $1400\,{}^\circ C$; it consists of 61% $CaAl_2Si_2O_8$ and 39% $CaMgSi_2O_6$. If the melt were more $CaMgSi_2O_6$-rich than this (composition x_1 for example), it would dissolve anorthite crystals and thereby increase the $CaAl_2Si_2O_8$ content until equilibrium was reached or until the anorthite present had all dissolved. If the liquid had the

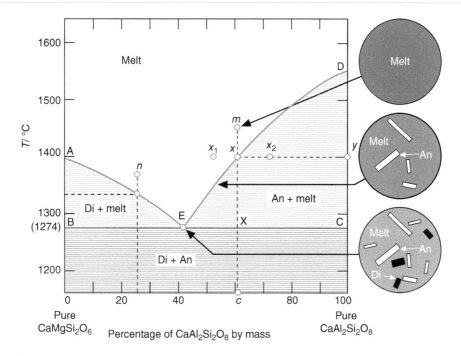

Figure 2.4 Melting relations in the **pseudo-binary** system $CaMgSi_2O_6$–$CaAl_2Si_2O_8$ at atmospheric pressure. The horizontal ruling represents two-phase fields: the solid–solid field is ruled more heavily than the solid–melt fields. 'Di' and 'An' refer to the phases diopside (composition $CaMgSi_2O_6$) and anorthite (composition $CaAl_2Si_2O_8$). This phase diagram is not strictly binary because small amounts of aluminium enter the pyroxene phase (see Morse, 1980, pp. 53–7, for details). The circular cartoons on the right illustrate what an experimental 'magma' might look like under the microscope at each stage.

composition x_2, it would precipitate anorthite, so reducing the $CaAl_2Si_2O_8$ content.

As the line DE shows, the composition of melt that can exist in equilibrium with anorthite (An) depends on the temperature. The corresponding line A–E shows that the same is true of the melts that can coexist with diopside (Di). The curve AED, the locus of the melt compositions that can coexist in equilibrium with either diopside or anorthite at different temperatures, is called the **liquidus**. All states of the system lying above it in Figure 2.4 consist entirely of the melt phase. The one point common to both limbs of the liquidus is E, which therefore represents the unique combination of melt composition and temperature at which all three phases are simultaneously at equilibrium. This condition is called a **eutectic**.

In applying the Phase Rule to Figure 2.4, we must recognize that a T–X diagram like this is no more than the end-view of more complex phase relations encountered in P–T–X space (cf. Figure 2.6a). By considering melting relations only at a single – in this case atmospheric – pressure, we are in fact artificially restricting the variance of each equilibrium. Any statements we

make about variance in this diagram relate only to an apparent variance F' where:

$$F' = F - 1 \qquad (2.10)$$

One may write the Phase Rule in terms of F' as follows:

$$\phi + F = \phi + (F' + 1) = C + 2$$

Therefore

$$\phi + F' = C + 1 \qquad (2.11)$$

This form of the Phase Rule, applicable to **isobaric** T–X (and, incidentally, **isothermal** P–X) phase diagrams, is sometimes known as the *Condensed Phase Rule*.

Point E	$\phi = 3$	(3 phases, Di + An + melt)
	$C = 2$	(2 components, $CaMgSi_2O_6$ and $CaAl_2Si_2O_8$)
	$3 + F' = 2 + 1$	
Therefore	$F' = 0$	an isobarically *invariant* equilibrium

The term *isobarically invariant* is jargon that reminds us that this equilibrium is invariant only so long as the pressure is held constant; if this constraint were to be

Box 2.3 Tie-lines and the Lever Rule

In *T–X*, *P–X* or *P–T–X* diagrams, tie-lines link together the compositions of two different phases that can coexist in equilibrium under specific conditions. Any composition lying between the ends of a tie-line must therefore represent a physical *mixture* of the two phases. From the position of that point on the tie-line, one can work out the relative proportions of the two phases in the mixture.

Figure 2.3.1a shows part of a phase diagram in which complete solid solution exists between two compounds, A and B (cf. Figure 2.5). The tie line *c–d* depicts equilibrium at temperature T_1 between a melt of composition *c* on the liquidus, and a solid solution of composition *d* on the solidus; both *c* and *d* are expressed in mass % B. Composition *x* lies in the *two-phase field* between *c* and *d*, and must signify a physical mixture of these two distinct phases. Let *C* and *D* represent the mass fractions (*i.e.* $C+D=1.00$) in which *c* and *d* are mixed to form *x*. We can then express the composition of *x* as a weighted average of *c* and *d*:

$$x = Cc + Dd \qquad (2.3.1)$$

Since $C = 1 - D$, this can be rewritten:

$$x = (1-D)c + Dd = c - Dc + Dd$$

Therefore $x - c = D(d - c)$

leading to $D = \dfrac{x-c}{d-c}$

Substituting $D = 1 - C$ into Equation 2.3.1, we can show in a similar fashion that:

$$C = \frac{x-d}{c-d} = \frac{d-x}{d-c} \ (\text{the change in sign cancels out})$$

The mass ratio in which *c* and *d* are present in *x* is therefore given by:

$$\frac{C}{D} = \frac{d-x}{d-c} \bigg/ \frac{x-c}{d-c} = \frac{d-x}{x-c} \ (\text{after cancelling denominators})$$

In other words : $C(x-c) = D(d-x)$ $\qquad (2.3.2)$

This useful equation is known as the **Lever Rule,** since it can also be applied to the 'lever effect' of the old-fashioned beam-balance (Figure 2.3.1b), in which the weight of a body *C* is inversely proportional to the distance from the fulcrum (*c–x*) at which it balances an opposing weight *D*:

$$\frac{\text{weight of } C}{\text{weight of } D} = \frac{x-d}{c-x}$$

Qualitatively, the closer the composition of a mixture plots (in composition space) to one of its constituents, the greater the percentage of that constituent in the mixture.

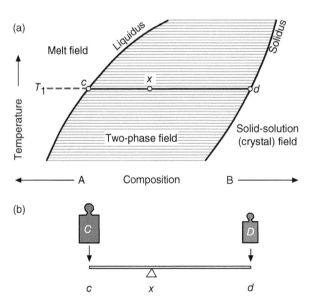

Figure 2.3.1 (a) Part of a phase diagram similar to Figure 2.5 to illustrate the Lever Rule. (b) The analogous geometry of the beam-balance.

relaxed, we would find that the eutectic was just one point of a univariant curve in $T–X–P$ space.

A eutectic is always an invariant point in any phase diagram. On the other hand, the one-phase 'melt' field has two degrees of freedom:

Point x_1	$\phi = 1$	(1 phase, melt)
	$C = 2$	(2 components, $CaMgSi_2O_6$ and $CaAl_2Si_2O_8$)
	$1 + F' = 2 + 1$	
Therefore	$F' = 2$	an isobarically divariant equilibrium

T and X have to be quantified to define fully the state of the system in this condition.

It would be natural to expect the fields ECD and ABE to be divariant as well, but the Phase Rule indicates otherwise. Consider the composition represented by point x_2. Neither the melt (at this temperature) nor anorthite can have this composition. The 'composition' x_2 only has meaning at 1400 °C when interpreted as the composition of a physical mixture of melt x and anorthite y. (The proportions of the two phases in this mixture can be worked out as explained in Box 2.3.) ECD and ABE are therefore two-phase fields. If $\phi = 2$ and $C = 2$, we cannot escape the conclusion that $F' = 1$. In other words, specifying temperature is sufficient to define the composition of all phases in equilibrium, or vice versa given that the pressure is already defined. Thus reaction-boundary lines (cf. Figures 1.3a and 2.1) are not the only manifestation of univariant equilibrium in phase diagrams; areas can also be univariant. Such fields arise in $T–X$ diagrams whenever two coexisting phases have different compositions. One can imagine them consisting of an infinity of horizontal tie-lines, as the horizontal ruling in Figure 2.4 is intended to symbolize.

$T–X$ diagrams are important in igneous petrology because they allow one to follow the evolution of melt composition with advancing crystallization in experimental and natural magmatic systems (at constant pressure). Imagine a melt m cooling from some temperature above the liquidus, say 1450 °C. The phase assemblage at this point consists of melt alone, as illustrated by the top 'microscope view' on the right. At first there will be no change other than a fall in temperature: we can imagine point m falling vertically through the 'melt' field. Arrival at the liquidus (point x) signals the first appearance of solid anorthite, which here begins to crystallize in equilibrium with the melt. Extraction of anorthite depletes the melt a little in the $CaAl_2Si_2O_8$ component, causing a shift in composition to the left in Figure 2.4. But the maintenance of univariant equilibrium demands that the melt composition should change in conjunction with falling temperature. As crystallization advances, therefore, the melt composition migrates steadily down the liquidus curve towards E, continuously crystallizing anorthite and changing composition. The phase assemblage here is shown in the middle roundel on the right, illustrating what might be seen in a quenched sample under the microscope.

On reaching the eutectic, the melt begins to crystallize diopside in addition to anorthite, as shown in the lowermost roundel. At this juncture, the melt composition becomes fixed, because anorthite and diopside crystallize in the same proportion as the $CaAl_2Si_2O_8$:$CaMgSi_2O_6$ ratio of the melt. The temperature also remains constant, because the eutectic is an invariant equilibrium (at least within the isobaric framework being considered): as long as three phases remain in equilibrium, neither melt composition nor temperature can change. Continued cooling in this context merely means the loss of heat (the latent heat of crystallization of diopside and anorthite) from the system at constant temperature and the formation of crystals at the expense of melt. Eventually the melt becomes exhausted, and invariant (Di + An + melt) equilibrium gives way to univariant (Di + An) equilibrium, allowing the temperature to resume its downward progress. The total solid assemblage will obviously now consist of 61% anorthite and 39% diopside (c in Figure 2.4).

The eutectic therefore represents the lowest projection of the melt field, the composition and temperature of the last melt to survive during the cooling of the system. Progressive crystallization of any melt in this system (with the special exceptions of pure $CaMgSi_2O_6$ and pure $CaAl_2Si_2O_8$) will lead its composition ultimately to the eutectic. This illustrates an important general principle of petrology: that the evolving compositions of crystallizing magmas of all types tend to converge upon one or two 'residual magma' compositions (a natural example being granite) at which the liquidus reaches its lowest temperature.

The eutectic also indicates the composition of the first melt to appear upon heating any mixture of diopside and anorthite (Box 2.4).

Box 2.4 Partial melting I: melting in the laboratory

Phase diagrams provide many insights into the important process of rock melting. Consider first the progressive melting of a mixture of diopside and anorthite, for example the mixture of composition c in Figure 2.4.

The mixture will simply heat up until the temperature reaches 1274 °C. At this point a melt of the eutectic composition E begins to form; this is the only melt composition that can be in equilibrium with diopside and anorthite, both of which are at this stage present in the solid mixture. Continued heating brings an increase in the proportion of melt at constant temperature, with no change of melt composition (invariant equilibrium) until the diopside disappears, having been entirely incorporated in the melt. (Anorthite has been dissolving too, but has not yet been used up.) At this stage, from the Lever Rule (Box 2.3), the ratio of melt:anorthite is XC:EX. Univariant equilibrium (An+melt) now obtains, and with increasing temperature the proportion of melt continues to rise, its composition proceeding up the liquidus curve as more and more anorthite dissolves in it. At x, the melt has the same composition as the solid starting mixture, and here the last remaining crystals of anorthite disappear. The system now enters the divariant melt field, where the temperature can continue rising without further change of state.

Similar principles govern the melting of solid-solution minerals. The olivine system shown in Figure 2.4.1 (analogous to Figure 2.5) provides an example relevant to basalt production by partial melting in the upper mantle (which consists largely of olivine). When olivine is heated up to the solidus (for example, point c_1), a small proportion of melt m_1 appears. The melt is much less magnesium-rich

than the olivine from which it is produced, a point of great petrological significance. Continued heating will cause the temperature to rise, the proportion of melt to increase, its composition to migrate up the liquidus curve towards m_2, and that of the remaining olivine crystals to migrate up the solidus towards c_2. The system would become completely molten at just over 1800 °C (m_3). Thus a gap exists between the temperature at which olivine begins to melt and that at which it becomes completely liquid (as in Figure 2.5). This gap, called the *melting interval*, is a feature of all minerals that exhibit solid solution. The everyday notion of a 'melting point' applies only to pure end-members, where the liquidus and solidus converge.

The complete melting of rocks like this only occurs in very unusual circumstances, like meteorite impacts. Generally, magmas are produced by a process of *partial melting*, in which temperatures are sufficient to melt a fraction of the source material but not all of it. Both in the olivine phase diagram and in actual rocks, partial melting will generate a melt less magnesium-rich (m_2) than the source material (c_1), leaving behind a refractory solid residue which contains more magnesium (c_2) than the source material prior to melting. The composition of both products – melt and residual solid – depends upon the degree (percentage) of melting, and therefore on the temperature attained.

We must be careful not to assume that the temperatures shown in the olivine diagram are necessarily characteristic of the upper mantle. We have seen that although pure anorthite (Figure 2.4) remains solid up to 1553 °C, a mixture of anorthite with diopside begins to melt below

---

Eutectics are common features in systems of this kind. They represent the general observation that *mixtures of minerals* (in other words, rocks) begin to melt at lower temperatures than any of the pure constituents (minerals) would on their own, just as a mixture of ice and salt conveniently melts at lower temperature than ice alone. This principle is widely used in industry when a flux is added to enable a substance to melt at a lower temperature than it would in the pure state (e.g. in soldering).

Crystallization in systems with solid solution

Diopside and anorthite belong to different mineral groups having different crystal structures, and the tendency for either to incorporate the constituents of the other into its crystal structure is negligible. But within many mineral groups it is common to find that crystal composition can vary continuously between one **end-member** composition and another. One can visualize one solid end-member 'dissolving' in the other to form

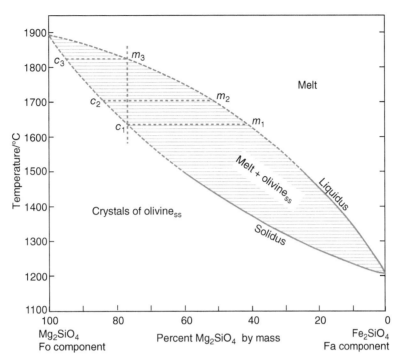

Figure 2.4.1 Melting relations in the olivine series at atmospheric pressure. This system has been determined experimentally only up to 1500 °C, and for pure forsterite. Interpolated boundaries are shown dashed.

1300 °C. The same is true of mantle rocks, which consist not of olivine alone (although it is the dominant constituent), but contain pyroxene and either garnet or spinel as well. Any such mixture will begin melting at temperatures below those at which the individual minerals would start to melt. Moreover, every mineral contributes to the composition of the melt right from the start, just as both anorthite and diopside contribute to the eutectic melt in Figure 2.4. It is a common misconception to believe that minerals in a rock will melt one after the other. A partial melt should be seen as a solution in which all the solid phases of the source rock are partly soluble.

The influence of pressure (depth) on melting is considered in Box 2.5.

a homogeneous crystal of intermediate composition. This phenomenon of miscibility in the solid state is referred to as *solid solution*.

Figure 2.5 shows the crystallization behaviour of a familiar example of such solid solution series, plagioclase feldspar (a solid solution between albite, $NaAlSi_3O_8$, and anorthite, $CaAl_2Si_2O_8$). Only one solid phase appears in the diagram.

Prominent in the diagram is a leaf-shaped feature, bounded by two curves depicting coexisting crystal and melt compositions as a function of temperature. Thus line *ab* in Figure 2.5 is a tie-line linking melt composition *a* to the plagioclase crystal composition *b* with which it is in equilibrium at that temperature. The curve through *a*, above which (in the 'melt' field) the system is entirely molten, is the **liquidus**; the curve through *b*, below which the system consists entirely of crystalline plagioclase, is called the **solidus**.

By applying the Condensed Phase Rule to Figure 2.5, we find that both the 'melt' and 'plagioclase' fields are divariant: $\phi = 1$, $C = 2$ ($NaAlSi_3O_8 + CaAl_2Si_2O_8$), therefore $F' = 2$. The 'melt + plagioclase' field is another

Box 2.5 Partial melting II: melting in the Earth's mantle

Figure 2.5.1 shows in simple terms how the **solidus** temperature of peridotite – the temperature at which it begins to melt – varies with depth in the mantle.[4] This is in essence a *P–T* diagram, drawn 'upside-down' in terms of pressure in order to depict temperature as a function of depth below the surface. The band between the solidus and liquidus lines, the 'melting interval', depicts the range of conditions under which partial melting (Box 2.4) of mantle peridotite can occur. These are the conditions necessary to produce basaltic magma.

The curve marked 'Oceanic geotherm' indicates how the ambient temperature in the upper mantle is believed to vary with depth beneath a typical sector of mid-ocean ridge. Note that the geotherm fails to reach the solidus at any depth. Why then should melting occur at all in the upper mantle?

To see why, we must recognize that the mantle is not a static body. Because the interior of the Earth is hot, the solid mantle undergoes continuous (although very slow) convective motion, with mushroom-like 'plumes' of buoyant hotter material ascending from below (e.g. beneath Hawaii), and dense colder material sinking down (e.g. cold oceanic lithosphere at subduction zones). In a convecting mantle, melting may arise purely as a consequence of upward motion. In Figure 2.5.1a, the solid peridotite at point X, for instance, can penetrate the solidus and begin to melt simply by migrating upward to lower pressures along the path X–Y. This process, known as **decompression melting**, is the primary cause of magma generation at mid-ocean ridges: plate forces continuously pull the lithospheric plates apart and thereby permit passive upwelling (Figure 2.5.1b) and melting of the underlying asthenosphere. No rise in temperature is required; indeed, the ascending material cools slightly as a result of the work it has to do in expanding (Box 1.1).

Mantle plumes, on the other hand, are sites of buoyant upwelling of deeper material that may be 150–300 °C hotter than the surrounding upper mantle. Melting in a plume is the combined effect of an elevated geotherm and decompression resulting from upwelling. As many plumes are located in intra-plate settings, the presence of thick cool lithosphere confines melting to deeper levels than beneath a mid-ocean ridge.

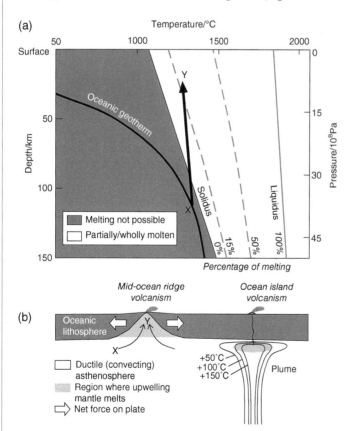

(a)

(b)

Figure 2.5.1 (a) Variation with pressure and depth of the solidus temperature of dry mantle peridotite. The heavy curve labelled 'oceanic geotherm' shows how ambient temperature beneath an oceanic ridge is believed to vary with depth. X–Y illustrates a passive upwelling path resulting from lithosphere extension. (b) Cartoon illustrating where decompression melting occurs beneath an extensional mid-ocean ridge and in an intraplate mantle plume head.

---
[4] In the absence of water.

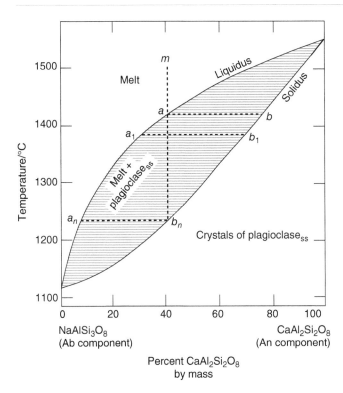

Figure 2.5 *T–X* diagram showing melting relations in plagioclase feldspars at atmospheric pressure. Horizontal ruling represents a two-phase field. The subscript 'ss' denotes that plagioclase is a solid solution (see text).

example of a univariant field (cf. Figure 2.4): $\phi = 2$, $C = 2$, therefore $F' = 1$. If equilibrium exists between melt and plagioclase, specifying T automatically defines the compositions of both phases. Conversely, knowing either phase composition defines the other composition and the temperature of equilibrium unambiguously. The line *ab* is one of an infinite series of such tie-lines traversing the two-phase field, as the horizontal ruling symbolizes.

Crystallization in this system leads to a series of continuously changing melt and solid compositions. Melt *m*, for example, will cool until it encounters the liquidus curve at *a*, where plagioclase *b* will begin to crystallize. Because *b* is more $CaAl_2Si_2O_8$-rich than *a*, its extraction will deplete the melt in $CaAl_2Si_2O_6$ and thereby enrich it in $NaAlSi_3O_8$; with continued cooling and crystallization the melt composition will migrate down the liquidus curve. The changing melt composition causes a corresponding evolution in the equilibrium composition of the plagioclase crystals. Not only will newly crystallized plagioclase be more albite-rich than *b*, but there will be a tendency for early-formed crystals, by

continuous exchange of Na, Ca, Al and Si, to re-equilibrate with later, more albite-rich fractions of the melt. Thus to maintain complete equilibrium as the melt evolves to point a_1 on the liquidus, all the crystals that have so far accumulated must adjust their composition to b_1 on the solidus. Such adjustment requires solid-state diffusion to and from the centre of each crystal, and is a slow process. Crystal growth during natural magma crystallization commonly proceeds too quickly to allow complete continuous equilibrium between crystals and magma: the surface layers readjust to changing melt composition, but the crystal interiors fall behind. The result is a compositional gradient between relatively anorthite-rich cores and relatively albite-rich margins of the plagioclase crystals, a phenomenon known among petrologists as **zoning** (Box 3.1). Zoning is seen in other mineral groups as well, notably pyroxene.

If crystallization proceeds slowly enough to permit continuous re-equilibration (an ideal situation known as *equilibrium crystallization*), the final melt has the composition a_2 and the end-product is a mass of crystals all having the composition b_n, the same as the initial melt composition *m*. Imperfect re-equilibration between crystals and the evolving melt, however, ties up a disproportionate amount of the anorthite component in the cores of early formed crystals owing to overgrowth or burial, and the melt can then evolve to compositions beyond a_n before it runs out, creating late zones of plagioclase more albitic than the original melt. This process, in which isolation of early-formed solids allows later melts to develop to more extreme compositions, is called *fractional crystallization*. Crystallization of natural melts in crustal magma chambers approximates closely to fractional crystallization, and this process contributes a lot to the chemical diversity of igneous rocks and magmas.

The solvus and exsolution

The final *T–X* section to be examined (Figure 2.6b) shows the phase relations of the alkali feldspars in the presence of water vapour (at a pressure of 2×10^8 Pa = 2 kbars). This diagram can be visualized as a cross-section of $P_{H_2O} - T - X$ space, coinciding with the plane in which $P_{H_2O} = 2 \times 10^8$ Pa (Figure 2.6a). One can speak of the diagram as an **isobaric** *T–X section* of phase relations in $P_{H_2O} - T - X$ space.

Box 2.6 Reaction points and incongruent melting

Every geology student knows that olivine and quartz are incompatible, and do not coexist stably in nature. (In fact, this is true only of magnesium-rich forsteritic olivines. Fayalite – Fe_2SiO_4 – is quite a common mineral in granites and quartz syenites.) How is this incompatibility expressed in a phase diagram?

The relevant part of the system Mg_2SiO_4–SiO_2 (omitting complications at the SiO_2-rich end) is shown for atmospheric pressure in Figure 2.6.1. In many respects it is similar to Figure 2.4. The difference is that, between Mg_2SiO_4 and SiO_2 along the composition axis, lies the composition of the pyroxene enstatite, $Mg_2Si_2O_6$. Consider the crystallization of melt composition m_1. On reaching the liquidus it will begin to crystallize olivine, whereupon further cooling and crystallization will lead the melt composition down the liquidus curve. On reaching R, the melt composition has become too SiO_2-rich (more so than enstatite) to coexist stably with olivine, which therefore reacts with the SiO_2 in the melt to form crystals of enstatite:

$$\underset{olivine}{Mg_2SiO_4} + \underset{melt}{SiO_2} \rightarrow \underset{pyroxene}{Mg_2SiO_6} \qquad (2.6.1)$$

(This symbolism does not mean that the melt consists of SiO_2 alone. Other components are present, but this reaction involves only the SiO_2 component.)

At R, the three phases are at equilibrium. Using the Condensed Phase Rule, it is clear that R is an **invariant point** like E. It is called a *reaction point*. Temperature and melt composition remain constant as the reaction proceeds (from left to right in reaction 2.6.1), until one or other phase is exhausted. In this case (beginning with m_1) the melt is used up first, and the final result is a mixture of olivine and enstatite: the melt never makes it to the eutectic. If, on the other hand, the initial melt had the composition m_2, more siliceous than enstatite, the reaction at R would transform all of the olivine into enstatite, with some melt left over. The disappearance of olivine releases the system from invariant equilibrium R, and the melt can proceed down the remaining liquidus curve, crystallizing enstatite directly until the eutectic is reached. The final result is a mixture of enstatite and silica (the high-temperature polymorph cristobalite). The proportions in the final mixture can be worked out by applying the Lever Rule to m_2.

During melting, this reaction relationship manifests itself as a phenomenon called **incongruent melting**. Pure enstatite, when heated, does not melt like olivine or anorthite but decomposes at 1557 °C to form olivine (less SiO_2-rich) and melt (more SiO_2-rich than itself), i.e. the reaction 2.6.1 run in reverse. The system is held in invariant three-phase equilibrium until the enstatite has been exhausted, then continues melting by progressive incorporation of olivine into the melt (cf. Figure 2.4).

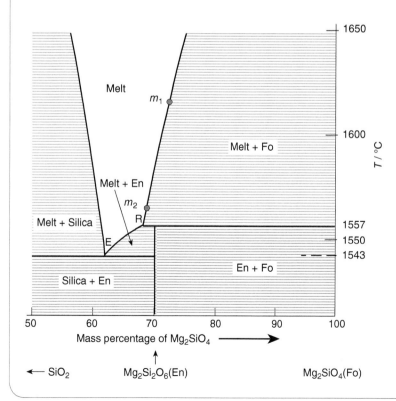

Figure 2.6.1 The Mg-rich part of the system Mg_2SiO_4–SiO_2 showing the reaction point R between Mg_2SiO_4 (forsterite) and SiO_2-rich melts (reaction 2.6.1).

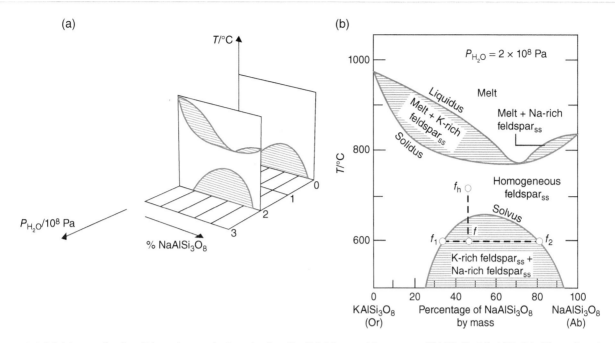

Figure 2.6 Melting and subsolidus phase relations in the alkali feldspars (the system $KAlSi_3O_8$–$NaAlSi_3O_8$). The subscript 'ss' denotes solid solution. (a) Perspective sketch of $P_{H_2O} - T - X$ space, showing the isobaric section at 2×10^8 Pa illustrated in (b). (b) Alkali feldspar phase relations at $P_{H_2O} = 2 \times 10^8$ Pa. Horizontal ruling represents two-phase fields.

The diagram shows the liquidus and solidus of the alkali feldspar series, which differ from those in Figure 2.5 only in that they fall to a minimum melting point in the middle of the series, rather than at one end. As a result we get *two* leaf-shaped fields instead of one. But interest is mainly in what happens in the *subsolidus region*. The 'homogeneous feldspar$_{ss}$' field immediately below the solidus means that here the end-members are completely miscible in the solid state: they form a complete solid solution in which any composition can exist as a single homogeneous phase. But at lower temperatures things get more complicated.

Beneath a boundary called the **solvus** a 'miscibility gap' appears. At these temperatures (for example 600 °C) the albite crystal structure is less tolerant of the $KAlSi_3O_8$ component (partly because of the large size of the potassium atom), and at a $KAlSi_3O_8$ content of about 20% ($f_2 = 80\%$ $NaAlSi_3O_8$) becomes **saturated** with it. Any $KAlSi_3O_8$ present beyond this limit is forced to exist as a separate $KAlSi_3O_8$-rich feldspar phase, whose composition can be found by extending a tie-line to the left-hand limb of the solvus, cutting it at f_1 (about 65% $KAlSi_3O_8$ or 35% $NaAlSi_3O_8$). This potassium feldspar is itself saturated with $NaAlSi_3O_8$.

A homogeneous alkali-feldspar solid solution such as f_h ceases to be stable as it cools through the solvus. At point f, for example, it is well inside the two-phase region. Such a point represents, at equilibrium, a mixture of *two* phases. The initially homogeneous feldspar therefore breaks down, or **exsolves**, into two separate phases f_1 and f_2. But solid-state diffusion is too slow to allow a cooling feldspar crystal to sort itself out into two separate crystals. The usual result of exsolution is a series of thin, lamellar domains of one phase enclosed within a host crystal of the other. The lever rule (Box 2.3) tells us that in the present example f_1 will be more abundant than f_2, and the cooling of crystal f_h will therefore produce a host crystal of composition f_1 containing *exsolution lamellae* of phase f_2. Such structures are characteristic of alkali feldspars, where they are known as **perthites**. Figure 2.7 shows a crystal of perthite viewed in a polarizing microscope configured to highlight this texture. The dark streaks are albite lamellae; the lighter host is orthoclase (divided into upper and lower portions that differ in shade owing to the **twinning** of the crystal). Exsolution textures analogous to perthite (although not given this name) are developed in some pyroxenes (Plate 3), owing to a similar miscibility gap between diopside ($CaMgSi_2O_6$) and enstatite ($Mg_2Si_2O_6$).

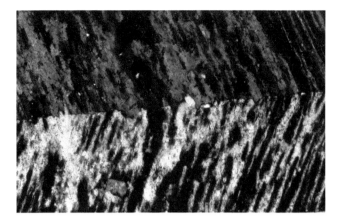

Figure 2.7 Part of a **twinned** crystal of **perthite** (orthoclase host enclosing darker **exsolution** lamellae of albite) seen under a polarizing microscope with crossed polars; width of field ~ 0.7 mm.

Application of the condensed phase rule (because P_{H_2O} is constant) to Figure 2.6b indicates that the sub-solvus region is a **univariant area**. In favourable circumstances, coexisting alkali feldspar (or pyroxene) compositions can be used to estimate the temperature of equilibration, a fact that has practical application in **geothermometry**.

Ternary phase diagrams

Merely to represent the range of compositions possible in a three-component system requires the use of the two dimensions of a piece of paper (Box 2.7). If we wish to represent comprehensively the phase relations of such a system over a range of temperature (in a form analogous to Figure 2.4), we would need to build a three-dimensional model. However, as a means of disseminating phase equilibrium data, such models would be less than convenient, so petrologists have devised various ways of condensing their content into two-dimensional form that can be printed on a page. Examples are shown in Figures 2.8 and 2.9, both of which relate to the crystallization of simple silicate melts, that is, to laboratory analogues of magmatic systems.

The base of any ternary phase diagram is an equilateral triangle, in which any composition in a ternary system can be plotted (Box 2.7). The temperature axis of the hypothetical model is constructed perpendicular to the plane of this triangle (see inset to Figure 2.8).

Ternary phase diagram with no solid solution

Figure 2.8 shows an example of a ternary phase diagram involving end-members that exhibit no mutual solid solution. The liquidus forms a 3-dimensional *surface* in this diagram, analogous to the liquidus *curve* in Figure 2.4. The surface takes the form of several curving 'hillsides' that meet along 'thermal valleys'. They can be portrayed either by a three-dimensional model constructed on a triangular base – see the inset in Figure 2.8 – or by plotting the model in the form of a two-dimensional 'map' with temperature 'contours' (isotherms) as shown in the main diagram. Note to begin with that the left-hand face of the model is equivalent to the Di-An binary phase diagram that has already been discussed in Figure 2.4, albeit shown here in a back-to-front sense. In the main diagram the liquidus surface consists of four sloping fields separated by gently curved boundaries. Each field represents a domain of melt composition within which one particular mineral crystallizes first (analogous to each curved line in Figure 2.4). The largest field is that in which forsterite is this 'liquidus phase'; the fields where diopside and anorthite crystallize first are significantly smaller.

Consider the course of crystallization followed by the initial melt composition labelled x. If at the outset it has a temperature above the liquidus (say 1500 °C), the first stage of its cooling path simply involves cooling to the liquidus temperature (about 1430 °C). There, because this composition lies within the forsterite field, olivine will begin to crystallize (see cartoon (a)). As no iron is present in this system, the olivine that forms will be pure forsterite (Mg_2SiO_4). Separating out forsterite in crystalline form depletes the remaining melt in the Mg_2SiO_4 component, a fact that we can represent on this diagram by drawing a line (dashed) from the Mg_2SiO_4 apex to the co-ordinates of x, then extending it beyond x (solid arrow). As cooling continues, the melt composition will migrate along this arrow, directly away from the forsterite apex. Eventually it reaches the boundary between the forsterite and diopside fields in Figure 2.8. At this boundary, called a *cotectic*, diopside begins to crystallize alongside forsterite. The crystalline extract that is removed from the melt from this point on will be a *mixture* of Di and Fo as in cartoon (b), the composition of which must lie somewhere on the bottom edge of the diagram (the domain of An-free Di–Fo mixtures), and extracting it will therefore drive the

Box 2.7 How a ternary diagram works

Three variables can be plotted in a two-dimensional graph if they add up to 100%. Any composition in a three-component system can therefore be represented in two dimensions, usually in the form of a *ternary diagram* plotted on special equilateral-triangle graph paper (Figure 2.7.1a). The user labels each apex with one of the components, as shown in Figure 2.7.1b.

Each apex represents 100% of the component with which it is labelled (the Di corner, for example, represents a composition consisting of 100% diopside). The side opposite represents compositions that are devoid of that component (in this case, a range of Ab–An mixtures). Lines parallel to this edge are contours representing different Di percentages from 0 (on the Ab–An edge) to 100% (at the Di apex).

To plot a composition such as Di 72%, Ab 19%, An 9% (note that the three co-ordinates must add up to 100%)[5], rule a line horizontally across the diagram at the position equivalent to 72% Di. Rule another line, parallel to the Di–An edge, at the position corresponding to 19% Ab (be sure to count this from the Di–An edge where Ab=0%). The intersection with the first line marks the composition being plotted. Note that only two readings need to be plotted; the third – being the difference from 100 – is not an independent variable, but it is useful to read it off the diagram (An=9%) to check that the point has been plotted accurately.

As with any plot showing mineral proportions, it is essential to indicate whether the numbers relate to mass % or molar proportions.

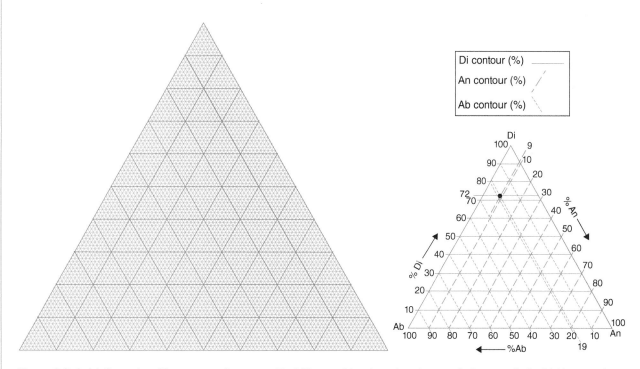

Figure 2.7.1 (a) Example of ternary graph paper with 100 mm sides (may be photocopied as needed). (b) How to plot a three-component mixture in a ternary diagram.

---

[5] The outcome of plotting three numbers that add up to *less than* 100% is a smaller triangle surrounding the target point, as illustrated in Gill (2010), Fig. B1, p. 363. In such cases, each value should be multiplied by 100/*Tot* (where *Tot* represents the initial total), thereby scaling them up to values that do sum to 100%.

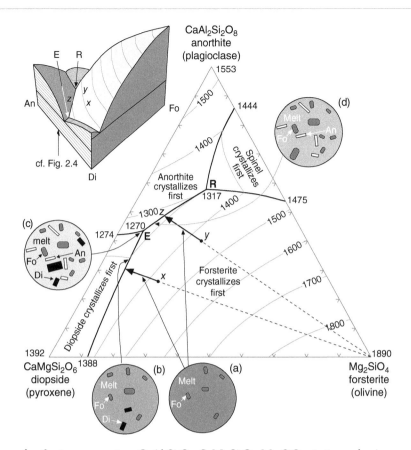

Figure 2.8 Phase diagram for the ternary system $CaAl_2Si_2O_8$–$CaMgSi_2O_6$–Mg_2SiO_4 at atmospheric pressure, after Osborne and Tait (1952). The liquidus surface is represented as isotherms (=temperature 'contours' graduated in °C). V-ticks mark 10% graduations in mass % of each component. 'x' and 'y' are illustrative initial melt compositions discussed in the text. Circular cartoons represent the 'phenocryst' mineralogy at key stages along each crystallization path, as seen under a microscope. E represents a **eutectic** point (see text) and R represents a *reaction point* (cf. Box 2.6). The inset sketch from Gill (2010) shows how the liquidus would appear as a 3D model (with temperature forming the vertical axis).

melt composition toward the An apex. In fact, because the Fo–Di boundary is a 'thermal valley', the melt composition migrates with falling temperature along the cotectic toward the point E – changing melt composition in any other direction would require a rise in its temperature. Continued removal of Di+Fo eventually drives the melt composition to the point E where the three fields meet. Here anorthite begins to crystallize alongside the forsterite and diopside (cartoon (c)).

Melt y will follow a different path as it crystallizes but will arrive at the same final melt composition. Initial crystallization of forsterite drives melt y to intersect the An-Fo cotectic, at which point anorthite will begin to crystallize alongside forsterite (see cartoon (d)). Diopside will only appear when the melt has reached point E (cartoon (c)). This point represents (a) the composition at which the liquidus surface reaches its lowest temperature, and (b) the composition toward which all melts

will converge as they crystallize, even melts whose initial compositions lie in the anorthite or diopside fields. It is called the *ternary eutectic* of this system.

Applying the Phase Rule to the composition y in Figure 2.8 at a temperature *above the liquidus* yields:

Point y	$\phi = 1$	melt alone
($T > T_{liquidus}$)		
	$C = 3$	$CaAl_2Si_2O_8 + Mg_2SiO_4 + CaMgSi_2O_6$
	$1 + F' = 3 + 1$	
Therefore	$F' = 3$	a trivariant equilibrium

The three degrees of freedom calculated here are (a) the two compositional coordinates required to define the position of y in the ternary diagram, plus (b) temperature, which (like point m in Figure 2.4) is unconstrained as long as the temperature remains above the liquidus.

When the temperature falls to meet the liquidus (about 1430 °C at point y), the variance of the equilibrium changes as olivine begins to crystallize:

Point y	$\phi = 2$	melt + forsterite
($T = T_{liquidus}$)		
	$C = 3$	$CaAl_2Si_2O_8 + Mg_2SiO_4 + CaMgSi_2O_6$
	$2 + F' = 3 + 1$	
Therefore	$F' = 2$	a divariant equilibrium

Melt composition y can coexist with olivine *only* at the liquidus temperature, so temperature (now dictated by that of the liquidus surface) ceases to be an independent variable and the variance decreases to 2.

When the melt evolves to point z, plagioclase begins to crystallize too:

Point z	$\phi = 3$	melt + forsterite + anorthite
	$C = 3$	$CaAl_2Si_2O_8 + Mg_2SiO_4 + CaMgSi_2O_6$
	$3 + F' = 3 + 1$	
Therefore	$F' = 1$	a univariant equilibrium

For this equilibrium to be maintained, the two compositional parameters can only vary in an *interdependent* way that confines z to the cotectic line, making this a univariant equilibrium.

When crystallization of forsterite and anorthite has driven the residual melt composition to point E:

Point E	$\phi = 4$	melt + forsterite + anorthite + diopside
	$C = 3$	$CaAl_2Si_2O_8 + Mg_2SiO_4 + CaMgSi_2O_6$
	$4 + F' = 3 + 1$	
Therefore	$F' = 0$	an isobarically invariant equilibrium

Here, with the melt in equilibrium with forsterite, anorthite and diopside, we have reached an invariant situation, the ternary analogue of the binary eutectic in Figure 2.4. Melt composition and temperature now remain fixed, and only the proportions of the various phases can vary: as heat is lost at a constant temperature of 1270 °C, melt crystallizes into forsterite, anorthite and diopside until no melt remains. This removes one of the four phases (melt) from consideration, leaving the three solid phases to continue cooling in a solid-state, univariant equilibrium.

A second invariant point R exists in this diagram, a reaction point similar to that discussed in Box 2.6. In the present context this complication can be ignored.

Ternary phase diagram with solid solution

Figure 2.9 shows another example of a ternary phase diagram relevant to magma crystallization. It incorporates along its edges two binary phase diagrams we have already discussed, Figures 2.4 and 2.5. The involvement of the plagioclase series (Figure 2.5) introduces solid solution, giving this diagram a different appearance to Figure 2.8. The gross features of the system can again be appreciated from a perspective sketch (Figure 2.9a) showing the topography of the liquidus surfaces as in Figure 2.8. In this diagram, however, these liquidus surfaces meet in a V-shaped low-temperature trough running out of the binary eutectic in the system $CaMgSi_2O_6$–$CaAl_2Si_2O_8$. The phase relations in this and the companion binary systems ($CaMgSi_2O_6$–$NaAlSi_3O_8$ and $CaAl_2Si_2O_8$–$NaAlSi_3O_8$) can be indicated on the vertical faces of the 'model'.

The main diagram (Figure 2.9b) is invaluable for examining the evolution of melt composition during crystallization, and considering the parallel magmatic evolution in real igneous rocks. The V-shaped valley divides the diagram into two fields, each labelled with the name of the solid phase that crystallizes first from melts whose compositions lie within that field. For example, a melt of composition a at 1300 °C will initially crystallize diopside. A line drawn from the $CaMgSi_2O_6$ apex to a, if extended beyond a, indicates the changes in melt composition caused by diopside crystallization. If the temperature continues to fall, the melt composition will eventually reach the boundary between the diopside and plagioclase$_{ss}$ fields (at point b), and here crystals of plagioclase begin to crystallize together with diopside. The boundary indicates the restricted series of melt compositions that can coexist with both diopside and plagioclase at the temperatures shown.

To work out the composition of the plagioclase that crystallizes from melt b requires the use of tie-lines. But one must remember that tie-lines are **isothermal** lines (since two phases in equilibrium must have the same temperature), and for this purpose it is appropriate to use a second type of diagram derived from the three-dimensional model. This is the *isothermal section* shown in Figure 2.9d. One can visualize this section as a horizontal slice through Figure 2.9a at a specified temperature (see Figure 2.9c). In principle an isothermal section can be drawn for any temperature for which phase

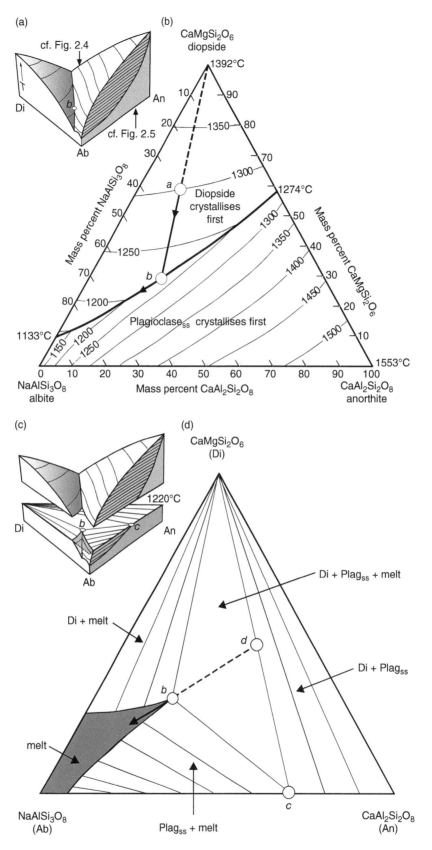

Figure 2.9 Various ways of representing crystallization in the **pseudo-ternary** system CaMgSi$_2$O$_6$–NaAlSi$_3$O$_8$–CaAl$_2$Si$_2$O$_8$. (a) A perspective sketch of the liquidus surface in three dimensions. Elevation and contours represent temperature. (b) A plan view of the liquidus surface, with topography shown by temperature contours (graduated in °C). (c) 3D sketch indicating the construction of the isothermal section shown in (d). (d) Isothermal section at 1220 °C. Tie-lines across the two-phase fields link coexisting phase compositions. The arrow (a tangent to the **cotectic** curve in Figure 2.9b) indicates the direction in which the melt composition b will evolve with further crystallization. This direction is controlled by the proportion in which diopside and plagioclase c crystallize (point d).

equilibrium data are available: this one relates to 1220 °C, the temperature of the liquidus at point *b* in Figure 2.9b. The area where the liquidus lies below the temperature of the section (shaded in Figure 2.9d) is a one-phase field where only melt is stable at 1220°C. The rest of the diagram can be regarded as the result of slicing the top off the solid model at this temperature (Figure 2.9c), revealing three 2-phase fields, each traversed by a family of tie-lines (summarizing the results of phase equilibrium experiments at 1220 °C). The composition of the plagioclase$_{ss}$ in equilibrium with the melt *b* at this temperature can be read from the diagram by following the tie-line from *b* to the $NaAlSi_3O_8$–$CaAl_2Si_2O_8$ edge of the diagram (point *c*).

Tie-line *b–c* forms one boundary of a *three-phase field* representing equilibrium between melt *b*, plagioclase *c* and diopside (composition $CaMgSi_2O_6$) at this temperature. Any point lying within this field signifies a physical mixture of these coexisting phases, the proportions of which could be worked out using the Lever Rule. Because all possible mixtures of diopside and plagioclase *c* lie to the right of melt *b* (along the line Di–c), the crystallization of these two minerals with cooling causes the melt composition to migrate leftwards, along the boundary – the **cotectic** – shown in Figure 2.9b. This direction is indicated by the arrow in Figure 2.9c. Point *d*, collinear with the arrow, indicates the proportion in which diopside and plagioclase (c) crystallize from the melt *b* (Exercise 5).

More detailed interpretation of such diagrams lies beyond the scope of this book. Further information can be found in the books by Morse (1980), Winter (2009) and Gill (2010).

Review

The great diversity of reactions and assemblages recorded in natural igneous and metamorphic rocks provides many avenues for investigating the conditions under which the rocks were formed. We have seen that, to analyse what such mineral assemblages mean in terms of pressure and temperature of formation, we can draw on two sorts of published experimental information. The primary source is the literature of *experimental petrology*, in which one can usually track down a number of phase diagrams relevant to the assemblages in a particular rock suite. Such diagrams, derived from well-established laboratory procedures

(Box 2.1) that are familiar to most petrologists, present phase equilibrium data in an easily understood form. But many diagrams of this kind refer to experiments on simple laboratory analogues rather than on the rocks themselves. The melts in Figure 2.9, for example, fall short of having true basaltic compositions, owing to the absence, among other things, of the important element iron. (One of the many consequences of this defect is that equilibria in Figure 2.9b are shifted to higher temperatures than would be found in a real iron-bearing basalt.) Thus diagrams like Figure 2.9, although invaluable for analysing general principles of phase equilibrium, do not reflect in quantitative detail the behaviour of more complex natural magmas and rocks. In some circumstances it can be helpful to carry out experiments on natural rock powders (see, for example, Gill, 2010, Figure 3.9) or comparable synthetic preparations, but the results cannot be directly displayed in simple phase diagrams and have less general application.

There are also useful applications in petrology for *thermodynamic* data (molar enthalpies, entropies and volumes). Using the Clapeyron equation and Le Chatelier's principle, we can predict certain features of phase diagrams without recourse to petrological experiment. Molar enthalpies and entropies of pure minerals are measured primarily by a completely different technique called *calorimetry*, involving the very accurate measurement of the heat evolved when a mineral is formed from its constituent elements or oxides. Such methods and data are less familiar to most geologists, and their successful application in solving petrological problems requires a command of thermodynamic theory beyond the scope of this book. Thermodynamics has, however, become one of the most versatile tools of metamorphic petrologists, enabling them to apply experimental data from simple synthetic systems to complex natural assemblages.

Further reading

Barker, A.J. (1998) *An Introduction to Metamorphic Textures and Microstructures*. Abingdon: Routledge.

Best, M.G. (2002) *Igneous and Metamorphic Petrology*. Oxford: Wiley-Blackwell.

Gill, R. (2010) *Igneous Rocks and Processes – a Practical Guide*. Chichester: Wiley-Blackwell.

Morse, S.A. (1980) *Basalts and Phase Diagrams*. New York: Springer-Verlag.

Yardley, B.W.D. (1989) *An Introduction to Metamorphic Petrology*. Harlow: Longman.

Winter, J.D. (2009) *Principles of Igneous and Metamorphic Petrology*, 2nd edn. Upper Saddle River, NJ: Pearson Education.

Sources of thermodynamic data for minerals

Berman, R.G. (1988) Internally-consistent thermodynamic data for minerals in the system $Na_2O–K_2O–CaO–MgO–FeO–Fe_2O_3–Al_2O_3–SiO_2–TiO_2–H_2O–CO_2$, *Journal of Petrology*, **29**, 445–522.

Holland, T.J.B. and Powell, R. (1998) An internally consistent thermodynamic data set for minerals of petrological interest. *Journal of Metamorphic Geology*, **16**, 309–43.

Exercises

2.1 Applying the Phase Rule to the reaction shown in Equation 2.5, discuss the variance of points X and Y in Figure 2.1.1a (Box 2.1). Identify the degrees of freedom operating at point X.

2.2 Why does ice floating on water tell us that the melting temperature of ice will be depressed at high pressures?

2.3 At atmospheric pressure (10^5 Pa), the following reaction occurs at 520 °C:

$$\underset{\substack{grossular \\ (a\ garnet)}}{Ca_3Al_2Si_3O_{12}} + \underset{quartz}{SiO_2} \rightarrow \underset{\substack{anorthite \\ (plagioclase)}}{CaAl_2Si_2O_8} + \underset{wollastonite}{2CaSiO_3}$$

Use the data below to plot a correctly labelled *P–T* diagram for pressures up to 10^9 Pa.

	Entropy S $JK^{-1} mol^{-1}$	Volume V $10^{-6} m^3 mol^{-1}$
Grossular ($Ca_3Al_2Si_3O_{12}$)	241.4	125.3
Quartz (SiO_2)	41.5	22.7
Anorthite ($CaAl_2Si_2O_8$)	202.7	100.8
Wollastonite ($CaSiO_3$)	82.0	39.9

2.4 Refer to Figure 2.5. Calculate the relative proportions of melt and crystals produced by cooling a melt of composition *m* to (a) 1400 °C, (b) 1300 °C, and (c) 1230 °C. What are the compositions of melt and plagioclase at these temperatures? (Assume that equilibrium is maintained throughout.)

2.5 Plot the following rock composition in the ternary system $CaAl_2Si_2O_8–CaMgSi_2O_6–Mg_2SiO_4$ (see Figure 2.8): plagioclase 42.5%, diopside 25.5%, nepheline ($NaAlSiO_4$) 15.0%, olivine 17.0%. (Note that nepheline does not appear in Figure 2.8.) What mineral would crystallize first from a melt corresponding to this composition?

2.6 Refer to Figure 2.9 and its caption. In what proportions must diopside and plagioclase crystallize from melt *b* to drive its composition along the cotectic curve (the arrow in Figure 2.9d)?

Calculate the compositions and proportions of the phases present in a solid mixture of composition *a* (Figure 2.9b). What would be the equilibrium assemblage for this mixture at 1220 °C? What would be the compositions and relative proportions of the phases present?

2.7 Figure 2.10 below shows the ternary eutectic in the system $CaMgSi_2O_6$ (Di)–Mg_2SiO_4 (Fo)–$Mg_2Si_2O_6$ (En) at a pressure of 20×10^8 Pa. To what depth in the mantle does this pressure correspond? What are the compositions (expressed as $Di_xFo_yEn_z$ where x, y and z are mass percentages) of the first melts to form as (a) the mixture M and (b) the mixture N are heated through the solidus?

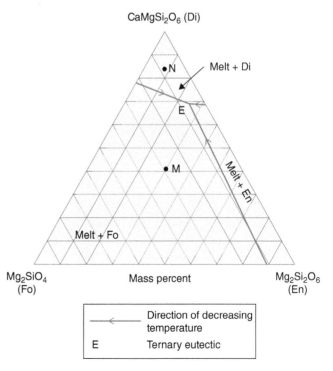

Figure 2.10 Simplified phase relations in the system $CaMgSi_2O_6$ (Di)–Mg_2SiO_4 (Fo)–$Mg_2Si_2O_6$ (En) at a pressure of 20×10^8 Pa (for Exercise 2.7).

3 KINETICS OF EARTH PROCESSES

In reading Chapter 2, it is easy to fall into the trap of supposing that reactions between minerals always proceed rapidly enough to reach chemical equilibrium during the time available, however short that may be. A little thought, however, indicates that this cannot be so. We have seen in Figure 1.3, for example, that aragonite is a **metastable** mineral at atmospheric pressure, and its survival in some outcrops of metamorphic rock crystallized at high pressure is a sign of *disequilibrium*: the rate of inversion to calcite has been too slow for the reaction to be completed before the processes of uplift and erosion exposed the rock at the surface. The same reasoning applies to occurrences in surface outcrops of the mineral sillimanite, which is stable only at elevated temperatures (Figure 2.1). Examination of igneous or metamorphic rocks in thin section often brings to light petrographic evidence of disequilibrium, in such forms as mineral zoning and corona structures (Box 3.1). From these examples it is clear that the *rate of progress* of geochemical reactions, and the way they respond to different conditions, are factors we cannot ignore.

The measurement and analysis of chemical reaction rates is called *chemical kinetics*, a science whose simpler geological applications are the subject of this chapter. Chemical kinetics provides the theoretical basis for understanding how (and why) reaction rates depend upon temperature, a matter of fundamental geological importance in view of the high temperatures at which magmas and metamorphic rocks crystallize. It also provides the algebra upon which radiometric (isotopic) dating methods are based (Box 3.2).

Chemical Fundamentals of Geology and Environmental Geoscience, Third Edition. Robin Gill.
© 2015 John Wiley & Sons, Ltd. Published 2015 by John Wiley & Sons, Ltd.
Companion Website: www.wiley.com/go/gill/chemicalfundamentals

Box 3.1 Disequilibrium textures

Mineral reactions that are unable to proceed to completion leave a rock in a state of chemical disequilibrum which, on the scale of a thin section, may be indicated by a variety of disequilibrium textures.

Coronas

Figure 3.1.1 shows a garnet crystal in a metamorphic rock (a metagabbro) which, as conditions changed during metamorphism, became unstable and began to react with adjacent quartz crystals to form a mixture of new minerals: plagioclase, magnetite (finely dispersed in the plagioclase) and pyroxene. The products of this metamorphic reaction are located at the periphery of the garnet crystal, in a zonal arrangement known as a **corona** structure. A corona records a mineral reaction that took place too slowly for it to proceed to completion (i.e. complete elimination of the garnet crystal) before changing conditions brought it to a halt. The rates of such reactions are controlled by diffusion rates and are therefore strongly temperature-dependent. What remains is a frozen-in disequilibrium texture.

It is often impossible (as here) to represent a corona reaction as a balanced chemical equation between the minerals observed, owing to the involvement of a fluid (or in some cases a melt) phase which can introduce and remove soluble reaction components without leaving any visible trace in Figure 3.1.1.

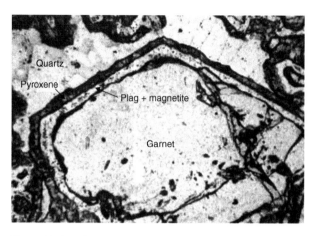

Figure 3.1.1 **Corona** structure rimming a six-sided garnet crystal in a metamorphosed gabbro. Opx and cpx stand for ortho- and clinopyroxene respectively. Width of field of view 4.5 mm. (Source: Carlson & Johnson, 1991, reproduced with permission of the Mineralogical Society of America.)

Reaction rims

Similar textures called **reaction rims** arise in various igneous rocks, due to reaction between early-formed crystals with later, more evolved melts. For example, olivine phenocrysts may become mantled by orthopyroxene owing to reaction with evolved melt richer in SiO_2 (Plate 1; see

Defining the rate of a reaction

It is easy to accept that some reactions proceed faster than others, but less easy to see how such differences can be expressed quantitatively. What precisely do we mean by the *rate* of a reaction?

Consider a simple chemical reaction, for example that between nitric oxide (NO) and ozone (a form of oxygen molecule comprising three oxygen atoms, O_3), two gaseous pollutants that occur in the troposphere as a result of the burning of fossil fuels. These **reactants** react with each other in equal molecular proportions to form the **products** nitrogen *di*oxide (NO_2) and ordinary oxygen (O_2), which are also gases:

$$\underset{gas}{NO} + \underset{gas}{O_3} \rightarrow \underset{gas}{NO_2} + \underset{gas}{O_2} \tag{3.1}$$

Imagine a laboratory experiment in which gaseous NO and O_3 are reacted together in a sealed vessel equipped with sensors that monitor the changing concentrations of NO, O_3, NO_2 and O_2 in the reaction vessel as the reaction progresses. (How these sensors work need not concern us.) The reaction consumes NO and O_3, whose concentrations (c_{NO} and c_{O_3}, each expressed in $mol\,dm^{-3}$) therefore decrease with time as shown in Figure 3.1. The concentrations of the products increase correspondingly as the reaction proceeds. The *rate* of the reaction at any stage is the gradient of the right-hand graph in Figure 3.1 at the moment concerned. Borrowing the symbolism of calculus (Appendix A):

$$rate = \frac{dc_{NO_2}}{dt} = -\frac{dc_{NO}}{dt} \tag{3.2}$$

also Box 2.6), or pyroxene may be rimmed by amphibole owing to reaction with hydrous late-stage melts at lower temperatures.

Zoning

In a rock that achieved complete chemical equilibrium between its phases at a given temperature, all mineral crystals would be homogeneous in composition. Igneous and metamorphic minerals are, however, quite commonly **zoned**. Zoning indicates that intracrystalline diffusion has failed to keep pace with changing external circumstances. Zoning in igneous minerals (see Plate 2) often reflects chemical evolution of the melt, with which only the rim of the growing crystal has maintained equilibrium (Figure 3.1.2). Zoning in igneous and metamorphic rocks may also be a response to changing physical conditions (*P*, *T*, etc.).

Exsolution

Perthites (Figure 2.7) and similar textures in pyroxenes (Plate 3) represent the solid-state decomposition of a homogeneous crystal into two immiscible phases (Figure 2.6). Intra-crystalline **exsolution** lamellae like these have a very large area of interface with the host crystal. The mismatch of structure across this interface generates a large positive interfacial energy, a situation undoubtedly less stable than equilibrium segregation into

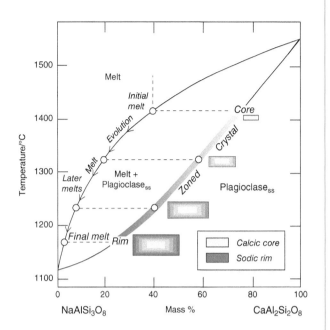

Figure 3.1.2 Formation of a zoned crystal of plagioclase through cooling and compositional evolution of melt.

separate crystals. The persistence of exsolution lamellae indicates that diffusion through the crystal was too slow to allow equilibration.

An example of how exsolution behaviour can be used to measure cooling rates is given in Box 3.5.

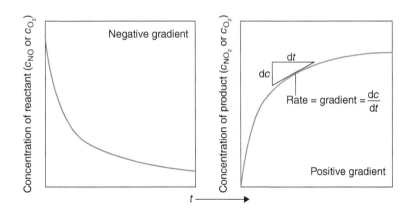

Figure 3.1 Composition–time curves for reaction 3.1.

Because one mole of NO_2 is *produced* for every mole of NO that is *consumed* by the reaction, the gradients of the left- and right-hand graphs in Figure 3.1 differ only in their sign (negative and positive respectively). This difference is represented by the minus sign in reaction 3.2.

Rate equation

If we repeated the experiment with double the concentration of NO present (c_{O_3} being initially the same as before), we would find that the initial rate is doubled.

Box 3.2 Kinetics of radioactive decay I: the Rb-Sr system

The radioactive isotope of rubidium, ^{87}Rb (Box 10.1), decays to the strontium isotope ^{87}Sr. The kinetics of this nuclear reaction can be treated in the same way as a chemical reaction:

$$^{87}Rb \rightarrow {}^{87}Sr + \beta^- + \bar{v}$$

$\quad$ 'parent' $\qquad$ 'daughter'
$\quad$ isotope $\qquad$ isotope

$\hfill (3.2.1)$

where the β-particle (β^-) and the antineutrino ($\bar{v}$) are released by the reaction.

The decay rate of any radioisotope is proportional to the number of the radioisotope ('parent') nuclei present in the sample (n_p) at the moment in question. This can be written as a rate equation:

$$\text{decay rate} = -\frac{dn_p}{dt} = \lambda n_p \hfill (3.2.2)$$

Because there is only *one* concentration term (n_p) on the right-hand side (unlike Equation 3.3), this is called a *first-order* reaction. λ is the rate constant analogous to k in Equation 3.3, but in this context it is called the *decay constant* (same algebra, different names).

The rate equation can be **integrated** to show how n_p varies with time:

$$n_p = n_p^0 \ e^{-\lambda t} \hfill (3.2.3)$$

where n_p^0 is the number of parent nuclei initially present (when $t = 0$). The decay profile for a short-lived isotope is shown in Figure 3.2.1a. If we transform both sides of Equation 3.2.3 into natural logarithms and rearrange things a bit, we get:

$$\ln\left(\frac{n_p^0}{n_p}\right) = \lambda t \hfill (3.2.4)$$

This equation is more useful because it is **linear** in relation to time (Figure 3.2.1b). For radioisotopes that decay rapidly, the decay constant can be determined by measuring the gradient of this graph (see Exercise 3.1).

The **half-life** $t_{\frac{1}{2}}$ of a radioisotope is the time it takes for n_p to decay to half of its original value ($n_p = \frac{1}{2} n_p^0$). Thus

$$\ln(2/1) = \lambda t_{\frac{1}{2}}$$

therefore $\quad t_{\frac{1}{2}} = 0.6931/\lambda \hfill (3.2.5)$

Because the decay of ^{87}Rb is very slow, the numerical value of λ is extremely small: 1.42×10^{-11} year^{-1}. During one year only about 14 out of every million million ^{87}Rb nuclei are likely to decay. The ^{87}Rb remaining in the Earth today has survived from element-forming processes (Chapter 11) that took place before the solar system was formed 4.6 billion years ago.

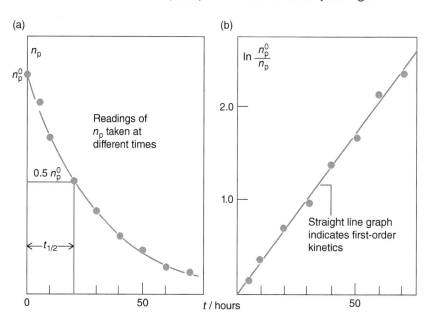

Figure 3.2.1 (a) The falling number n_p of radionuclei in a sample plotted against time. (b) The function $\ln[n_p^0/n_p]$ plots as a straight line against time; the gradient equals λ.

If we double the initial concentrations of both NO and O_3, we find the initial rate is quadrupled. This suggests that the rate is related to reactant concentrations:

$$\text{rate} = -\frac{dc_{NO}}{dt} = k \cdot c_{NO} \cdot c_{O_3} \qquad (3.3)$$

Equation 3.3 is called the *rate equation* for this reaction. Because it contains *two* concentration terms (c_{NO} and c_{O_3}), the reaction is said to have *second-order* kinetics. The constant k, whose numerical value is specific to this reaction (and to the temperature at which the experiment is run), is called the *rate constant*. The equation predicts that as the reactants are used up the rate will decline, which is consistent with the flattening out of the slopes in Figure 3.1.

The process of radioactive decay can be analysed in a similar manner (Boxes 3.2 and 3.3).

Heterogeneous reactions

Reactions like 3.1 that take place within a single phase (in this case a homogeneous gas mixture) are called **homogeneous reactions**. Nearly all reactions of geological significance, on the other hand, are **heterogeneous reactions,** involving the participation of two or more phases (minerals, melts, solutions …). Because they require the migration of components across the *interface* dividing one phase from another, the formulation of rate equations for heterogeneous reactions is much more complicated than for homogeneous reactions.

The most obvious consequence of involving two phases in a reaction is that the surface area of their interface becomes a variable in the rate equation. Interfacial surface area is determined chiefly by particle size. The surface area of a cube 1 cm across is 6 cm² (six sides each of 1 cm² area). Cutting the cube in half in each direction produces eight cubes, each 0.5 cm across and each having a surface area of $6 \times 0.5^2 = 1.5$ cm². The total volume of all the cubes together is unchanged (1 cm³) but the total surface area has increased from 6 cm² to $8 \times 1.5 = 12$ cm². Dividing the original cube into 1000 cubelets each of 0.1 cm size would increase the total area to 60 cm², while reducing to particle sizes equivalent to silt and clay sediments would increase their surface area to 3000 and 60,000 cm² respectively. Particle or crystal size, because it determines the area of contact between phases, has a profound effect on the rate of a heterogeneous reaction. This is one reason why diesel fuel injected as a fine spray into an engine reacts explosively with air, whereas the bulk liquid burns much more slowly.

The condition of the interface is also a very important factor. The rate of inversion of aragonite to calcite, for example, is greatly accelerated by the presence of traces of water along the grain boundaries. Surface chemistry has many important applications in the chemical industry and in mineral processing (for example, the use of a frothing agent to optimize the separation of ore minerals by flotation).

Mechanical factors come into play as well. When a solid dissolves in still water, the aqueous phase surrounding it becomes locally saturated, and this impedes further solution until diffusion has distributed the dissolved species more evenly. Dissolution of sugar in coffee can therefore be accelerated by the use of a teaspoon to promote homogenization, and natural forms of agitation can be correspondingly effective in the marine environment. Experiments show that the rate at which calcite dissolves in water can be represented like this:

$$\text{Rate} = kA\alpha^{\frac{1}{3}} \left\{ K^0 - \left(c_{Ca^{2+}}\right)^{\frac{1}{2}} \left(c_{CO_3^{2-}}\right)^{\frac{1}{2}} \right\} \qquad (3.4)$$

The c terms refer to concentrations of ions in solution, K^0 and k are constants, A is the total surface area of the calcite phase present, and α is the experimental stirring rate (which appears as the cube root for reasons that need not concern us). No doubt the effect of natural wave-agitation is still more complicated. This equation illustrates how rapidly the complexities multiply when even the simplest heterogeneous reactions are studied kinetically.

Temperature-dependence of reaction rate

Everyday experience tells us that chemical reactions, whether homogeneous or heterogeneous, *speed up* as the temperature is raised. Epoxy adhesives cure more quickly in a warm oven. Conversely, the very fact that we use refrigerators and freezers to preserve food indicates that biochemical reactions *slow down* at lower temperatures. Quantitatively the temperature effect which these examples illustrate is quite pronounced: many laboratory reactions roughly double

Box 3.3 Kinetics of radioactive decay II: the U–Th–Pb system

Each of the naturally occurring isotopes of the trace elements uranium (^{235}U and ^{238}U) and thorium (^{232}Th) decays through a complex series of intermediate radioactive nuclides to an isotope of lead (Pb). This is illustrated for the decay of ^{238}U to ^{206}Pb in Figure 3.3.1; similar flow diagrams can be drawn for the decay of ^{235}U to ^{207}Pb and for ^{232}Th to ^{208}Pb. Together these decay chains form the basis of U–Th–Pb radiometric dating.

Despite its complexity, the overall decay process in Figure 3.3.1 conforms to first-order kinetics, because the first step in the process (to ^{234}Th, a short-lived radioactive isotope of thorium) happens to be the slowest. The kinetic complexities of the subsequent branching decay series are immaterial because the rate of the whole process is controlled by this one *rate-determining step*, just as the flow of water from the end of a hose can be controlled by adjusting the tap supplying it. This phenomenon is not limited to radioactive decay: the kinetics of some complex chemical reactions are also controlled by a slow, rate-determining step.

For every uranium or thorium nucleus that decays to lead within the Earth, between 6 and 8 alpha particles are released. By capturing electrons the **α-particles** become ^4He atoms, which form the bulk of the helium flux escaping from the Earth's interior.

All but one of the nuclides involved in Figure 3.3.1 are radioactive solids, likely to be retained within the mineral hosting the original U. The sole exception is ^{222}Rn, one of the isotopes of the inert gas radon, whose mobility presents an environmental hazard in areas underlain by high-U-Th rocks like granites, as discussed in Box 9.9.

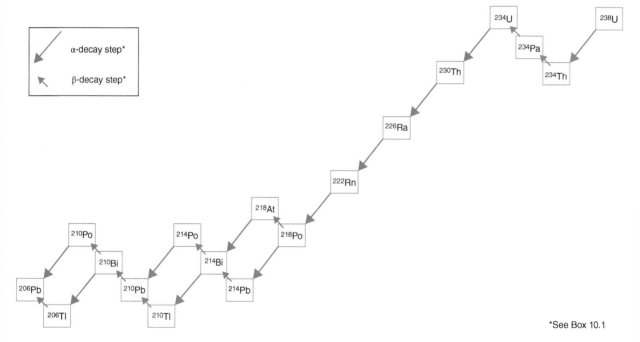

Figure 3.3.1 The chain of radioactive decay steps by which ^{238}U (also written 'uranium-238') decays to ^{206}Pb (lead-206).

their reaction rates when the temperature is raised by just 10°C (see Exercise 3.2 at the end of this chapter). The temperature-dependence of reaction rates is particularly significant for geological processes, whose environments can vary in temperature over many hundreds of degrees.

Most reaction rates vary with temperature in the manner shown in Figure 3.2a. The Swedish physical chemist Svante August Arrhenius[1] showed in the late 1880s that this behaviour could be represented

---

[1]   Arrhenius is also notable for being the first scientist to postulate (in 1896) a climatic 'greenhouse effect' arising from the presence of CO_2 in the atmosphere, and for recognizing even in the nineteenth century that mankind's burning of fossil carbon contributed to global warming.

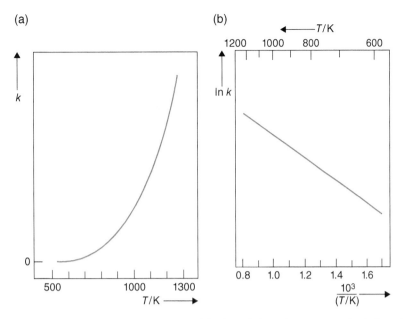

(a)

(b)

Figure 3.2 Variation of the rate constant with temperature. (a) k plotted directly against T (in kelvins). (b) k plotted against reciprocal temperature. T^{-1} has been multiplied by 10^3 to give a more convenient number scale. The reciprocal temperature scale is graduated directly in kelvins along the top margin. (This axis shows the recognized way of writing $10^3/T$ with T expressed in kelvins).

algebraically by expressing the rate constant k for a reaction in terms of an exponential equation:

$$k = Ae^{-E_a/RT} \qquad (3.5)$$

This has become known as the *Arrhenius equation* and it has a number of important applications in geology. R is the **gas constant** ($8.314\,\mathrm{J\,mol^{-1}K^{-1}}$) and T represents temperature *in **kelvins***. A and E_a are constants for the reaction to which the rate constant refers (varying from one reaction to another). A is called the pre-exponential factor, and it has the same units as the rate constant (which depend on the order of the reaction concerned). The constant E_a is called the *activation energy* of the reaction and has the units $\mathrm{J\,mol^{-1}}$.

Chapter 1 showed that physical and chemical processes often encounter an energy 'hurdle' that impedes progress from the initial, high-energy (less stable) state to the lower-energy configuration in which the system is stable (Figure 1.4). The hurdle is illustrated for a hypothetical chemical reaction in Figure 3.3. It arises because the only pathway leading from reactants to products necessarily passes through a higher-energy *transition state*. For a reaction involving the rupture of one bond and the establishment of a new one, this involves the formation of a less stable intermediate molecular species

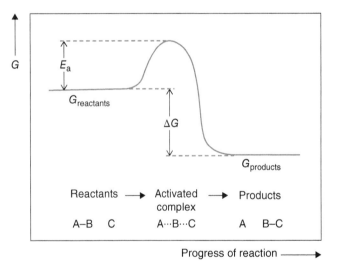

Figure 3.3 Energetics of a hypothetical reaction $AB + C \rightarrow A + BC$. The vertical axis represents the free energy of the system. E_a is the activation energy, which is released again (except in the case of endothermic reaction, when only a part is released) on completion of the molecular reaction.

(the *activated complex* in Figure 3.3). The activation energy E_a in the Arrhenius equation may be visualized as the 'height' of the hurdle (in free energy units) relative to the initial reactant assemblage. Box 3.4 explains how the activation energy can be understood on the atomic scale.

Box 3.4 What activation energy means on the atomic scale

In a chemical reaction (Figure 3.3):

$$AB + C \rightarrow A + BC$$

the established A–B bond must be weakened (stretched) before a new bond (B–C) can begin to form. The energetics of the A–B bond are shown in Figure 3.4.1a. The process begins with AB in its most stable configuration (internuclear distance = r^0_{A-B}). Energy is required to stretch the A–B bond to the stage when formation of the B–C bond becomes an equally probably outcome (i.e. sufficient to form the activated complex A $\cdots$ B $\cdots$ C); this energy input constitutes the activation energy E_a (Figures 3.3 and 3.4.1c). The whole reaction from AB to BC can be visualized by considering the energy–distance curves of both molecules, Figures 3.4.1a and b, 'back to back' as in Figure 3.4.1c. Note that the AB bond need not be completely broken before the BC molecule can begin to form.

The explanation of the activation energy in reactions between ionic compounds is slightly different, but is still associated with the need to disrupt one arrangement of atoms or ions before another more stable arrangement can be adopted.

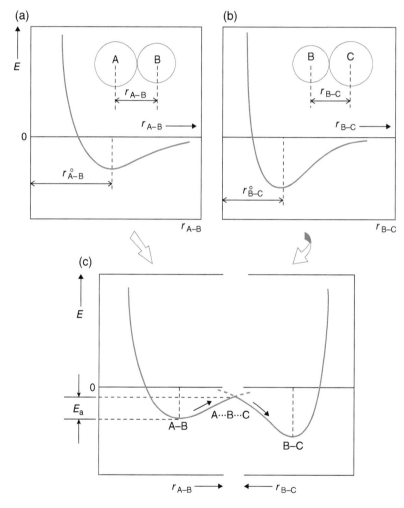

Figure 3.4.1 (a) and (b). Bond energy versus nuclear separation r_{A-B} and r_{A-B} for molecules AB and BC. (c) Bond energy and activation complex in the transition from AB + C to A + BC.

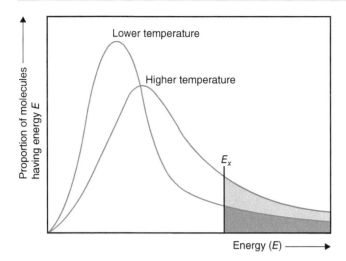

Figure 3.4 Molecular kinetic-energy distribution for two temperatures. The shaded areas show the portions of each distribution that lie above a specified energy E_x. This proportion is greater at the higher temperature.

For a collision between reactant molecules to lead to the formation of product molecules, it must derive sufficient thermal energy from the participants to generate the activated complex. From theoretical calculations one can predict the kinetic energy distribution among reactant molecules at a given temperature T: the results are illustrated for two alternative temperatures in Figure 3.4. One can show that the proportion of molecular encounters involving kinetic energies greater than some critical threshold energy E_x (shaded areas in Figure 3.4) is given by:

$$\text{Proportion of molecular collisions exceeding } E_X \propto e^{-E_x/RT} \qquad (3.6)$$

The term on the right is called the *Boltzmann factor*. It appears in the Arrhenius equation as a measure of the proportion of the reactant molecule collisions that possess sufficient energy (i.e. greater than E_a) at the temperature T to reach the transition state and thereby to complete the reaction. The form of the Boltzmann factor shows that an increase of temperature will shift the energy distribution towards higher energies (Figure 3.4), so that a greater proportion of reactant molecules can collide with energies exceeding E_a and surmount the energy hurdle, like water flowing over a weir. In other words the reaction rate will increase. But reducing the temperature will inhibit the reaction, because $-E_a/RT$ becomes a larger negative number

and makes the Boltzmann factor and therefore k smaller.

The value of the activation energy tells us how sensitive the rate of a reaction will be to changes in temperature, and measuring this temperature-dependence provides the means by which E_a can be determined experimentally. (Because it is not a net energy change between reactants and products like ΔH, E_a cannot be measured calorimetrically.) Transforming both sides of Equation 3.5 into natural logarithms ('ln' – see Appendix A), one gets:

$$\ln k = -\frac{E_a}{R} \cdot \frac{1}{T} + \ln A \qquad (3.7)$$

which has a **linear** form: $y = m \cdot x + c$ (see Appendix A). Thus if $\ln k$ ('y') is plotted against $1/T$ ('x', T being in kelvins), a straight line should be observed as shown in Figure 3.2b and discussed in Appendix A. The gradient of this line is $-E_a/R$. The numerical value of the activation energy can therefore be established by repeating the rate experiment at a number of pre-determined temperatures, and plotting the rate constants obtained in the form shown in Figure 3.2b. Such a diagram is called an *Arrhenius plot*. An Arrhenius plot for reaction 3.1 is illustrated in Figure 3.5a. A geochemical example is shown in Figure 3.5b. See also Exercise 3.2 at the end of the chapter.

Photochemical reactions

Is a rise in temperature the only mechanism by which reactant molecules can surmount the activation energy barrier? A number of gas reactions known to take place in the stratosphere show that energetic **photons** from the Sun provide an alternative means of energizing molecules into reacting. One example is the formation of stratospheric ozone (O_3):

$$O_2 \xrightarrow{\text{UV photon}} O\bullet + O\bullet \qquad (3.8)$$

The intense flux of solar ultraviolet (UV) radiation – see Box 6.3 – in the upper atmosphere causes a small proportion of oxygen molecules to disintegrate into separate oxygen atoms which, by virtue of their disrupted chemical bonds (represented by the symbol $O\bullet$), are highly reactive: they effectively sit on top of the energy barrier shown in Figure 3.3. Collision of either

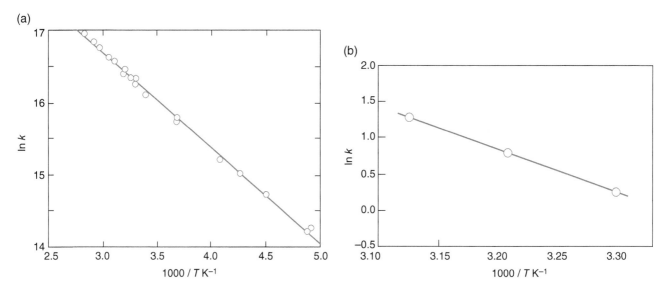

Figure 3.5 (a) Arrhenius plot for reaction 3.1. The rate constant k is expressed in $dm^3 mol^{-1} s^{-1}$. (b) Arrhenius plot for the photochemical oxidation of pyrite after here the rate k is expressed in units of $\mu mol\ m^{-2} min$. Note that temperature is expressed in kelvins (K) in each case. (Sources: After Zhang (2008). © Princeton University Press, based on data of Borders & Birks (1982); Schoonen *et al.* (2000). Reproduced with the authors' kind permission.)

of these oxygen **free radicals** with another oxygen molecule generates an ozone (tri-oxygen) molecule, as shown in reaction 3.9.

$$O\cdot + O_2 \rightarrow O_3 \qquad (3.9)$$

Chemical reactions like 3.8 that are initiated by energetic photons are described as **photodissociation** reactions. In kinetic terms they differ from conventional reactions in that their rate constants depend not on reactant temperature (as in Equation 3.5) but on the UV photon flux to which the reactants are exposed.

Solar UV radiation in a different wavelength range stimulates a reverse reaction:

$$O_3 \xrightarrow{\text{second UV photon}} O_2 + O\cdot \qquad (3.10)$$

This is the key stratospheric reaction that strongly attenuates the incoming UV flux in this wavelength range and protects living things on the Earth's surface from the most damaging effects of solar UV radiation. This filter is highly efficient: we often refer to the 'ozone layer' protecting us, yet remarkably (in view of the vital protection all living things on Earth derive from it) the steady-state ozone concentration in the stratosphere rarely exceeds 5 **ppm**.

Ozone also forms photochemically in the lower troposphere, but here the intensity of solar UV is much lower, and ozone formation can only occur when **catalysed** by nitrogen oxides arising from traffic pollution (a factor in the infamous 'photochemical smog' that afflicts many of the world's sunniest cities today, especially those at higher altitude like Mexico City).

Diffusion

Any process that requires the input of thermal energy to surmount an energy hurdle – a *thermally activated process* – will show an Arrhenius-type temperature dependence (Equation 3.5 and Figure 3.2). This property is characteristic of a number of physical processes in geology in addition to geochemical reactions.

When a component is unevenly distributed in a phase – whether solid, liquid or gas – so that its concentration in one part is higher than in another, random atomic motions tend to 'even out' the irregularities over time, leading to a net migration, called **diffusion**, of the component 'down' the gradient to regions of lower concentration. Given sufficient time, diffusion leads eventually to a homogeneous distribution (labeled t_∞ in Figure 3.6a). In fact, atoms diffuse

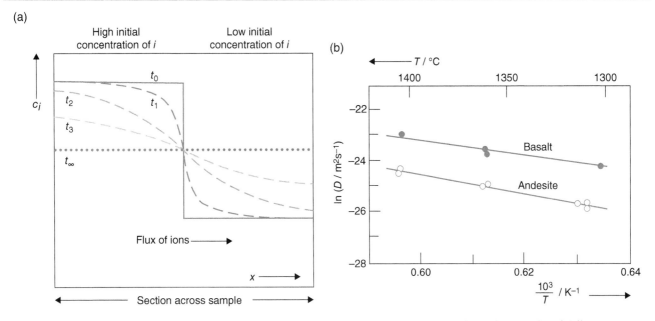

Figure 3.6 (a) The diffusion of ions in response to a non-uniform concentration distribution. (b) Arrhenius plot of diffusion coefficients for the diffusion of cobalt ions in silicate melts. The activation energies measured from the slopes of the two graphs are:

basalt: 220 kJ mol^{-1}
andesite: 280 kJ mol^{-1}

The units of D are m^{2}s^{-1}. Temperature is expressed in kelvins (K). (Source: Adapted from Lowry *et al.* (1982). Reproduced with permission from Springer Science and Business Media.)

randomly even through a homogeneous substance, but only where a concentration gradient exists is a *net* flow of chemical components observed.

Imagine a crystal with an abrupt internal discontinuity in the concentration of a component i, as shown in Figure 3.6a. Suppose the crystal is maintained at a constant temperature T, high enough for solid-state diffusion to occur at a significant rate. If we were to measure the distribution of component i on several successive occasions t_1, t_2, t_3, we would see the development of a progressively smoother concentration profile, leading eventually to a uniform distribution of i (at t_∞ in Figure 3.6a). These changes point to a net flux of component i from left to right through the plane of the original discontinuity. The magnitude of the flux in moles per second will depend on the surface area of this interface. One therefore expresses the flux f_i as the amount of component i (in **moles**) that migrates through a unit area of the plane per second, so the units are 'moles per square metre per second' (mol m^{-2}s^{-1}). Common sense suggests that the flux will depend upon the steepness of the concentration gradient $\dfrac{dc_i}{dx}$; so no net flux will occur if the concentration is uniform and the gradient zero, whereas a high flux will occur if the gradient is steep. The 19th century German physicist Adolf Fick backed up this hunch theoretically in 1855, showing that:

$$f_i = -D_i \frac{dc_i}{dx} \qquad (3.11)$$

This equation is known as *Fick's First Law of Diffusion*. The negative sign indicates that the direction of the net flux f_i is *down* the concentration gradient (i.e. towards the right in Figure 3.6a). The constant D_i is called the *diffusion coefficient* for the species i in the crystal concerned (at a given temperature). The units of concentration c_i are 'moles per cubic metre' (mol m^{-3}), so the concentration gradient dc_i/dx will have units of moles per cubic metre *per metre* in the x direction ([mol m^{-3}] m^{-1} = mol m^{-4}) from which it is simple to show that the units of D_i must be m^{2}s^{-1} (see Exercise 3.5).

Many experimental determinations of diffusion coefficients have been made for various elements in a range of silicate materials at various temperatures. Figure 3.6b shows how the measured diffusion coefficient for cobalt (D_{Co}) changes with temperature and composition in basalt and andesite melts. The data are presented in the form of an Arrhenius plot, in which they lie on straight lines. Thus one can express the temperature dependence of D in terms of an equation similar to Equation 3.5:

$$D = D_0 e^{-E_a/RT} \tag{3.12}$$

or in log form:

$$\ln D = -\frac{E_a}{R} \cdot \frac{1}{T} + \ln D_0 \tag{3.13}$$

These equations are identical to Equations 3.5 and 3.7, except that the pre-exponential factor is written D_0. Thus, although we think of diffusion as a physical phenomenon, it resembles a chemical reaction in being governed by an activation energy E_a. Like a person caught up in a dense crowd, the diffusing atom or ion must jostle and squeeze its way through the voids of the melt (Chapter 9), and pushing through from one structural site to the next presents an energy hurdle which only the more energetic ('hotter') atoms can surmount.

Measured diffusion coefficients differ from one element to another (Co as opposed to Cr, for example) for diffusion in the same material at the same temperature. Figure 3.6b shows that the diffusion coefficient for one element in a melt will also vary with the composition of the melt (Box 9.2).

Solid-state diffusion

From the point of view of diffusion, a crystalline solid differs from a melt in several important respects. Atoms diffusing through a silicate melt encounter a continuous, **isotropic** medium (D is independent of direction) that is a relatively disordered structure. Most crystalline solids, on the other hand, are polycrystalline aggregates that offer two routes for diffusion: within and between crystals.

Volume (intra-crystalline) diffusion
Volume diffusion through the three-dimensional volume of the constituent crystals is similar in general terms to diffusion through a melt. The main difference

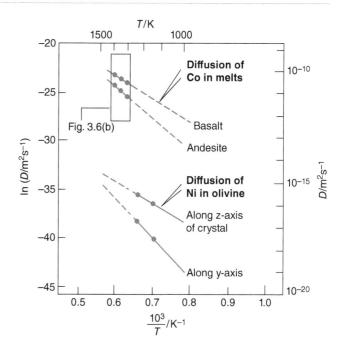

Figure 3.7 Comparison of volume diffusion rates in silicate melts and olivine crystals. (Source: Data from Henderson 1982).

is that the crystals have more closely packed, ordered atomic structures than melts (Box 8.3), and diffusion through them is much slower (lower D). Figure 3.7 contrasts the diffusion behaviour of similar metals in melts and in olivine crystals (in which the diffusion coefficients are lower by a factor of about 10^5). Notice that diffusion in olivine takes place more readily along the crystallographic z-axis than along the y-axis. Crystallographic direction is an important factor in diffusion through **anisotropic** crystals, reflecting the internal architecture of the crystals (Chapter 8).

Grain-boundary (inter-crystalline) diffusion
Grain-boundary diffusion exploits the structural discontinuity between neighbouring crystals as a channel for diffusion. This is much more difficult to quantify because the rate of diffusion depends upon:

(a) the grain size of the rock: in a fine-grained rock, the total area of grain boundaries is larger in relation to the total volume of the rock, and diffusion will be easier;
(b) the microscopic characteristics of the grain boundaries, e.g. the presence or absence of water.

Grain-boundary diffusion appears to be much more rapid in most circumstances than volume diffusion. Grain boundaries provide the main conduit through which volatile species penetrate the volume of a rock during metamorphism and hydrothermal alteration. Open fractures and faults do not occur in the deep crust owing to the high temperature and confining pressure. Under these conditions fluid migration is concentrated along *shear zones*, where deformation has led to a local reduction of grain size and often to a strong **foliation**, both of which promote diffusion and fluid movement because of the increase in grain-boundary area.

Box 3.5 illustrates how diffusion studies can be applied to estimate the cooling rates experienced by iron meteorites.

Melt viscosity

The flow of silicate melts is another kind of geological process where the rate varies strongly with temperature. Like treacle, silicate melts become less viscous as the temperature is raised.

A liquid flows in response to shear stress applied to it, which generates a velocity gradient in the liquid. The velocity of water flowing in a pipe of radius r, for example, increases from zero at the wall to a maximum v at the centre, a mean gradient of v/r. For most liquids the velocity gradient dv/dz is proportional to the applied shear stress σ:

$$\frac{dv}{dz} = \frac{\sigma}{\eta} \qquad (3.14)$$

The parameter η in Equation 3.14 is called the *viscosity* of the liquid. Because viscosity measures *resistance* to flow, it is like an inverse rate constant: a *low* viscosity indicates a runny liquid capable of rapid flow, whereas a *high* value signifies a 'stiff' or viscous one that flows only slowly.

As to units of measurement, shear stress σ is measured in $Pa = N\,m^{-2}$ (Appendix A, Table A2), and dv/dz is measured in $(m\,s^{-1})\,m^{-1} = s^{-1}$. Rearranging Equation 3.14 gives $\eta = \sigma/dv/dz$, so the units in which viscosity is measured are $Pa/(s^{-1}) = Pa\,s$. To provide a feel for how widely viscosities vary in nature, Table 3.1 provides some illustrative values.

Table 3.1 Illustrative viscosity values (in Pa s) for everyday liquids and for dry silicate melts

Substance	Viscosity at room temperature (~25 °C)	Melt viscosity above liquidus temperature
Water	0.001	
Motor oil SAE 50	0.5	
Egg white	2.5	
Treacle	20	
Basalt melt (dry)*		7.5–150
Smooth peanut butter	250	
Glazier's putty	~10^5	
Rhyolite melt (dry)*		~10^{10}

*Melt viscosity increases if suspended crystals or bubbles are present.

The viscosity of an erupted lava affects the style of a volcanic eruption, and its value depends upon both melt composition and temperature. Siliceous lavas like rhyolite have viscosities several orders of magnitude higher than basalts at their relevant liquidus temperatures (Table 3.1), as explained in Box 8.3. Figure 3.8 shows both this compositional effect and the temperature variation. By analogy with Figures 3.5 and 3.6b, viscosity is plotted in a logarithmic form that varies linearly with inverse temperature. Each of these straight lines can be described by an equation involving an activation energy E_a:

$$\frac{1}{\eta} = \frac{1}{\eta_0} e^{-E_a/RT} \qquad (3.15)$$

or using natural logarithms : $\ln\left(\frac{1}{\eta}\right) = -\frac{E_a}{R}\frac{1}{T} + \ln\left(\frac{1}{\eta_0}\right)$

$$(3.16)$$

The resemblance of Equation 3.16 to Equation 3.13 tells us that the flow of a silicate melt – like diffusion – requires atoms or molecules to jostle past each other, surmounting energy hurdles as they do so. The collective effect of these energy hurdles is expressed in the flow activation energy E_a, the value of which can be determined by measuring the gradient of each line ($= -E_a/R$). The activation energies for the flow of silicate melts (which Figure 3.8 shows to be, like viscosities themselves, higher for siliceous melts than for low-silica melts) are generally in

Box 3.5 Diffusion and cooling rate: applications to meteorites

At temperatures below 900°C, the continuous high-temperature solid solution between metallic iron and nickel divides into two solid solutions, separated by a two-phase region below a solvus (Figure 3.5.1a). Accordingly many iron meteorites exhibit complex exsolution intergrowths of two separate Fe–Ni alloys, kamacite and taenite, developed during cooling through this solvus (Figure 11.1.1).

Consider the cooling of a crystal of alloy X which at nearly 900°C is a homogeneous metallic solid solution. At about 700°C the alloy encounters the solvus, below which taenite becomes supersaturated with Fe, and expels it in the form of platelets of a separate kamacite phase, whose composition is indicated by drawing a **tie-line** to the other limb of the solvus (Ni ~ 6%). The upper inset shows a profile of the Ni content in a representative cross-section of the crystal (on the scale of a few mm) at 680°C: slender lamellae of low-Ni kamacite have appeared in the initially homogeneous taenite.

The Fe–Ni solvus is unusual in that both limbs slope in the same direction (cf. Figure 2.6). As the temperature falls, therefore, *both* immiscible phases grow more Ni-rich.

The lever rule (Box 2.3) indicates that their relative proportions must also change: the middle and lower insets show that kamacite lamellae grow at the expense of taenite, forming ever-thicker plates.

This process relies on the diffusion rates of Fe and Ni atoms in the two phases. Laboratory experiments show that diffusion rates are slower in taenite than in kamacite, and thus Ni is expelled from kamacite more rapidly than it can diffuse into the interior of the adjacent taenite crystal. Consequently the Ni distribution in taenite lamellae develops an M-shaped profile, with Ni concentrated at the edges. The more rapidly the meteorite is cooled, the more pronounced is the central dip in the Ni profile. Calculations based on diffusion profiles allow estimation – from the shape of the M-profiles in iron meteorites – of the cooling rates they experienced during the early development of the solar system (Figure 3.5.1b). These estimates, commonly between 1°C and 10°C per million years, suggest that iron meteorites are derived from relatively small parent bodies (diameter < 400 km); large planetary bodies would cool more slowly (Hutchison, 1983).

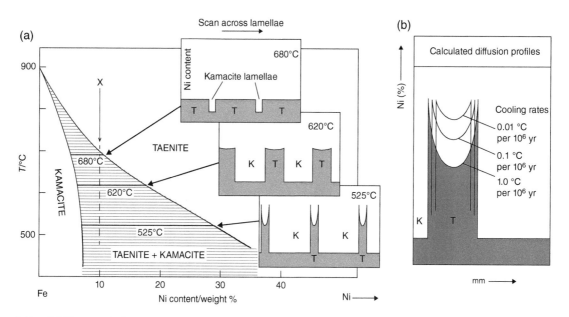

Figure 3.5.1 (a) Phase relations for Fe–Ni alloys showing the subsolidus fields of kamacite and taenite and the 2-phase field (ruled) between them (see Goldstein and Short, 1967). (b) Calculated diffusion profiles across a taenite platelet as a function of cooling rate, based broadly on Wood (1964). (Source: Adapted from Hutchison (1983)). See http://www.psrd.hawaii.edu/April07/irons.html

the same range as for diffusion (see captions to Figures 3.6 and 3.8).

The deformation of crystalline rocks, although a much more complicated process, shows a similar dependence on temperature:

$$\text{strain rate} = \left(Ae^{-E_a/RT} \right)\sigma^N \qquad (3.17)$$

where σ is again shear stress, and N is a constant depending on details of the deformation process (its value is typically about 8). The factor in brackets is analogous to inverse viscosity. It is not given that name because σ, instead of appearing in linear form, is raised to the power N. Activation energies are nonetheless in the same range as those for viscous flow of melts.

Persistence of metastable minerals: closure temperature

An equilibrium Earth would be extremely boring (e.g. there would be no life, no O_2 in the air, no plate tectonics, etc.). Disequilibrium is what makes the Earth so diverse and interesting. Without kinetic barriers there would be no geochemists to study kinetics or science, because all human beings, and in fact all life forms, 'should burst into flames'! (Zhang, 2008)

Paradoxically, chemical equilibrium in geological processes would be of less interest to the petrologist were it not for the intervention of disequilibrium. Consider an argillaceous rock undergoing metamorphic recrystallization in conditions corresponding to the kyanite–sillimanite phase boundary (Figure 2.1). Once equilibrium is achieved, kyanite and sillimanite will coexist stably in the rock. If, following uplift and erosion, the rock is found at the surface still containing kyanite and sillimanite, their coexistence and textural relationship will testify to the high-temperature conditions under which the rock crystallized and will give the petrologist an indication of the conditions of metamorphism. Yet the current state of the rock is plainly one of disequilibrium, as sillimanite is now well outside its stability field; it remains only as a **metastable** relic of a former state of equilibrium.

Metamorphic equilibrium seems generally to keep pace with changing conditions more effectively in prograde reactions (those involving a temperature increase) than during the waning stages of regional metamorphism. One reason is that volatile constituents such as

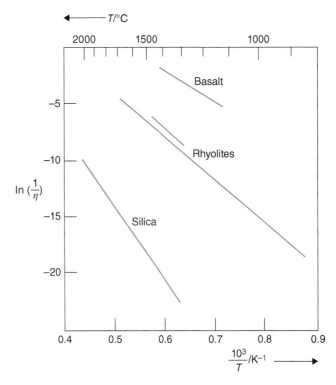

Figure 3.8 Arrhenius plot of viscosities (in $N\,s\,m^{-2}$) for some silicate melts; Viscosity is the inverse of the flow rate-constant so it has been plotted here in reciprocal form. Temperature is expressed in kelvins (K). The slope of the graphs increases with silica content, reflecting an increase in activation energy as follows:

Basalt melt:	$230\,kJ\,mol^{-1}$
Rhyolite melt:	$350\,kJ\,mol^{-1}$
Pure silica melt:	$500\,kJ\,mol^{-1}$

(Source: data from Scarfe (1978) and other authors cited therein).

water vapour, whose presence accelerates recrystallization, will be more abundant during the prograde stage owing to the dehydration reactions taking place. Once volatiles have been lost from the system, they are no longer available to facilitate the retrograde (falling-temperature) stages of metamorphism that follow. In the absence of volatiles it is likely that temperature asserts the dominating influence on metamorphic reaction rates.

As metamorphic or igneous rocks cool, they pass into a temperature range of increasing kinetic paralysis, in which reaction rates fall behind changing circumstances (as illustrated in Box 3.5). Eventually a temperature is reached (depending on the reaction being considered) at which diffusion rates become substantially slower that the cooling rate of the rock, and reaction has effectively ground to a halt. This is called the **blocking** or **closure temperature**.

The closure temperature is an important concept in geochronology. In the potassium–argon dating method (Chapter 10) the age of a rock is established by measuring the minute amount of ^{40}Ar (argon is a gas) that has accumulated in a potassium-bearing mineral from the decay of ^{40}K. The method therefore depends critically on the ability of the relevant mineral grains to imprison this intra-crystalline argon component that, being a gas, tends to escape. At temperatures above the closure temperature for argon diffusion, the ^{40}Ar atoms diffuse to grain boundaries and escape. A K–Ar age determination on an igneous rock therefore records not the age of intrusion or eruption, but the date when the rock had cooled to temperatures low enough for the rate of diffusion of ^{40}Ar out of the grains of the potassium mineral to be insignificant. A later phase of metamorphism in which the rock is again heated above the closure temperature would discharge the ^{40}Ar accumulated up to that point, and the isotopic clock would thereafter record a 'metamorphic age', indicating the time when the rock body had again cooled below the closure temperature.

Review

A chemical reaction can be visualized in free energy terms as a journey leading from one valley, the domain of the reactants, to another where the product species form the dominant population (Figure 3.3). The route from one valley to the other leads across a high pass in free energy space, the 'transition state', and reactant molecules that cannot summon sufficient energy to traverse this high ground will not reach their 'destination'. The key factor is the availability of energy: if energy is plentiful, the traffic over the pass will be heavy (the reaction rate will be high); if energy is hard to come by, many reactant molecules will be forced to turn back before attaining the pass.

What sources of energy drive geochemical reactions? For reactions taking place in the Earth's interior, reactants must rely on the kinetic energy possessed by their constituent atoms and molecules. Accordingly the temperature plays a crucial role in determining the rates of geochemical reactions, as enshrined in the Arrhenius equation. Reactions involving silicate minerals are characterized by large activation energies, and therefore only at high temperatures is there a significant population of reactant 'molecules' possessing the kinetic energy required to surmount the activation energy hurdle (Figure 3.4). Disequilibrium textures in many rocks (Box 3.1) testify to the rapid slowing-down of chemical reactions as temperature falls, and at surface temperatures *disequilibrium* is the rule rather than the exception: the persistence of Fe^{2+} minerals on the Earth's surface, where atmospheric oxygen makes Fe^{3+} the stable form of iron, is one obvious example. Both the flow of silicate melts (Figure 3.8) and diffusion are also strongly temperature-dependent (Figures 3.6 and 3.7), suggesting they too involve an activation step. The marked slowing-down of diffusion with falling temperature, and its effective cessation at the closure temperature, are essential requirements for radiometric dating.

Many reactions in the atmosphere, however, rely on another energy source, the Sun. Solar photons, particularly those of UV wavelengths, are energetic enough to tear apart chemical bonds in molecules like O_2, O_3, NO_2, H_2O_2, $CFCl_3$ and HCHO (formaldehyde). The result of such **photodissociation** reactions is often the formation of **free radicals** such as $O\cdot$, $HO\cdot$, $H\cdot$, $HO_2\cdot$, $NO_3\cdot$ $Cl\cdot$, $ClO\cdot$ and $HCO\cdot$. Free radicals, possessing unpaired electrons (represented by the symbol '$\cdot$'), are highly reactive: through photodissociation they have already reached the energy 'pass' and thus they are capable of initiating gas reactions in a manner unrestrained by considerations of activation energy. For example:

$$CH_4 + HO\cdot \rightarrow H_3C\cdot + H_2O \qquad (3.18)$$

This reaction incidentally illustrates the key role of the hydroxyl free radical $HO\cdot$ as an atmospheric cleansing agent, removing many forms of pollution (including the potent greenhouse gas methane, CH_4) from the air we breathe.

Further reading

Atkins, P.W. and de Paula, J. (2009) *Elements of Physical Chemistry*, 5th edn. Oxford: Oxford University Press.

Krauskopf, K.B. and Bird, D.K. (1995) *Introduction to Geochemistry*, 3rd edition. New York: McGraw-Hill.

Zhang, Y. (2008) *Geochemical Kinetics*. Princeton, NJ: Princeton University Press.

Exercises

3.1 The table below gives the Geiger counter count-rate at various times during an experiment on radioactive iodine-131 (^{131}I). Treating the decay of ^{131}I as a chemical reaction in which the count rate may be taken as being a measure of the concentration of ^{131}I, confirm that the reaction is first-order and determine the decay constant. From the linear graph you have drawn, calculate the half-life of ^{131}I (Box 3.2).

Time (hours)	Counts (sec^{-1})
0	18,032
25	16,410
50	15,061
100	12,590
200	8789
300	6144
400	4281
500	3002

3.2 There is a rule of thumb that says that the rate of many room-temperature chemical reactions approximately doubles when the temperature is increased by 10°C. Calculate the activation energy for a reaction that exactly conforms to this relationship. (Gas constant $R = 8.3143\,\mathrm{J\,K^{-1}\,mol^{-1}}$.)

3.3 Use the viscosity measurements given below to verify that the Arrhenius equation is applicable to the flow of a silicate melt, and determine the activation energy. (*Hint*: viscosity is the *resistance* to flow, to which flow rate is inversely proportional.)

Temperature (°C)	Viscosity of rhyolite melt (Ns m^{-2})
1325	2042
1345	1585
1374	1097
1405	741

(Gas constant as given in Exercise 3.2.)

3.4 Calculate the half-life of ^{87}Rb ($\lambda_{^{87}\mathrm{Rb}} = 1.42 \times 10^{-11}\,\mathrm{year^{-1}}$). What percentage of the ^{87}Rb incorporated in the Earth 4.6×10^{9} years ago has now decayed to ^{87}Sr?

3.5 Demonstrate that the units of diffusion coefficient D_i in Equation 3.11 are $\mathrm{m^2\,s^{-1}}$, when concentration c_i is expressed in $\mathrm{mol\,m^{-3}}$ and the net flux f_i is expressed in $\mathrm{mol\,m^{-2}\,s^{-1}}$.

4 AQUEOUS SOLUTIONS AND THE HYDROSPHERE

The great importance of water and aqueous solutions on the surface of the Earth barely needs pointing out. Water is the principal agent of erosion and of the transportation of eroded materials, either by mechanical or chemical means. The world's oceans are the primary medium for sedimentation, they act as a global chemical repository for many substances of geological significance, and they play a crucial part in moderating the climate and supporting life on the planet (not to mention the role water plays in living things). Water also has important functions in the Earth's interior: ore transport, rock alteration and metamorphism all involve the migration of hot aqueous fluids through the crust. The chemistry of aqueous solutions – the subject of this chapter – is a vital factor in all these geological processes.

A solution consists of two kinds of constituent which are given confusingly similar names:

- **solute** refers to the dissolved species (e.g. the salt in a saline solution); a solution may contain several solutes, such as when sodium chloride *and* potassium nitrate are dissolved in the same solution. Seawater is a complex example of such a mixed solution.
- **solvent** refers to the medium in which the solutes are dissolved. *Aqueous* solutions are those in which water is the solvent (Latin *aqua*: water).

The basic jargon of solution chemistry is reviewed in Appendix B and key terms are defined in the Glossary. Some of the important properties of water as a solvent are reviewed in Box 4.1.

Chemical Fundamentals of Geology and Environmental Geoscience, Third Edition. Robin Gill.
© 2015 John Wiley & Sons, Ltd. Published 2015 by John Wiley & Sons, Ltd.
Companion Website: www.wiley.com/go/gill/chemicalfundamentals

Box 4.1 The special properties of water

Water is so commonplace that one tends to overlook how unusual its properties are in comparison with other liquids. The water molecule is bent (Figures 4.1.1 and 7.9a,b), with both hydrogen atoms occurring on the same side. As oxygen attracts electrons more strongly than hydrogen, a slight excess of electron (negative) charge $\delta-$ gathers at the oxygen side of the molecule, leading to a corresponding deficiency (a net positive charge $\delta+$) on the two hydrogen atoms at the other side (Figure 4.1.1). This *polarity* is responsible for most of the properties peculiar to water:

(a) Molecules of water in the liquid and solid states are loosely bonded to each other by an electrostatic attraction between the negative end of one molecule and the positive end of a neghbouring one. This *hydrogen bonding* (Chapter 7) lies behind many of water's unique properties.

(b) Hydrogen bonding gives liquid water a very high **specific heat**, a high **latent heat of vaporization** (because water molecules are more difficult to separate and disperse into a gas phase), and an unusually large liquid-state temperature range (100 °C). These thermal properties of water give it a heat-exchange capacity of great climatic significance on the Earth, as the moderating effect of the oceans on seaboard climate illustrates. It takes only 2.5 m depth of ocean water to match the heat capacity of the entire atmospheric column above.

(c) Hydrogen bonding in the liquid state is also responsible for making liquid water denser than ice at temperatures close to 0 °C (for the consequences of which, see Box 2.2). Water therefore *expands* on freezing, leading to the important erosional processes of frost-shattering and frost-heaving (and the familiar fact that ice floats on water). Water also has an unusually high surface tension, resulting in strong capillary penetration of pore waters.

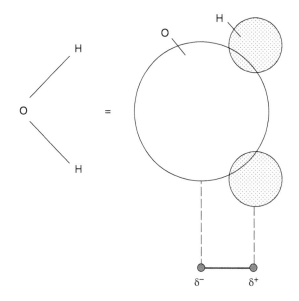

Figure 4.1.1 The geometry and polarity of the water molecule.

(d) The polar nature of the water molecule makes water an attractive environment for ions, and gives water its unique capacity as a solvent of ionic compounds. Polar water molecules, attracted by the electrostatic field of each ion, cluster in a loose layer around it, aligning their polarity according to the ion's charge. This association of water molecules around a dissolved ion (an *ion dipole interaction* – Chapter 7) is called **hydration**. The combined electrostatic attraction of the polar water molecules lowers the ion's potential energy and stabilizes it in solution. Many ionic compounds remain hydrated when they crystallize from solution, forming hydrated salts such as the mineral gypsum, $CaSO_4 \cdot 2H_2O$. Water is thus the prime example of a *polar solvent*.

(e) Conversely, water is much less effective as a solvent of non-polar organic substances.

Ways of expressing the concentrations of major constituents

Solutions

The composition of a solution can be expressed by stating the **concentration** of each of the solute species present (the amount of solvent present is so nearly constant that it is rarely of interest). There are several ways of expressing concentration. The obvious one is to state the mass of each solute per unit volume of solution,[1] in units such as $g\,dm^{-3}$, $mg\,dm^{-3}$ or $\mu g\,dm^{-3}$. But in considering the properties of the solution and the reactions

---

1 The **SI** unit of volume used in solution chemistry is the cubic decimetre (dm^3), which is equal to a litre (L). See Appendix A.

that take place in it, it is the number of **moles** of each major solute present that matters, rather than the weight. Dividing the concentration C of each compound in the solution (expressed in $g\,dm^{-3}$) by its **relative molecular mass** M_r gives the **molarity**, c:

$$c = C/M_r \text{ mol dm}^{-3} \qquad (4.1)$$

Aqueous geochemists measure solute concentrations using a slightly different quantity called the **molality**, m. The molality m_i of species i in an aqueous solution is the number of moles of i *per kilogram of water* (solvent, not solution). There are sound practical and thermodynamic reasons, however, for expressing molality in a **dimensionless** form (one in which the units of measurement cancel out). For this purpose we use the **activity** a_i, which for **dilute solutions** is numerically equal to the molality:

$$a_i = m_i/m^{\ominus} \qquad (4.2)$$

where $m^{\ominus}$ represents what is called the standard molality, equal to $1\,mol\,kg^{-1}$. This artifice removes the necessity of associating units with complex formulae like Equation 4.11.

Solids

In geology, one often deals with heterogeneous reactions between aqueous solutions and solid or gaseous phases. The concentration of a component in a solid phase can be expressed in a variety of ways, such as the mass percentage of each element, or the mass percentage of the oxide of each element (Box 8.4). When considering equilibrium between solids and solutions, however, we want to express solid compositions in **molar** terms, of which **mole fraction** (X) is the most commonly used form.

Consider a sample of olivine that has been found to contain 65.0% by mass of Mg_2SiO_4 (the forsterite end-member) and 35.0% of Fe_2SiO_4 (the fayalite end-member). To calculate the mole fractions of these components (assuming no other components are present), one must first establish their relative molecular masses M_r from the following **relative atomic masses**:

$$Mg = 24.31 \quad Si = 28.09 \quad Fe = 55.85 \quad O = 16.00$$

The relative molecular mass (RMM) of forsterite is therefore $(2 \times 24.31) + 28.09 + (4 \times 16.00) = 140.71$, and

therefore 1 mole of Mg_2SiO_4 will weigh $140.71\,g$. The RMM of fayalite is $(2 \times 55.85) + 28.09 + (4 \times 16.00) = 203.79$. Thus $100\,g$ of the olivine will contain:

(a) $65.0\,g$ of $Mg_2SiO_4 = 65.0/140.71 = 0.4619$ moles Mg_2SiO_4;
(b) $35.0\,g$ of $Fe_2SiO_4 = 35.0/203.79 = 0.1717$ moles Fe_2SiO_4.

The mole fraction of Mg_2SiO_4 in this olivine is therefore:

$$X_{Mg_2SiO_4} = \frac{\text{moles of } Mg_2SiO_4}{\text{total moles}} = \frac{0.4619}{0.4619 + 0.1717} = 0.7290$$

$$(4.3a)$$

Likewise the mole fraction of Fe_2SiO_4 in the olivine is:

$$X_{Fe_2SiO_4} = \frac{\text{moles of } Fe_2SiO_4}{\text{total moles}} = \frac{0.1717}{0.4619 + 0.1717} = 0.2710$$

$$(4.3b)$$

In this case, checking whether the mole fractions add up to 1.0000 is a useful way to confirm that no arithmetical error has been made. Note that a mole fraction is a **dimensionless number**.

Gases

A gas is a high-**entropy** state of matter that expands to fill the volume available (Box 1.3). The concentrations of individual gaseous species in a gas mixture can be expressed in three alternative ways:

- **mole fraction** X_i, expressed either as a fraction of 1.0000 as above for solids, or in molar **ppm** (moles of component i per 10^6 moles of gas mixture);
- volume per cent (Figure 9.3) or parts per million by volume (ppmv = number of m^3 of pure gas i present in $10^6\,m^3$ of the gas mixture);
- the **partial pressure** p_i of component i present in the gas mixture (in Pa – see Appendix D).

These three measures are related through the *Ideal Gas Law*:

$$PV = nRT \qquad (4.3c)$$

where P is the pressure exerted by the gas mixture on the walls of its container, V is the volume of the container, n is the total number of moles of gas present in the container, R is the **gas constant**, and T is the temperature in **kelvins**. Both the volume (at constant pressure) and the pressure (at constant volume) of a gas are proportional to the *number of moles* present n, regardless of their mass (which is why mass units are not used for reporting gas composition). For an ideal gas mixture at atmospheric pressure ($\approx 10^5$ Pa), the following *numerical* relationships hold:

$$\text{Mole fraction of component } i = \frac{\text{volume percent of } i}{100}$$
$$= \frac{\text{partial pressure of } i \text{ in Pa}}{10^5 \text{ Pa}}$$

Equilibrium constant

In this chapter we are concerned with many reactions in solution that can proceed in either direction, depending on the circumstances. Consider a reaction:

$$A + B \rightarrow C + D \tag{4.4}$$
reactants   *products*

By analogy with Equation 3.5, we would expect the rate of this reaction in solution to be represented by:

$$\text{Rate} = k_{AB} a_A a_B \tag{4.5}$$

where a_A and a_B are the activities of reactants A and B in the solution (Equation 4.2), and k_{AB} is the rate constant for the A + B reaction. The presence in solution of C and D, produced by this 'forward' reaction, is likely to initiate a *reverse reaction* that regenerates A and B:

$$A + B \leftarrow C + D \tag{4.6}$$

$$\text{Rate of reverse reaction} = k_{CD} \, a_C \, a_D \tag{4.7}$$

This tells us that as the products of the forward reaction (C and D) build up in the solution, the reverse reaction will speed up, and eventually a *steady state* will be reached in which the rate of the forward reaction is matched by that of the reverse reaction, and hereafter no *net* change in activities takes place. This state of *equilibrium* is written:

$$A + B \rightleftharpoons C + D \tag{4.8}$$

The relative proportions of A, B, C and D present in solution when this steady state has been achieved are summarized by the **equilibrium constant** K:

$$K = \frac{a_C a_D}{a_A a_B} \tag{4.9}$$

The value of K is a constant characteristic of a particular equilibrium at a specified temperature. It indicates at which point on the path from '100% reactants' (A + B) to '100% products' (C + D) the reaction settles into equilibrium. The system, if disturbed, will always readjust to the same position of equilibrium as defined by Equation 4.9. For example, if more A is added to the equilibrated solution represented by Equation 4.8 (i.e. if a_A is increased), A and B will react together to a greater extent, thereby reducing a_B (and the new a_A) and increasing a_C and a_D. These individual activity adjustments occur in such a way that the ratio $a_C a_D / a_A a_B$ resumes the original value, K, the value it had before the disturbance occurred. Altering the temperature, however, will generally cause the value of K to change (Box 4.2).

The algebraic form of the equilibrium constant in equations like 4.9 reflects the nature of the reaction equation. If we were to look at a more complicated equilibrium, for example one in which b molecules of species B react with c molecules of species C like this:

$$bB + cC \rightleftharpoons dD + eE + fF \tag{4.10}$$

the equilibrium constant would normally have the form:

$$K = \frac{a_D^{\,d} \cdot a_E^{\,e} \cdot a_F^{\,f}}{a_B^{\,b} \cdot a_C^{\,c}} \tag{4.11}$$

Solubility and the solubility product

To see how an equilibrium constant works, consider the solubility of various substances in water. The **salt** calcium fluoride, CaF_2 (which, in crystalline form, is the familiar vein mineral *fluorite*), is only very slightly soluble in cold water: only 0.017 g of CaF_2 will dissolve in 1 kg of water at 25 °C, corresponding to a molality of 0.00022 mol kg^{-1}. (The relative molecular mass of CaF_2 is $40.08 + (2 \times 19.00) = 78.08$.) This quantity is the *solubility* of CaF_2 at this temperature. Adding further solid CaF_2 to the system will cause no increase in the concentration in solution, no matter how long we wait. Such a solution, which has reached equilibrium with solid

Box 4.2 The effect of temperature on equilibrium constants

Like the equilibria considered in Chapter 2, the positions of aqueous equilibria vary with temperature (and to a much smaller extent with pressure). An equilibrium constant is therefore only constant for a particular set of physical conditions. The temperature variation is important, considering the wide range of temperatures at which aqueous solutions are encountered in the geological world (from 0 °C to hundreds of degrees).

Raising the temperature of a system in equilibrium will alter the equilibrium constant in a direction that favours the higher enthalpy side of the equilibrium (Le Chatelier's principle – Chapter 2). Consider for example the solubility of a salt like $BaSO_4$:

$$\underset{\text{Solid}}{BaSO_4} \rightleftharpoons \underset{\text{Saturated solution}}{Ba^{2+} + SO_4^{2-}} \qquad (4.2.1)$$

This reaction has a positive ΔH (the reaction is endothermic). As enthalpy is higher on the right-hand side, Le Chatelier's principle (Chapter 2) indicates that a rise in temperature will increase the equilibrium constant (Equation 4.17), shifting the equilibrium to the right, in accordance with the everyday experience that salts are more soluble in hot water than cold.

During the 1980s a programme of exploration in the Pacific Ocean using the manned submersible *Alvin* located active, chimney-shaped hydrothermal vents, which discharge jets of water at temperatures up to 350–400 °C.[2] Such vents have since been identified on mid-ocean ridges in all of the major oceans. The hot, mildly acid, oxygen-free fluids discharged from them are saturated with metal sulfides and other salts, which are immediately precipitated upon coming into contact with cold, neutral, oxygen-bearing seawater (see Figure 4.2.1). The fine sulfide precipitates form dense black clouds in the water, so

Figure 4.2.1 A black smoker issuing from a chimney in the Logatchev Hydrothermal Field of the Mid-Atlantic Ridge. (Source: Reproduced with permission of GEOMAR Helmholtz Centre for Ocean Research, Kiel, Germany.)

these vents have become known as 'black smokers' (Edmond and von Damm, 1983).

Gases, on the other hand, become *less* soluble at higher temperatures. The expulsion of dissolved CO_2 as water is heated causes the deposition of the familiar 'fur' inside kettles and heating pipes, particularly in **hard**-water areas. 'Hard' waters contain calcium bicarbonate which, with the expulsion of dissolved CO_2 as the water is heated, forms the less soluble calcium carbonate:

$$\underset{\substack{\text{hard water} \\ \text{(cold)}}}{Ca^{2+} + 2HCO_3^-} \rightarrow \underset{\substack{\text{solid} \\ \text{'fur'}}}{CaCO_3} + \underset{\substack{\text{water} \\ \text{(hot)}}}{H_2O} + \underset{\substack{\text{expelled} \\ \text{gas}}}{CO_2} \qquad (4.2.2)$$

Some solids also have inverse solubility/temperature relationships, for example, the anhydrite ($CaSO_4$) from which black smoker chimneys are initially built. Anhydrite is deposited from seawater as it is heated by the hot vent.

---

[2] The highest vent discharge temperature so far discovered is 464 °C (Koschinsky *et al.*, 2008).

CaF_2 left over, is said to be **saturated** with this component.

When dissolved in aqueous solution, calcium fluoride – being an ionic compound (Chapter 7) – is completely dissociated into calcium and fluoride **ions** (Ca^{2+} and F^- respectively). The calcium atom loses two electrons and becomes a doubly charged positive ion (called a **cation**). Each fluorine atom (of which there are twice as many as calcium) captures one extra electron to become a singly charged negative ion or **anion**. The dissolution of solid fluorite can therefore be written as a chemical reaction leading to an equilibrium:

$$\underset{\text{solid}}{CaF_2} \rightleftharpoons \underset{\text{solution}}{Ca^{2+} + 2F^-} \tag{4.12}$$

(Because that is a **heterogeneous equilibrium** like those considered in Chapter 2, one needs to specify in which phase each reactant or product resides when writing the reaction.) As dissolution proceeds, the concentration of Ca^{2+} and F^- ions in solution builds up, and increasingly the ions will react with each other to produce solid CaF_2 again (the reverse reaction). Saturation of the solution is an equilibrium in which the gross rate of dissolution of solid CaF_2 is equalled by the rate of precipitation from solution, so that no further *net* change is observed. One can formulate an equilibrium constant for reaction 4.12 (cf. Equation 4.11):

$$K_{CaF_2} = \frac{a_{Ca^{2+}} \cdot \left(a_{F^-}\right)^2}{X_{CaF_2}} \tag{4.13}$$

Because the solid phase is pure CaF_2, the **mole fraction** $X_{CaF_2} = 1.00$. Thus:

$$K_{CaF_2} = a_{Ca^{2+}} \cdot \left(a_{F^-}\right)^2 \tag{4.14}$$

where $a_{Ca^{2+}}$ and a_{F^-} are the activities of Ca^{2+} and F^- ions in a solution saturated with CaF_2.

This kind of equilibrium constant provides an alternative way of expressing the solubility of a slightly soluble salt (in this case CaF_2) in water. It is called the *solubility product* of CaF_2. It conveys the same information as the solubility, but in a different – and more versatile – form.

The value of K_{CaF_2} can be calculated from the solubility data given above. Reaction 4.12 tells us that one mole of solid CaF_2 dissolves (in sufficient water) to yield one mole of Ca^{2+} ions and two moles of F^-ions. Thus if 0.00022 moles of CaF_2 can saturate 1 kg of water at $25\,°C$, the activities of Ca^{2+} and F^- in the saturated solution will be 0.00022 and 0.00044 respectively.

Therefore:

$$\begin{aligned} K_{CaF_2} &= a_{Ca^{2+}} \cdot \left(a_{F^-}\right)^2 \\ &= 0.00022 \times (0.00044)^2 \\ &= 4.26 \times 10^{-11} \end{aligned}$$

Given the small size of this number, it is often more convenient to write K_{CaF_2} in logarithmic form. Since $\log(4.26 \times 10^{-11}) = -10.37$:

$$\log_{10} K_{CaF_2} = -10.4 \tag{4.15a}$$

which is equivalent to writing $K_{CaF_2} = 10^{-10.4}$ (4.15b)

or, by analogy with **pH** notation (Appendix B),

$$pK_{CaF_2} = 10.4 \tag{4.15c}$$

Solubility products are available in published tables (see references at the end of this chapter). Like all equilibrium constants, they vary with temperature (Box 4.2).

If, in a solution of CaF_2, the observed ion activity product

$$a_{Ca^{2+}} \cdot \left(a_{F^-}\right)^2$$

has a value (e.g. 10^{-12}) that is numerically smaller than the solubility product K_{CaF_2} for the appropriate temperature, the solution is not saturated with CaF_2. Any solid CaF_2 introduced will be unstable, and will tend to dissolve. But if circumstances produce an ion activity product greater than $10^{-10.4}$ (say 10^{-8}) at $25\,°C$, the solution has become **supersaturated** with CaF_2, and it will precipitate solid CaF_2 until the activity product has fallen to the equilibrium value indicated by the solubility product.

Table 4.1 shows the solubility products for a few important minerals. More data relevant to geochemistry are given by Krauskopf and Bird (1995).

Interaction between ionic solutes: the common-ion effect

We have been concerned up to now with solutions containing only a single salt (CaF_2). Natural waters, however, are mixed solutions containing many salts, and in such circumstances the question of solubility ceases to be a simple matter, because the solubility of CaF_2, for example, is now affected by contributions of Ca^{2+} and F^- from other salts present which contain these ions (such as $CaSO_4$ and NaF respectively).

Consider the salt barium sulfate, $BaSO_4$, which forms the mineral barite (another vein mineral). On dissolving in water, $BaSO_4$ ionizes as follows:

$$\underset{\text{crystal}}{BaSO_4} \rightleftharpoons \underset{\text{solution}}{Ba^{2+} + SO_4^{2-}} \tag{4.16}$$

Table 4.1 Solubility products of various **salts** (tabulated as pK ($=-\log_{10}K$) for $25\,°C$). Names of corresponding minerals are given in brackets

Halides			Carbonates*		
$PbCl_2$	4.8	(i.e. $K=10^{-4.8}$)	$CaCO_3$	8.3	(calcite)
BaF_2	5.8		$BaCO_3$	8.3	
$CuCl$	6.7		$FeCO_3$	10.7	
$AgCl$	9.7		$MgCO_3$	6.5	
CaF_2	10.4	(fluorite)			
			Sulfides		
Sulfates			PbS	27.5	(galena)
$BaSO_4$	10.0	(barite)	HgS	53.3	
$CaSO_4$	4.5	(anhydrite)	ZnS	24.7	(sphalerite)
$PbSO_4$	7.8				
$SrSO_4$	6.5		Phosphate		
			$Ca_5(PO_4)_3F$	60.4	(fluorapatite)

*Solubility dependent on pH and H_2CO_3 concentration.

The solubility product for $BaSO_4$ at $25\,°C$ is (from Table 4.1):

$$K_{BaSO_4} = a_{Ba^{2+}} \cdot a_{SO_4^{2-}} = 10^{-10.0} \tag{4.17}$$

The corresponding calcium salt $CaSO_4$ is more soluble at $25\,°C$

$$K_{CaSO_4} = 10^{-4.5} \tag{4.18}$$

Suppose now we mix equal volumes of:

(a) a saturated $BaSO_4$ solution ($BaSO_4$ activity 10^{-5}), and

(b) a $CaSO_4$ solution with an activity of $0.001 = 10^{-3}$ (which the reader can easily confirm is undersaturated with $CaSO_4$).

Mixing these solutions dilutes both $BaSO_4$ and $CaSO_4$ by a factor of two (the same amount of each salt now dissolved in twice the volume of water). We might therefore expect the mixed solution to be less than saturated with $BaSO_4$. But consider the new ion activities of the individual ions:

$$a_{Ba^{2+}} = 0.5 \times 10^{-5}$$
$$a_{Ca^{2+}} = 0.5 \times 10^{-3}$$
$$a_{SO_4^{2-}} = \underset{\substack{contribution \\ from\,BaSO_4}}{0.5 \times 10^{-5}} + \underset{\substack{contribution \\ from\,CaSO_4}}{0.5 \times 10^{-3}}$$
$$= 0.505 \times 10^{-3}$$

Because of the contribution from the $CaSO_4$ solution, the SO_4^{2-} activity is now much higher (50 times) than it was in the $BaSO_4$ solution. Calculating the ion activity product for $BaSO_4$ in the mixed solution gives:

$$a_{Ba^{2+}} \cdot a_{SO_4^{2-}} = 0.5 \times 10^{-5} \times 0.505 \times 10^{-3}$$
$$= 10^{-8.6}$$

This is considerably greater than the solubility product of $BaSO_4$ at $25\,°C$. In spite of the two-fold dilution of Ba, the additional concentration of sulfate ions has made the solution supersaturated with $BaSO_4$, and we can expect precipitation of $BaSO_4$ to occur until the activity product has been reduced to the equilibrium value 10^{-10}.

This unexpected outcome has arisen because $BaSO_4$ and $CaSO_4$ share an ionic species in common, the sulfate ion SO_4^{2-}. Had the second solution (b) consisted of calcium chloride $CaCl_2$, not $CaSO_4$, no extra sulfate ions would have been introduced, and no $BaSO_4$ would have precipitated. The precipitation of $BaSO_4$ by the addition of $CaSO_4$ is an example of the *common-ion effect*. The same outcome could be obtained by adding $BaCl_2$ solution instead of $CaSO_4$ (in which case Ba^{2+} would have been the common ion).

Barite deposits are found on the sea floor at places where barium-containing hydrothermal fluids (in which sulfur is present only as sulfide species like S^{2-}, H_2S and HS^-, and not as sulfate) emerge into sulfate-bearing seawater, a natural example of precipitation due to the common-ion effect.

Note that the simple quantitative concept of solubility applicable to single-salt solutions ceases to have any meaning in natural waters, where the activity of each ionic species can include contributions from a number of dissolved salts. This stresses the value of expressing solubility in the form of an equilibrium constant, the solubility product K. To summarize:

Ion activity product > K means a supersaturated solution;

= K means a saturated solution in equilibrium with solid phase;

< K means an undersaturated solution.

Other kinds of equilibrium constant

Solubility of a gas

Water can dissolve gases as well as solids. A geologically important example is carbon dioxide (CO_2), which on dissolving forms a **weak acid** called *carbonic acid* (H_2CO_3). Water in equilibrium with carbon dioxide in the air is therefore always slightly acidic, a property relevant to many chemical weathering processes.

The dissolving of CO_2 in water can be written as a chemical reaction:

$$\underset{gas}{CO_2} + \underset{solution}{H_2O} \rightleftharpoons \underset{solution}{H_2CO_3} \qquad (4.19)$$

Using the appropriate ways of recording concentrations in these phases, the equilibrium constant is:

$$K_{H_2CO_3} = \frac{a_{H_2CO_3}}{p_{CO_2} \cdot X_{H_2O}} \approx \frac{a_{H_2CO_3}}{p_{CO_2}} \qquad (4.20)$$

since the mole fraction of water X_{H_2O} in dilute solution is very close to 1.00. p_{CO_2} is the **partial pressure** of CO_2 in the atmosphere, which in normal air at the present time[3] is equal to 39.1 Pa (=391 ppmv).

This equilibrium constant, although relating to the solubility of a solute species, is quite different in form from the solubility product of Equation 4.14. The reason is that the behaviour of CO_2 in a solution, as reaction 4.19 shows, is unlike that of ionic compounds, and this is reflected in the mathematical form of an equilibrium constant.

Increasing the air pressure or the concentration of CO_2 in air would shift equilibrium 4.19 to the right, increasing the solubility of CO_2 in water, as expressed by $a_{H_2CO_3}$. Opening a can of fizzy drink, on the other hand, releases the gas pressure inside, and because the dissolved CO_2 has suddenly become supersaturated it forms bubbles of the gas phase, a process similar to the formation of vesicles in a molten lava.

Most gases (including CO_2) are less soluble in hot water than cold (Box 4.2).

Dissociation of weak acids

Equation 4.19 is in fact a slight simplification. The carbonic acid term (H_2CO_3) on the right-hand side actually represents the sum of *three* dissolved carbonate species occurring in natural aqueous solutions: $H_2CO_3^0$, HCO_3^- and CO_3^{2-}. (The superscripts 0, – and 2– merely indicate the charge – or lack of it – on each of these molecules.) In what relative proportions do these carbonate species occur?

Carbonic acid is an example of a weak acid (Appendix B), meaning that, unlike familiar strong laboratory acids such as HCl (hydrochloric acid) which dissociate (ionize) completely in aqueous solution, carbonic acid dissociates into ions only to a small extent. This takes place in two successive stages:

1. $\underset{\substack{carbonic\\acid}}{H_2CO_3} \rightleftharpoons H^+ + \underset{\substack{bicarbonate\\ion}}{HCO_3^-} \qquad K_1 = \frac{a_{H^+} \cdot a_{HCO_3^-}}{a_{H_2CO_3}} = 10^{-6.4}$

$$(4.21)$$

2. $\underset{bicarbonate}{HCO_3^-} \rightleftharpoons H^+ + \underset{carbonate}{CO_3^{2-}} \qquad K_2 = \frac{a_{H^+} \cdot a_{CO_3^{2-}}}{a_{HCO_3^-}} = 10^{-10.3}$

$$(4.22)$$

It is these dissociation reactions, and the accompanying release of hydrogen H^+ ions, that make aqueous solutions of CO_2 slightly acidic (Appendix B). Indeed, knowing the equilibrium constants K_1 and K_2 for reactions 4.21 and 4.22, it is possible to calculate the **pH** (acidity) of pure water that has equilibrated with atmospheric carbon dioxide (Exercise 4.3).

K_1 and K_2 represent a class of equilibrium constant known as *dissociation constants*. One finds that, with all **polyprotic** acids such as H_2CO_3 (Appendix B), K_1 is much larger than K_2. The acidity that arises from dissolution of CO_2 in pure water is therefore almost entirely due to reaction 4.21 alone.

---

[3] The annual average value of p_{CO_2} has risen from 31.5 Pa in 1960 to 39.1 Pa today (Figure 9.2).

But the pH of natural waters can vary widely (Figure 4.1). Box 4.3 shows how pH affects the relative stability and occurrence of the three natually occurring forms of dissolved carbonate.

The carbonic acid present in natural waters determines whether they will dissolve carbonates (limestone) or precipitate them:

$$CaCO_3 + H_2CO_3 \underset{\text{deposition}}{\overset{\text{weathering}}{\rightleftharpoons}} Ca^{2+} + 2HCO_3^-$$ (4.23)

solid   solution   bicarbonate
solution

Physical conditions influence this equilibrium chiefly through the amount of dissolved CO_2 (as H_2CO_3). A local increase in atmospheric CO_2 content (as occurs in soil pore spaces owing to the oxidation of organic matter, for example), or an increase in total pressure leads to a higher H_2CO_3 concentration which 'shifts the equilibrium to the right' (meaning more $CaCO_3$ is dissolved to form bicarbonate).[4] An increase in temperature, on the other hand, expels dissolved CO_2 (Box 4.2 and previous section) and so makes $CaCO_3$ less soluble.

---

[4]   Some spring waters, having become saturated with calcium carbonate through equilibration with limestone at depth (i.e. under elevated pressure), discharge CO_2 during ascent and deposit at the surface a distinctively porous form of limestone called *tufa*.

Box 4.3 Speciation of carbonic acid: the influence of pH

Equations 4.21 and 4.22 suggest that potentially H_2CO_3, HCO_3^- and CO_3^{2-} might *all* coexist in solution. Furthermore, since a_{H^+} appears in both equilibrium constants, their relative proportions must vary with pH. That being so, which of these three carbonate species would predominate in which part of the pH range? We can answer that question by rearranging equations 4.21 and 4.22 as follows:

1. $\dfrac{a_{H_2CO_3}}{a_{HCO_3^-}} = \dfrac{a_{H^+}}{K_1} = \dfrac{10^{-pH}}{K_1}$ (4.3.1)

2. $\dfrac{a_{HCO_3^-}}{a_{CO_3^{2-}}} = \dfrac{a_{H^+}}{K_2} = \dfrac{10^{-pH}}{K_2}$ (4.3.2)

since $pH = -\log_{10}(a_H^+)$ and therefore $a_{H^+} = 10^{-pH}$ (see Appendix A).

These rearranged equations tell us that the activity ratios $\dfrac{a_{H_2CO_3}}{a_{HCO_3^-}}$ and $\dfrac{a_{HCO_3^-}}{a_{CO_3^{2-}}}$ are pH-dependent. We can see from Equation 4.3.1 that H_2CO_3 will be most abundant at low values of pH (= high values of 10^{-pH}), i.e. in acid solutions – as shown in Figure 4.3.1 – whereas Equation 4.3.2 tells us that CO_3^{2-} will be found only in alkaline solutions (high pH = low values of 10^{-pH}). To calculate $a_{HCO_3^-}$ as a percentage of $\left(a_{H_2CO_3} + a_{HCO_3^-}\right)$:

$$\frac{a_{H_2CO_3} + a_{HCO_3^-}}{a_{HCO_3^-}} = \frac{a_{H_2CO_3}}{a_{HCO_3^-}} + 1 = \frac{10^{-pH}}{K_1} + 1$$

Taking the inverse and multiplying by 100%:

$$100 \times \left(\frac{a_{HCO_3^-}}{a_{HCO_3^-} + a_{H_2CO_3}}\right) = 100 \times \left[\frac{10^{-pH}}{K^1} + 1\right]^{-1} \%$$ (4.3.3)

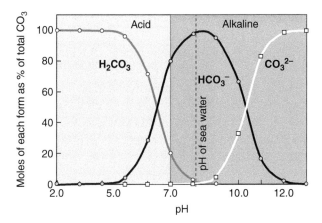

Figure 4.3.1 The relative abundance and stability ranges of undissociated carbonic acid (H_2CO_3), bicarbonate ion (HCO_3^-) and carbonate ion (CO_3^{2-}) – all represented as mole % of total carbonate present – as a function of pH. (Note that $a_{CO_3^{2-}}$ is negligible at low pH when H_2CO_3 is present and $a_{H_2CO_3}$ is insignificant at high pH when CO_3^{2-} is present.)

Figure 4.3.1 shows how the abundance of each species varies with pH. The term *speciation* is often used to describe the identity and relative abundance of the different chemical forms that an element or compound may adopt under different conditions (for example, at various pH values).

Similar plots (called Bjerrum plots)[5] can be drawn for other **polyprotic** acids present in seawater – such as boric acid (H_3BO_3), phosphoric acid (H_3PO_4) and silicic acid (H_2SiO_3) – which, like H_2CO_3, occur in a number of pH-dependent dissociated forms (see Libes, 2009, for Bjerrum plots of these acids).

---

[5]   After Danish chemist Niels Bjerrum.

Warm surface ocean water is saturated or oversaturated with dissolved carbonate. The solubility of CO_2 and $CaCO_3$ increases with depth, however – partly because deep ocean water is colder – and one can identify a *carbonate compensation depth* or *lysocline* below which ocean water becomes undersaturated with carbonate. This change is encountered about 3 km down in the Pacific and about 4.5 km down in the Atlantic. It has been estimated that 80% or more of the calcareous shell material precipitated near the surface is redissolved during or after settling to the deep ocean floor.

Non-ideal solutions: activity coefficient

All of the solutes considered so far have been only slightly soluble in water (their solubility products have been very small numbers). The discussion has in effect been limited to very **dilute** solutions, in which the ions are so dispersed that electrostatic attraction and repulsion between them (called 'ion–ion interactions') can be ignored. The behaviour of ionic species in such solutions can be accurately expressed, as we have seen, in terms of equilibrium constants that involve only the concentrations of the species of interest. Solutions sufficiently dilute to comply with this simple model of behaviour are called **ideal solutions**.

Geochemical reality is less straightforward, however. Most natural waters are complex, multi-salt solutions, whose properties may be far from ideal. Although we may direct our attention to species like $BaSO_4$ that occur only in low concentrations, the solutions in which they are dissolved will typically contain larger amounts of other more soluble salts such as chlorides or bicarbonates. In these stronger solutions, all ions present, including the dispersed Ba^{2+} and SO_4^{2-} ions, will experience many ion–ion interactions, which impede their freedom to react. In particular, Ba^{2+} and SO_4^{2-} ions, tangled up electrostatically with other types of ion, will be less likely to meet and react with each other than if dissolved at the same concentration in pure water. This depression of reactivity is a function of the *total salt content* of the solution. When trying to understand equilibria between specific ionic species in such *non-ideal solutions*, account must therefore be taken of the concentrations of all solutes present, not just the species of interest.

Consider the solubility products of $BaSO_4$, measured (a) in pure water and (b) in 0.1 molal[6] NaCl solution.

_____

[6] That is, a solution in which the molality of NaCl is 0.1.

$$K_{BaSO_4} = a_{Ba^{2-}} \cdot a_{SO_4^{2-}} = \frac{m_{Ba^{2-}}}{m^\ominus} \cdot \frac{m_{SO_4^{2-}}}{m^\ominus} \qquad (4.24)$$

(a) pure water :   solubility product $= 1.0 \times 10^{-10}$
$$(4.25a)$$

(b) NaCl solution :   solubility product $= 7.5 \times 10^{-10}$
$$(4.25b)$$

Notice that substantially more $BaSO_4$ will dissolve in the non-ideal saline solution (b) than in the same amount of pure water. The reason is that the reverse reaction responsible for restricting solubility:

$$\underset{solid}{BaSO_4} \leftarrow \underset{solution}{Ba^{2+} + SO_4^{2-}}$$

is inhibited by the non-ideal, ion–ion interactions that Ba^{2+} and SO_4^{2-} ions experience in the presence of abundant Na^+ and Cl^- ions.

To be dealing with equilibrium 'constants' that vary according to the nature of the host solution is unacceptable. The problem can be overcome by redefining activity so that it serves as a measure of 'effective concentration', incorporating the reduction of ionic reactivity in stronger solutions. A complete definition of activity is therefore:

$$a_{Ba^{2-}} = \gamma_{Ba^{2-}} \cdot \frac{m_{Ba^{2-}}}{m^\ominus}$$
$$a_{SO_4^{2-}} = \gamma_{SO_4^{2-}} \cdot \frac{m_{SO_4^{2-}}}{m^\ominus} \qquad (4.26)$$

γ is the Greek letter gamma, and the functions $\gamma_{Ba^{2-}}$ and $\gamma_{SO_4^{2-}}$ are called the *activity coefficients* for Ba^{2+} and SO_4^{2-} in the saline solution. An activity coefficient is simply a variable factor that expresses the degree of non-ideality of the solution (for an ideal solution it equals 1.00). Although it may look like a 'fiddle factor', the activity coefficient has some foundation in solution theory, and its value can in many cases be predicted with acceptable accuracy.

Expressing the equilibrium constant in Equation 4.24 in terms of the newly defined activities:

$$K_{BaSO_4} = a_{Ba^{2-}} \cdot a_{SO_4^{2-}} = \gamma_{Ba^{2-}} \cdot \gamma_{SO_4^{2-}} \left\{ \frac{m_{Ba^{2+}}}{m^\ominus} \cdot \frac{m_{SO_4^{2-}}}{m^\ominus} \right\} \quad (4.27)$$

This equation remains true for all circumstances. The solubility product (which by definition relates to the observed concentrations, $m_i/m\ominus$, not effective concentrations a_i) is the quantity in brackets above, and this can vary in non-ideal solutions according to the values of the activity coefficients.

Ionic strength

The activity coefficient γ_i for a particular species i depends upon the concentrations of all the solutes present in the solution. How is this overall 'strength' of the solution to be expressed?

Because departures from ideality arise from electrostatic interaction between ions, it is logical to devise a parameter that combines the amount of each type of ion present in solution and the charge on the ion. This is accomplished by the **ionic strength,** introduced by the American chemists G.N. Lewis and M. Randall in 1921. The ionic strength I of a solution is given by the formula

$$I = \tfrac{1}{2}\sum_i m_i z_i^2 \qquad (4.28)$$

Different values of the integer subscript i (1, 2, 3, etc.) identify in turn the various ionic species present in the solution. m_i is the molality of ionic species i (which can be established from a chemical analysis of the solution) and z_i is the charge on the ion concerned, expressed as a multiple of the charge on an electron. The summation symbol Σ_i means adding together the $m_i z_i^2$ terms for all the values of i (i.e. for every ion species in solution).

Consider a solution in which NaCl is present at a molality of $0.1\,\text{mol}\,\text{kg}^{-1}$ and BaF_2 has a molality of $0.005\,\text{mol}\,\text{kg}^{-1}$ (i.e. below saturation). The ionic strength I of this solution is given by:

Values in equation	What they represent
$I = \tfrac{1}{2}\cdot\big[0.1\times 1^2$	$\tfrac{1}{2}\cdot\big[m_{Na^+}\cdot\big(z_{Na^+}\big)^2$
$+\,0.1\times 1^2$	$m_{Cl^-}\cdot\big(z_{Cl^-}\big)^2$
$+\,0.005\times 2^2$	$m_{Ba^{2+}}\cdot\big(z_{Ba^{2+}}\big)^2$
$+\big(2\times 0.005\big)\times 1^2\big]$	$m_{F^-}\cdot\big(z_{F^-}\big)^2\big]$
$= 0.115\,\text{mol}\,\text{kg}^{-1}$	

(4.29)

Why does the charge of each ion z_i appear in this formula as z_i^2 rather than just z_i? A rigorous explanation would require a digression into electrostatic field theory (Atkins and de Paula, 2009), but we can see that the appearance of z_i^2 is plausible in the following way. The force attracting ion i to one of its oppositely charged neighbours is proportional to z_i: if $z_i = 2$, the ion will be twice as strongly attracted at a given distance as it would be if $z_i = 1$. But the stronger attraction will tend to draw ion i closer to the neighbour, increasing the attraction still further, making it more than twice as strong as that felt by a singly charged ion. This relationship is represented more faithfully by $m_i z_i^2$ than $m_i z_i$.

Natural waters span a considerable range of ionic strength, as these representative figures show:

	$I/\text{mol}\,\text{kg}^{-1}$
River water	<0.01
Seawater	0.7
Brines	1–10

Natural waters

There is no universal theoretical treatment capable of predicting non-ideal behaviour across the whole range of ionic strengths given above. The degree of interionic attraction and repulsion and their influence on the properties of a solution change considerably as the ionic strength increases. It will be helpful to divide the spectrum of natural waters into smaller ranges of ionic strength, for which different assumptions and approximations apply.

River water ($I < 0.01\,\text{mol}\,\text{kg}^{-1}$): Debye-Hückel Theory

Table 4.2 shows the principal dissolved constituents of average river water. The reader can confirm that its ionic strength is about $0.002\,\text{mol}\,\text{kg}^{-1}$. Solutions as dilute as this exhibit the weakest ionic interactions, because the ions are widely separated from each other. There is nevertheless a tendency for each ion to attract a diffuse, continually changing jumble of oppositely charged ions around itself, aptly described by chemists as the *ionic atmosphere* of the ion concerned. Around each cation there is a slight statistical preponderance of anions, and vice versa. Like the phenomenon of hydration (Box 4.1), this tenuous ion–ion association is sufficient to depress the free energy of the ions in solution, making them less likely to take part in chemical reactions such as precipitation. The extent of this non-ideality can be estimated using a simple equation derived by physicists P.J.W. Debye and E. Hückel in 1923, from a consideration of the free-energy change associated with the electrostatic properties of the ionic atmosphere:

$$\log_{10} \gamma_i = -Az_i^2 I^{\frac{1}{2}} \qquad (4.30)$$

where γ_i is the activity coefficient of ionic species i; z_i is the charge on ion i (± 1, 2, 3, etc.); A is a constant which is characteristic of the solvent ($A = 0.509\,kg^{1/2}\,mol^{-1/2}$ for water at 25 °C); and I is the ionic strength of the solution. This equation is known as the *Debye-Hückel equation*, which works accurately for non-ideal solutions of ionic strength up to $0.01\,mol\,kg^{-1}$. Conveniently, most fresh waters fall into this category (broadly $I < 0.01\,mol\,kg^{-1}$).

For univalent ions in Table 4.2:

$$A = 0.509\ kg^{\frac{1}{2}} mol^{-\frac{1}{2}}$$

$$z_i = \pm 1, \qquad \text{therefore } z_i^2 = 1$$

$I = 0.0021\,mol\,kg^{-1}$ for average river water.
Therefore $I^{\frac{1}{2}} = 0.046\ mol^{\frac{1}{2}}\ kg^{-\frac{1}{2}}$
Thus $\log \gamma_i = -0.0234$
and so $\gamma_i = 0.95$

Table 4.2 Composition of average river water

Ion	Concentration (ppm = mg kg⁻¹)	Molality m_i (10^{-3} mol kg⁻¹)
HCO_3^-	58.3	0.955
Ca^{2+}	15.0	0.375
Na^+	4.1	0.274
Cl^-	7.8	0.220
Mg^{2+}	4.1	0.168
SO_4^{2-}	11.2	0.117
K^+	2.3	0.059

Thus in river water the behaviour of univalent ions (Na^+, K^+, HCO_3^- and Cl^-) is only marginally non-ideal:

$$a_i = 0.95\ m_i / m^{\ominus}$$

and assuming ideal behaviour would introduce an error of only 5% for each of these ions.

The appearance of z_i^2 in Equation 4.28 (for reasons similar to those explained in the preceding section) suggests that divalent ions will show a larger departure from ideality. For $z_i = 2$, $z_i^2 = 4$ and:

$$\log \gamma_i = -0.0937$$

and therefore $\gamma_i = 0.81$

The activity of each divalent ion will therefore be about 20% below its molality. Thus even at an ionic strength as low as $0.002\,mol\,kg^{-1}$, river water is perceptibly non-ideal. For example, from equations 4.29 and 4.17 we would expect that sparingly soluble species like $BaSO_4$ would be about 25% more soluble than in pure water.

In addition to dissolved material and sediment, rivers transport some products of weathering and erosion in the form of colloidal suspension (Box 4.4).

Seawater (I = 0.7 mol kg⁻¹)

Seawater, as the principal medium of sediment deposition and the ultimate sink for the dissolved products of erosion and anthropogenic pollution, is geologically the most important category of natural water. Analyses show that it has remarkably constant composition across the world. Confining attention to the open oceans, both the **salinity** (the total salt content) and the concentration ratios between elements vary by less than 1%. In enclosed basins the composition of seawater may vary more widely owing to evaporation or freshwater runoff.

Table 4.3 shows the global average composition of seawater. Calculation of the ionic strength, assuming all the constituents shown are fully ionized, gives $0.686\,mol\,kg^{-1}$. This is well outside the range of composition to which Debye-Hückel theory is applicable.

The population of the ionic atmosphere around an ion in fresh water is essentially transient: ions are too dispersed for permanent associations between ions to

Box 4.4 Colloids

Colloids consist of ultrafine particles (usually much smaller than 1 μm) of one phase dispersed metastably in another. Most colloids fall into one of three categories:

Sol: Solid particles dispersed in a liquid, like the clay particles suspended in river water. Certain types of sol 'set' into a turbid, semi-solid form called a **gel** (e.g. gelatin).

Emulsion: One liquid dispersed in another (e.g. milk).

Aerosol: Liquid or solid particles dispersed in a gas (e.g. smoke and fog in the troposphere; desert dust; sulfuric acid/sulfate aerosol in the stratosphere[7] resulting from a major volcanic eruption).

---

[7] Because such aerosols reduce the intensity of solar radiation reaching the surface, deliberate injection of sulfate aerosol into the stratosphere has been suggested as a 'geoengineering' strategy to counteract anthropogenic climate change.

Because of their enormous surface area, the chemistry of colloids is dominated by the surface propeties of the colloidal particles. They possess a surface charge owing to the adhesion of ions. When dispersed in a solution of low ionic strength, interparticle repulsion prevents coagulation into larger particles. In a strong aqueous solution, however, a dense ionic atmosphere forms around the particles, they repel each other less effectively, and they therefore aggregate into larger, more stable particles. This process, seen for example when lemon juice is added to milk, is called *flocculation*. Much of the silting that occurs in estuaries is due to flocculation of colloidal clay particles when the river water in which they are suspended mixes with seawater.

Mineral and organic matter colloids play an important part in determining the cation exchange capacity of soils (Box 8.2).

Table 4.3 Principal ionic constituents of seawater

Ion	Concentration (ppm = $mg\,kg^{-1}$)	Molality m_i ($10^{-3}\,mol\,kg^{-1}$)	% free ion (calc)*	γ_i measured[†]
Cl^-	19,011	535.5	100	–
Na^+	10,570	459.6	99	0.70
Mg^{2+}	1271	53.0	87	0.26
SO_4^{2-}	2664	27.8	54	0.07
Ca^{2+}	406	10.2	91	0.20
K^+	380	9.7	99	0.60
HCO_3^-	121	2.0	69	0.55
Br^-	66	0.8	–	–
CO_3^{2-}	18	0.3	9	0.02

*See Box 4.4. Percentage for Cl^- is assumed.
†See Berner (1971), Table 3.6.

be of any significance. In stronger solutions like seawater, however, certain ions associate on a more permanent and specific basis. For example, Mg^{2+} and HCO_3^- (bicarbonate) ions are abundant enough for a significant proportion of them to combine to form the **ion pair** $MgHCO_3^+$.

About 19% of the bicarbonate in seawater is thought to be present in this *associated* form (Box 4.5). Thus, instead of considering just two ionic species:

$$Mg^{2+} \qquad HCO_3^-$$

a realistic chemical model of seawater must distinguish three ionic species:

$$Mg^{2+} \qquad HCO_3^- \qquad MgHCO_3^+$$

as separate chemical entities. Ions can also associate by forming coordination **complexes** (Chapters 7 and 9) in which the cohesive force resembles a covalent bond rather than an ionic one. Complex formation is particularly prevalent among the transition metals.

The extent of ion pairing in seawater is shown in Box 4.5.

pH of seawater: carbonate equilibria and buffering

If we wish to adjust the **pH** of an aqueous solution in the laboratory, we may add a small amount of a **strong acid** such as hydrochloric acid (HCl) or a strong alkali such as sodium hydroxide (NaOH). 'Strong' in this context means that the acid or base is one that is completely ionized in solution (Appendix B), and therefore a small addition delivers a large dose of H^+ or OH^- respectively to the solution it is added to.

Sea water, however, is devoid of strong acids and bases, and its pH is controlled instead by the dissociation behaviour of **weak acids**. The most abundant weak acid in the oceans is carbonic acid, whose partial

Box 4.5 The extent of ion-pairing in seawater

An ordinary chemical analysis of seawater, like that in Table 4.3, does not identify the actual species (ion pairs, complexes or free ions) in which the element occurs. For example, magnesium could be present in seawater in any of several alternative forms.

Mg^{2+} $MgSO_4^0$ $MgCO_3^0$ $MgHCO_3^+$
free ion   ← ion pairs/complexes →

(The superscript ^{0} signifies the zero charge on a neutral dissolved species.) Table 4.3 gives no clue as to how important each of these species might actually be, but clearly it must be true that:

$$m_{Mg}(total) = m_{Mg^{2+}} + m_{MgSO_4^0} + m_{MgCO_3^0} + m_{MgHCO_3^+} \quad (4.5.1)$$
$$= 53 \times 10^{-3} \, mol \, kg^{-1}$$

This is an example of a *mass balance equation*.

The stability of an ion pair or complex is measurable in terms of its *dissociation constant*. For example, the dissociation reaction

$$MgHCO_3^+ \rightleftharpoons Mg^{2+} + HCO_3^- \quad (4.5.2)$$

has a dissociation (equilibrium) constant given by:

$$K_{MgHCO_3^+} = \frac{a_{Mg^{2+}} \cdot a_{HCO_3^-}}{a_{MgHCO_3^+}} \quad (4.5.3)$$

Laboratory experiment shows that:

$$K_{MgHCO_3^+} = 10^{-1.16} \quad \text{at } 25°C$$

Figure 4.5.1 shows the molal proportions of principal cations and anions in seawater. The detached segments show the proportion of each ion involved in ion pairing (ignoring less significant pairings such as $CaHCO_3^+$). The remaining part of each segment represents the proportion left as free ions.

It is clear that a high proportion of certain anions like SO_4^{2-} are bound up in ion pairs. More than 40% of the sulfate ion in seawater appears to be paired – in more or less equal proportion – with Na^+ and Mg^{2+}; only 55% of the sulfate present exists as free ions. No more than 10% of the CO_3^{2-} in seawater appears to be present as free ions. As one would expect, pairing is more prevalent among divalent ions.

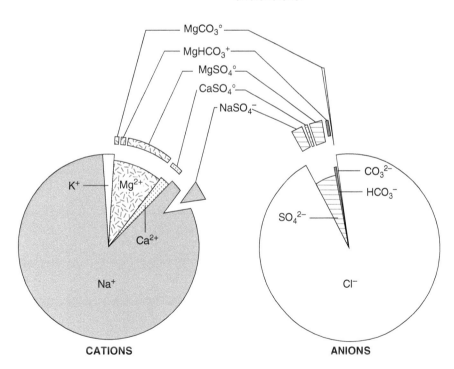

Figure 4.5.1 Pie charts showing the molal proportions of principal cations, anions and ion pairs in seawater.

dissociation into bicarbonate (HCO_3^-) and carbonate (CO_3^{2-}) ions has already been considered in equations 4.21 and 4.22. The fact that weak acids like carbonic acid are only partially dissociated gives them a capacity to stabilize a solution against pH change. This capacity is referred to as **buffering**.

Imagine two beakers on a laboratory bench, one containing a litre of distilled water and the other a litre of seawater. Upon adding a few drops of a strong acid like HCl to the distilled water beaker, while observing the change in pH using a pH-meter (Appendix B), we would find that the pH falls from about 7 to perhaps 2 or 3, consistent with the increase in a_{H^+} that we would expect to accompany the addition of strong acid.

The pH of the seawater, on the other hand, although the beaker is treated exactly the same way, would be found to *remain close to its initial value of 8.1*. The explanation lies in Equation 4.21.[8] The momentary increase in a_{H^+} that occurs when the strong acid is added raises the *ion activity product* $a_{H^+} \cdot a_{HCO_3^-}$ of the seawater, boosting the ratio $\dfrac{a_{H^+} \cdot a_{HCO_3^-}}{a_{H_2CO_3}}$ well above its equilibrium value ($K_1 = 10^{-6.4}$). This causes the reverse reaction in Equation 4.21 to accelerate, driving the equilibrium back to the left. In other words, the added H^+ reacts with HCO_3^- to form additional H_2CO_3, and this reaction mops up much of the H^+ added and restores the pH to close to its original value. The *new* a_{H^+} actually remains slightly higher (and the pH slightly lower) than the original value, but by no means as high (or the pH as low) as it would have become had HCO_3^- not been present.

This buffering capacity explains why seawater samples across the globe have pH values lying within narrow limits of 8.1 to 8.3. Because the dissolved inorganic carbon (often abbreviated to 'DIC') in the oceans represents a very large reservoir of carbonate, the capacity of this buffer system is huge. Other weak acids present in seawater, such as boric (H_3BO_3) and phosphoric (H_3PO_4) acids, also have a buffering effect but – being less abundant – their contribution is much smaller.

Many other compositional aspects of seawater are regulated by chemical reactions. If all of the Mg^{2+} delivered by rivers were to accumulate in the oceans, for example, seawater would be many times richer in Mg^{2+}

than it actually is. The present level in the oceans is moderated by exchange reactions between seawater and ocean-floor basalts, which remove Mg^{2+} into minerals like chlorite which form as alteration products of the original **ferromagnesian** minerals present in basalt. The oceans do not simply accumulate in solution all of the solute delivered by continental runoff; the concentrations of many elements in seawater are subject to complex regulatory mechanisms, in many of which biota also play an important role.

Brines and hydrothermal fluids ($I > 1.0 \, mol \, kg^{-1}$)

Near-surface ground waters in the continental crust are largely of **meteoric** origin; that is they are derived ultimately from atmospheric precipitation. They often have compositions of low ionic strength not very different from river water (Table 4.2), although depending to some extent on the type of rock through which they have flowed. In coastal areas there may also be a component of seawater present. In deeply buried sedimentary rocks, however, the pore waters are **connate** in origin, originating as seawater trapped during accumulation of the sediment. Drilling shows that such waters – oilfield brines, pore waters, 'formation water' and so on – are highly saline (Table 4.4), having remained in contact with the host lithologies at elevated temperatures for millions of years.

Hot, hypersaline aqueous fluids are important in another context: the transport of metals and their

Table 4.4 Ground water and brine

Ion	Ground water in Mississippi sandstone at depth 40 m* (ppm)	Oilfield brine, Mississippi, at depth 3330 m[†] (ppm)
Cl^-	4.4	158,200
Na^+	60	59,500
K^+	4.1	538
Ca^{2+}	44	36,400
Mg^{2+}	11	1730
Fe^{2+}	1.3	298
SO_4^{2-}	22	310
HCO_3^-	327	–
Zn	0	300

*From Todd and Mays (2006).
[†]From Barnes (1979), Table 1.1.

---

8   In principle, equilibrium 4.22 contributes to the buffering effect too, but at pH $\approx$ 8 the concentration of CO_3^{2-} present is too low (Figure 4.3.1) for it to have significant impact.

emplacement in mineral veins and ore bodies. The compositions and temperatures of ore-forming hydrothermal fluids can be established from the study under the microscope of the *fluid inclusions* left behind in the ore and gangue minerals (Box 4.6). It is clear that simple ionic solubilities of ore minerals (solubility products; Table 4.1) are many times too low to explain the dissolved-metal concentrations actually measured in such fluids, which are often in the 100–500 ppm range for metals like copper (Cu) and zinc (Zn). The discrepancy reflects the dramatic increase in the solubility of such metals brought about by **complexing** in highly saline hydrothermal fluids (see Chapter 7, 'The co-ordinate bond').

Determining the chemical forms – or **speciation** – of metals dissolved and transported in the hydrothermal fluids that circulate in the Earth's crust is an important step in understanding the formation of **hydrothermal** ore deposits. The complexes formed depend on the temperature and pH of the solution, and on the **ligands** present. The most important ligands in typical ore-forming fluids are probably Cl^-, H_2S and HS^-. The sulfide ion itself (S^{2-}) is unlikely to be an important ligand in the neutral or mildly acidic fluids believed to be typical of ore-forming systems.

Because ore-forming fluids are known to be highly saline, chloride **complexes** dominate the hydrothermal transport of many important metals. Thus lead (Pb) may be present as $PbCl^+$ or $PbCl_4^{2-}$ depending on the salinity of the host solution:

$$Pb^{2+} + Cl^- \rightarrow PbCl^+ \quad \text{low salinity solution} \quad (4.31)$$

$$Pb^{2+} + 4\,Cl^- \rightarrow PbCl_4^{2-} \quad \text{high salinity solution} \quad (4.32)$$

These species enable solutions in contact with galena (PbS) to contain as much as 600 parts per million of lead (by weight), whereas pure water saturated with PbS contains only 4×10^{-9} ppm.

Experiments and calculations suggest that the principal zinc species in hot saline fluids is neutral $ZnCl_2^0$, that silver is transported chiefly as $AgCl_2^-$, and tin as $SnCl^+$. Copper forms the complex $(CuCl)^0$, which dominates its hydrothermal solution chemistry above 250 °C, but at lower temperatures the bisulfide complex $Cu(HS)_3^{2-}$ may play a significant role.

The stability of such complexes is very sensitive to the temperature of the fluid, to its pH and to the salinity. It follows that if a sulfide-bearing hydrothermal fluid experiences a significant change in any of these variables, there may be a drastic reduction in the solubility of certain metals present, leading to their deposition as sulfide ore. This can happen if there is a drop in temperature (Box 4.2), mixing with other solutions of higher pH or lower salinity, or reaction with wall rocks.

Oxidation and reduction: *Eh*-pH diagrams

Two components – protons (H+ ions) and electrons – are omni-present in the natural environment; their activities can be measured electrometrically as pH and redox potential Eh; these activities may be plotted one against the other in an Eh–pH diagram. The total area of such a diagram contained by (field) measurements delineates the natural (aquatic) environment. (Baas Becking et al., 1960)

Oxidizing power and acidity[9] are the two most important parameters of any sedimentary environment, jointly delineating the limits of stability of minerals that are found there. They are expressed in terms of the **redox potential** (*Eh*–see Box 4.7) and the pH (Appendix B) values of solutions with which the minerals coexist. Where several alternative minerals may crystallize depending on the conditions, it is logical to map their stability fields on *Eh*-pH diagrams. As an illustration, Figure 4.1a shows the stability fields of various copper minerals, together with relevant dissolved Cu species (shaded field), that are stable in the presence of water, chloride, a sulfur and carbon dioxide.

What do the sloping lines at the top and bottom of this diagram signify? Above the top line lie conditions so strongly oxidising as to cause water to decompose and liberate oxygen:

$$H_2O \rightleftharpoons \tfrac{1}{2}O_2 + 2H^+ + 2e^- \quad (4.33)$$

As the forward reaction does not take place in nature, the line represents a formal *upper* limit for *Eh* values found in natural waters. This upper boundary is inclined because the value of *Eh* required for the forward reaction to occur depends on the pH (since H^+ is among the products).

---

[9] Oxidizing power can be thought of as the capacity to *remove* electrons from atoms, while reducing power reflects the power to *donate* electrons. Acidity, on the other hand, reflects the capacity of a solution to *donate* hydrogen ions (H+ ions=protons).

Box 4.6 Fluid inclusions in minerals

During the growth of a crystal from a hydrothermal fluid, it is common for the growing lattice to enclose and trap a minute volume of fluid, accidentally preserving a small sample of it for subsequent microscopic examination. Such *fluid inclusions* (Figure 4.6.1), which vary in size from less than 1 μm to more than 100 μm, can be used to estimate the temperature and composition of the original fluid.

The post-enclosure evolution of a saline fluid inclusion is shown in Figure 4.6.2, which resembles the phase diagram for pure water in Box 2.2. The saline fluid occupies a constant-volume enclosure (neglecting thermal contraction of the crystal) and therefore cooling causes the fluid pressure to fall from A to B along a constant-volume path called an *isochore*. At B the fluid becomes saturated with vapour, and after some supercooling a bubble nucleates at B_1. As the inclusion cools along the liquid/vapour phase boundary to room temperature (R), thermal contraction of the liquid allows the bubble to grow. The liquid may become saturated with one of its solutes (at D, say), so that a *daughter crystal* of halite (Figure 4.6.1) or some other salt nucleates at D_1 and grows during subsequent cooling.

Using a microscope whose stage is specially equipped to heat the sample, the geologist can recreate and observe the cooling process in reverse, and the temperature at which the bubble just disappears indicates approximately the temperature at which the crystal and its inclusions were formed. The 'homogenization temperature' measured is actually T_B, but this is a useful minimum estimate of T_A. Measurement of the original pressure by other means allows T_A to be determined more accurately.

Like the salt used to melt ice on the roads in winter, the salinity of the fluid depresses the freezing point relative to that of pure water. The phase relations of the system

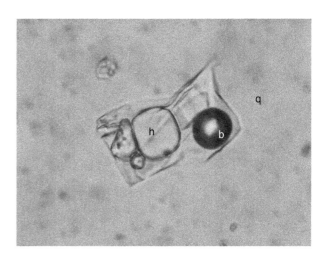

Figure 4.6.1 Thin section view of a fluid inclusion (about 30 μm across) from a quartz vein in the Dartmoor Granite, Devon, England. h = halite crystal, b = gas bubble, q = host quartz crystal. The gas bubble boundary appears more prominent (in optical terms, it has higher *relief*) because of the high contrast in refractive index between the gas and the surrounding liquid. (Source: Photo courtesy of Dr D. Alderton.)

water–salt resemble those of Di–An (Figure 2.4). Modern microscope 'heating stages' are also equipped to cool samples as far as −180 °C, making it possible to determine the temperature at which ice crystals first appear upon cooling or disappear upon warming up, from which the salinity of the fluid can be estimated.

Fluid inclusions provide a very powerful means of exploring the physical and chemical properties of hydrothermal systems. More information can be found in Rankin (2005) and the book by Samson *et al.* (2003).

The lower diagonal line marks the onset of reducing conditions powerful enough to reduce water to hydrogen:

$$H_2O + H^+ + 2e^- \rightleftharpoons H_2 + OH^- \tag{4.34}$$

The occurrence of reduced hydrogen H_2 in aqueous systems is very rare, so this line approximates to the *lower* boundary for *Eh* in the hydrosphere.

Between these two diagonal boundaries lies the 'water window' that embraces all natural aquatic environments.

It is useful to compare Figure 4.1a with Figure 4.1b, which summarizes the *Eh*–pH ranges of waters from a variety of aquatic environments. The dots show the range of the large number of water analyses compiled by Baas Becking *et al.* (1960). Waters that circulate in contact with oxygen in the atmosphere and remain well

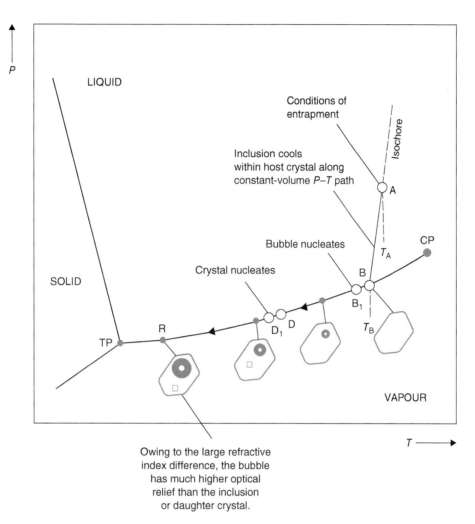

Figure 4.6.2 The *P–T* path of fluid inclusion formation.

aerated are oxidizing and lie in the upper, 'oxic' part of the 'water window'. Ferrous minerals coexisting with them (Box 4.7) are eventually oxidized to red ferric minerals such as hematite (Fe_2O_3) or goethite (hydrated Fe_2O_3), and iron-bearing rocks like some shales and sandstones therefore weather a reddish colour. Stagnant or waterlogged environments – particularly those rich in organic matter – tend to be strongly reducing (lying in the lower 'anoxic' part of the water window). Weathering

in these conditions, found below the water table, produces grey or green surfaces characteristic of ferrous-iron silicate minerals. Magnetite ($Fe_3O_4 = FeO.Fe_2O_3$) and sulfide minerals are also stable in such environments.

Eh–pH diagrams serve a similar purpose for the sedimentary geochemist as *P–T* or *T–X* phase diagrams (Chapter 2) do for 'hard-rock' petrologists, helping us to reconstruct from the minerals contained in an actual rock the conditions under which it formed. Many

Box 4.7 Oxidation and reduction ('redox') reactions: *Eh*

In forming stable compounds with other elements, iron may adopt one of two **oxidation states.** A *ferrous* atom has donated or shared *two* of its electrons in forming bonds with other atoms (Chapter 7) and is said to be ***di*valent,** or in oxidation state II. Silicate minerals containing ferrous ions are commonly grey or green in colour (e.g. olivine, Box 9.10). Alternatively an iron atom may commit *three* electrons to bonding, forming a ***tri*valent** species (*ferric iron*, oxidation state III). Many compounds of ferric iron have a red or orange colour (e.g. rust). These states of iron can be symbolized Fe(II) and Fe(III). This notation recognizes that iron in these oxidation states may not *necessarily* take the form of **ions** Fe^{2+} and Fe^{3+}: this depends on the type of bonding involved, as explained in Chapter 7.

Pure metallic iron is said to be in oxidation state 0, written Fe(0). Oxidation states of metals are discussed in more detail in Chapter 9.

A reaction that causes iron to *increase* the number of its electrons committed to bonds with other atoms is called an **oxidation** reaction, as it increases the oxidation state of iron. An example is the weathering of the iron-rich olivine fayalite (a ferrous compound):

$$2Fe_2SiO_4 \ + \ O_2 \ + \ H_2O \ \rightarrow \ 2Fe_2O_3 \ + \ 2H_2SiO_4 \quad (4.7.1)$$

olivine	air	water	hematite	dissolved
			(ferric oxide)	form of SiO_2

Oxidation of an ion (e.g. oxidizing Fe^{2+} to Fe^{3+}) equates to the *removal* of one or more electrons.

Reduction is the opposite process, for example when an oxidized element receives back some or all of its bonding electrons, illustrated by the formation of **native** iron when a basaltic lava comes into contact with bituminous shale:

$$2FeMgSiO_4 \ + \ C \ \rightarrow \ 2Fe \ + \ 2MgSiO_3 \ + \ CO_2 \quad (4.7.2)$$

'olivine' component in melt : Fe(II)	carbon content of shale	'native' iron : Fe(0)	'pyroxene' component in melt	gas

Reaction 4.7.2 also illustrates an extremely important geochemical property of organic matter (containing C, H and O) and elemental carbon: their capacity to act as a natural *reducing agent* in a wide variety of geological circumstances.

Oxidation and reduction ('**redox**') reactions are important for many other elements that exist in nature in more than one oxidation state. Carbon, for example, can exist as the pure element [C(0) in diamond or graphite], or combined with oxygen ('oxidized carbon', C(IV) as in calcium carbonate, $CaCO_3$). In organic compounds, however, it is combined with hydrogen, an element of lower electronegativity (Figure 6.3), and in this

---

examples of *Eh*–pH diagrams are given in the classic book by Garrels and Christ (1965) and in a recent compilation by the Geological Survey of Japan (2005).

Case study – arsenic in Bangladesh groundwater and drinking water

Arsenic is a ubiquitous (trace) *element found in the atmosphere, soils and rocks, natural waters and organisms. … Most environmental arsenic problems are the result of mobilisation under natural conditions* (Smedley and Kinniburgh, 2002)

Bangladesh is a small, densely populated country located on a flood-prone delta where two major rivers

rising in the Himalayas and Tibet – the Ganges and Brahmaputra – discharge into the Indian Ocean. Population pressures mean that surface water supplies are often too polluted microbiologically to provide safe drinking water (see Box 4.8). Groundwater development for water supply purposes has therefore been actively encouraged by government and international agencies as a means of reducing the incidence of waterborne diseases. Millions of wells have been drilled, mostly into alluvial aquifers 10–60 m deep, although in the south, where these shallow aquifers are too saline for human consumption, deeper boreholes abstract from aquifers at depths between 100 and 150 m. However, drawing water from these deeper aquifers – in fluvial sediments rich in organic matter – introduces

state [C(–IV) as in methane, CH_4] it is referred to as 'reduced carbon'.

Sulfur behaves in a rather similar fashion: depending on the environment, it may occur:

(a) in the elemental state S(0) as the 'native' sulfur often deposited around volcanic vents;
(b) as 'oxidized sulfur' in sulfur dioxide, SO_2 [S(IV)], and the sulfate ion, SO_4^{2-} [S(VI)];
(c) as 'reduced sulfur' in H_2S [hydrogen sulfide, S(–II)] and in metal sulfides like galena (PbS).

The oxidation state of sulfur, like iron (Equation 4.7.2), can be changed by interaction with organic matter. In the **diagenesis** of sediment accumulating on the ocean floor, for example, dissolved sulfate present in pore water (Table 4.3) may be reduced by organic matter in the sediment to form crystals of pyrite (FeS_2).

The stability of minerals containing these oxidation states depends on the external conditions to which they are exposed. Environments open to the atmosphere are intrinsically oxidizing, because oxygen is good at drawing electrons away from metal atoms, and such conditions stabilize oxygen-bearing minerals like sulfates, hematite (Fe_2O_3) and cuprite (Cu_2O). Environments dominated by organic matter, on the other hand, or those cut off from the atmosphere (below the water table, for example), tend to be reducing. These are conditions favourable to the formation of sulfides and other oxygen-poor minerals.

These relationships are illustrated by the downward progression from Cu_2O through Cu metal to Cu_2S minerals in Figure 4.1a.

The relative oxidizing or reducing character of a natural solution, determining the stability of minerals that coexist with it, is expressed in terms of its **redox** (or **oxidation**) potential, Eh, expressed in volts (V) or millivolts (mV). In practical terms, the Eh value of an aqueous environment (such as open seawater, lake sediment or bogwater) is measured by inserting a platinum electrode into the solution and reading the voltage it develops relative to a reference electrode.[10] High values represent oxidizing conditions, whereas low or negative values signify reducing environments (Figure 4.1b).

To quote Krauskopf and Bird (1995):

Redox potential in many ways is analogous to pH. It measures the ability of an environment to supply electrons to an oxidizing agent, or to take up electrons from a reducing agent, just as the pH of an environment measures its ability to supply protons (H^+ ions) to a base or to take up protons from an acid.

_____

[10] The symbol Eh derives from the fact that the measured potential E of the platinum electrode is by convention expressed relative to a reference electrode called a *hydrogen electrode*.

a new problem recognized only recently: dangerously high levels of easily mobilized arsenic.

Arsenic is a metalloid infamous for its toxicity. Its inorganic solution geochemistry is summarized in Figure 4.2. Most toxic trace metals occur in solution as simple cations (like Cd^{2+}, Cu^{2+}, Pb^{2+}, Zn^{2+}) that are soluble only in acid solution; in neutral and alkaline waters these metals tend to precipitate out (as illustrated by the stability fields of crystalline Cu minerals shown in Figure 4.1a) or **adsorb** on to solid mineral surfaces. The aqueous chemistry of arsenic, on the other hand, is dominated by **oxy-anions** that remain soluble across the pH and Eh range (Figure 4.2), including As(III) arsenites and As(V) arsenates. Unlike Figure 4.1a, all the species shown in Figure 4.2 are dissolved forms

of As. Arsenic differs from other oxy-anion-forming elements like Cr, V, Mo and P in remaining soluble even under reducing (low Eh) conditions, a fact that lies behind arsenic's environmental mobility.

The high-As aquifers in southern Bangladesh consist of micaceous sands, silts and clays, capped by impervious clay that inhibits entry of air. Isolation from the atmosphere, combined with an abundance of solid organic matter, leads to highly reducing conditions in these deeper aquifers that favour selective mobilization of As. The aquifer sediments here are not particularly As-rich (typically 10–30 ppm), but arsenic is known to adsorb on to hydrous iron oxide minerals present as ubiquitous coatings on sedimentary grains, in which As is present at higher levels up to 500 ppm

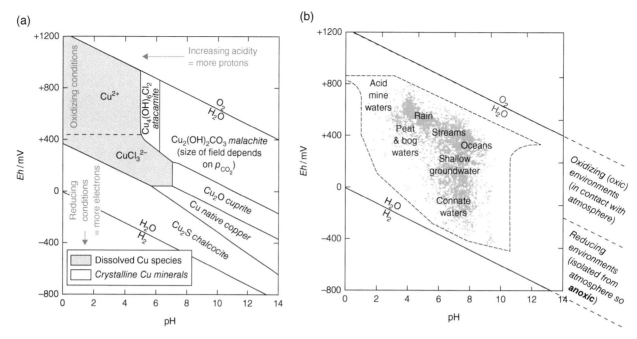

Figure 4.1 *Eh*–pH diagrams, showing (a) approximate stability fields of copper minerals and dissolved spacies (shaded area) in equilibrium with water, S^{2-}, Cl^- and CO_2^{2-} (simplified from Rose (1976, 1989) using malachite data from Vink (1986) and Garrels and Christ (1965)); (b) approximate *Eh*-pH ranges of some natural aquatic environments. Dots and the enclosing dashed line delimit the range of water analyses observed in nature, from the compilation by Baas Becking *et al.* (1960).

Box 4.8 Drinking water quality

What factors determine whether water is safe to drink? Four categories of water contamination pose a potential threat to human health:

- Microbiological contaminants, including viruses and bacteria (e.g. fecal coliform bacteria caused by inadequate sewage treatment of discharges upstream).
- Inorganic chemical solutes[11] such as nitrate and phosphate (which are usually traceable to fertilizer use) and heavy metals.
- Organic chemical contaminants such as industrial solvents, pesticides and pharmaceuticals.
- Radiological contaminants such as radon (Box 3.3) and radioiodine (^{131}I).

In developed countries, water utilities bear responsibility for regularly monitoring the concentrations of such con-taminants in their public water supplies. When contaminant concentrations are found to exceed prescribed limits (e.g. **WHO**, 2011), a utility must shut off the affected source from the supply network, or blend in water from other sources to bring the concentrations down below the prescribed limit.

It is good practice to monitor a wider range of water quality indicators of concern to the comsumer, including physical attributes like transparency, taste, odour and colour, as well as technical chemical parameters such as pH, **hardness**, dissolved oxygen content, and *chemical oxygen demand* (COD – a measure of the total oxidizable organic compounds present).

---

[11] Some inorganic solutes occur naturally and may, like Ca^{2+} and Mg^{2+}, be beneficial to health.

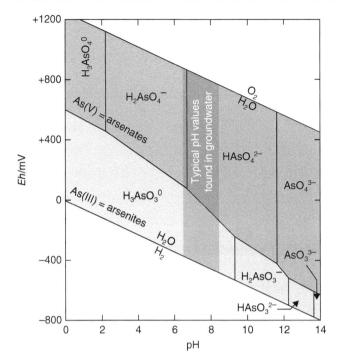

Figure 4.2 *Eh*–pH diagram showing the stability ranges of dissolved forms of arsenic. Arsenic exists in aqueous solution in one of two **oxidation states** (see Box 4.7): As(III) in the lower, light orange field, and As(V) in the darker field above. In each field, dissolved arsenic occurs as a number of pH-dependent **oxy-anions**, some of which incorporate hydrogen (cf. bicarbonate HCO_3^-). The darker vertical band illustrates the limited range of pH values characteristic of most groundwater samples. (Source: Adapted from Smedley & Kinniburgh (2002). Reproduced with permission of Elsevier.)

(Nickson *et al.*, 2000). When reducing conditions develop in the aquifers after burial, these iron coatings dissolve and liberate Fe and As into the groundwater.

It is estimated that 57 million people, mostly in SE Bangladesh, are drinking water with As levels higher than the current recommended **WHO** safe upper limit of $10\,\mu g\,dm^{-3}$; 30 million of these even depend on water with As levels above $50\,\mu g\,dm^{-3}$ (Smedley and Kinniburgh, 2002). In the short term, partial decontamination is possible by thoroughly aerating drinking waters drawn from the contaminated wells: this causes re-oxidation and precipitation of the dissolved Fe and coprecipitation of some of the As from solution.

Still deeper wells (>150 m in depth) yield water that has acceptably low As levels and may provide a longer-term answer to SE Bangladesh's chronic water supply problems.

Further reading

Albarède, F. (2009) *Geochemistry, an Introduction*, 2nd edition. Cambridge:Cambridge University Press.

Atkins, P.W. and de Paula, J. (2009) *Elements of Physical Chemistry*, 5th edition. Oxford: Oxford University Press.

Barrett, J. (2003) *Inorganic Chemistry in Aqueous Solution*. Cambridge: Royal Society of Chemistry.

Drever, J.I. (1997) *Geochemistry of Natural Waters*, 3rd edition. Upper Saddle River, NJ: Prentice Hall.

Edmond, J.M. and von Damm, K. (1983) Hot springs on the ocean floor. *Scientific American 248* (April), 78–93.

Krauskopf, K.B. and Bird, D.K. (1995) *Introduction to Geochemistry*, 3rd edition. New York: McGraw-Hill.

Libes, S.M. (2009) *An Introduction to Marine Biogeochemistry*, 2nd edition. San Diego:Academic Press.

Rankin, A. (2005) Fluid inclusions. In *Encyclopedia of Geology* (eds Selley, R.C., Cocks, L.R.M. and Plimer, I.R.) p. 253–60. Amsterdam: Elsevier.

Smedley, P.L. and Kinniburgh, D.G. (2002) A review of the source, behaviour and distribution of arsenic in natural waters. *Applied Geochemistry 17*, 517–68

UNICEF (2008) Arsenic mitigation in Bangladesh. Available at www.unicef.org/bangladesh/Arsenic.pdf

Walther, J.V. (2008) *Essentials of Geochemistry*, 2nd edition. Sudbury, MA: Jones and Bartlett Publishers.

Exercises

4.1 Calculate the solubility (in $mol\,kg^{-1}$) of $BaSO_4$ at 25 °C in:
 (a) pure water;
 (b) water containing $10^3\,mol\,kg^{-1}$ dissolved $CaSO_4$.

4.2 Calculate how many grams of CaF_2 will dissolve in 1 kg of pure water at 25 °C. (Solubility product of CaF_2 25 °C = $10^{-10.4}$; Ca = 40; F = 19.)

4.3 Calculate the activity of carbonic acid in rainwater that has equilibrated with air:
 (a) with a nominal pre-industrial $P_{CO_2}{}^{air}$ = 28.0 Pa,
 (b) with the actual mean value of $P_{CO_2}{}^{air}$ = 31.5 Pa measured in 1960, and
 (c) with the actual mean value of $P_{CO_2}{}^{air}$ = 39.1 Pa measured in 2012.[12]

Calculate the pH values of these solutions (considering only reaction 4.21) and estimate the change in rainwater pH over the last 50 years. (Equilibrium constant $K_{H_2CO_3}$ (reaction 4.20) = $0.31 \times 10^{-6}\,mol\,kg^{-1}$ Pa^{-1} at 25 °C.)

---

[12] www.esrl.noaa.gov/gmd/ccgg/trends/global.html

5 ELECTRONS IN ATOMS

Why does a geologist need to understand atoms?

We now turn to examine the behaviour of matter on the much smaller scale of individual atoms and molecules, which are about 10^{-10} m in size (Box 5.1). The sub-microscopic atomic world seems at first sight to have little bearing on everyday geology, which people usually associate with events on a much grander scale, such as earthquakes and volcanic eruptions. However, as the next few chapters show, many important properties of geological materials can be related to the types of atoms of which they are made and the chemical bonds that hold these atoms together. The geometry of a lava flow, for instance, depends upon the viscosity of the lava: basaltic lava of low viscosity tends to spread out (if erupted on land) to form tabular flows which may flow a long distance and become quite thin; volcanoes constructed chiefly of basalt flows, such as the Hawaiian islands, are therefore characterized by very gentle slopes. Dacite or rhyolite lava, on the other hand, has a much higher viscosity and tends to form a bulbous, steep-sided lava dome directly above the vent. Lava viscosity is determined by the bonding structure linking individual atoms together, and as we shall see in Chapter 7 this varies according to the dominant types of atom present. For example, the atomic characteristics of silicon (Si) are such that melts rich in SiO_2 (dacite and rhyolite, 65–75% SiO_2) are much more viscous than silica-poor lavas (basalt, 45–52% SiO_2). Atomic interactions on the subnanometre scale thus exert a direct influence on the shape of geological structures some 10^{13} times larger.

The properties that geological materials inherit from their atomic constitution also have important economic applications. One geophysical approach to mineral exploration is to measure the electrical conductivity of the ground at a series of points in a

Chemical Fundamentals of Geology and Environmental Geoscience, Third Edition. Robin Gill.
© 2015 John Wiley & Sons, Ltd. Published 2015 by John Wiley & Sons, Ltd.
Companion Website: www.wiley.com/go/gill/chemicalfundamentals

promising area. The chemical bonding in many sulfides makes them behave in some respects like metals, giving them a greater capacity to conduct electricity than the silicate rocks that host them (Chapter 7). This property can be exploited by the geophysicist to track down undiscovered ore bodies. Once again we see the important influence of atomic characteristics on the macroscopic properties of minerals and other geological materials. A basic grasp of the nature of atoms and chemical bonding is therefore directly relevant to the work of the ordinary geologist.

Box 5.1 Units of atomic size

The size of an atom may be measured in nanometres (nm) or picometres (pm):

$1 \, nm = 10^{-9} \, m$ or one thousand-millionth of a
    $metre = 1000 \, pm$.
$1 \, pm = 10^{-12} \, m$ or one million-millionth of a metre.

Most atoms and ions have diameters in the range 0.1–0.3 nm or 100–300 pm.

The nanometre is the currently recognized SI unit for atomic size (Appendix A). The traditional unit for atomic dimensions was the 'Ångstrom unit' ($1 \, Å = 0.1 \, nm = 10^{-10} \, m$) named after Swedish spectroscopist Anders Jonas Ångström.

The atom

Every atom consists of two parts:

(i) a nucleus at the centre, containing nearly all of the mass of the atom but accounting for only one ten-thousandth of its diameter;
(ii) a family of electrons gathered around the nucleus, forming a three-dimensional 'cloud' that makes up the volume of the atom.

The basic facts about these two atomic components are given in Table 5.1. The nucleus represents an extraordinarily dense state of matter, and it accommodates the atom's positive charge (which is proportional to Z, the number of protons it contains). Protons in such close proximity exert a powerful electrostatic repulsion on each other, but the nucleus is held together by a still greater force called the **strong force** (Box 11.2).

In geochemistry we are more interested in the negatively charged electrons gathered around the nucleus, trapped in the electrostatic pull of its positive charge. The number of electrons in a neutral atom is equal to the number of protons in the nucleus. These captive electrons provide the means by which atoms can associate and bond together. They are the currency of chemical reactions, being exchanged or shared whenever atoms interact and form bonds. It is therefore to the behaviour of electrons in atoms that the present chapter is devoted.

Table 5.1 Basic facts about atoms

	Nucleus	Electron cloud
Approximate size	$10^{-14} \, m$	$10^{-10} \, m$
Electrostatic charge	Positive	Negative
Constituent particles	Protons and neutrons*	Electrons
Approximate mass of individual particles	$1.7 \times 10^{-27} \, kg$ (≈ 1800 electron masses)	$9 \times 10^{-31} \, kg$
Relative numbers of particles	In most types of nucleus, there are slightly more neutrons than protons.	The number of electrons in a neutral atom is equal to Z, the number of protons*.
Approximate density of matter ($kg \, dm^{-3} = g \, cm^{-3}$)	$10^{12} \, kg \, dm^{-3}$	$1 \, kg \, dm^{-3}$

*Protons carry one unit of positive charge and neutrons are electrically neutral. Electrons each have one unit of negative charge. Z is called the **atomic number**.

Those reading this chapter for the first time may find that Box 5.4 provides a useful short cut.

The mechanics of atomic particles

Mechanics is the science of bodies in motion. It describes the motion of anything from billiard balls to space satellites and planets. Except for the twentieth-century contributions of relativity and quantum mechanics, the basic rules of mechanics have been known for more than two centuries, since the days of Isaac Newton. His contribution was of such importance that today we often refer to the classical mechanics of the macroscopic world as 'Newtonian mechanics'.

It is natural to expect the movement of electrons around the nucleus in an atom to conform to the principles of Newtonian mechanics, like the planets in their motion around the Sun. The attractive forces in these two cases, although different in kind – electrostatic and gravitational respectively – conform to the same inverse square 'law' (Appendix A, equations A8 and A9), so one would expect the mathematical analysis to be the same. Nevertheless, it has been clear for more than 80 years that classical mechanics is not wholly applicable to physics on the atomic and sub-atomic scale. Other influences seem to be at work which, although having little obvious impact on the macroscopic world, are paramount in the physics of the atom.

Perhaps the most radical departure from everyday experience is the notion that atomic particles such as electrons possess some of the properties of waves, a suggestion first made by the French physicist Louis de Broglie in 1924. (Readers unfamiliar with the physics of waves may find Box 5.2 helpful.) De Broglie's idea was

---

Box 5.2 What is a wave?

A wave describes a periodic disturbance in the value of some physical parameter. When a ripple travels across the surface of an otherwise still pond, the physical parameter being disturbed is the elevation of the pond surface: a twig floating on the surface will be seen to bob up and down as the ripple passes by, indicating that the pond surface is periodically displaced from its equilibrium position. A wave is periodic in both time and space: the twig bobs up and down v times per second (we call v the frequency of the wave, measured in units of s^{-1}; v is the Greek letter 'nu'), but if we take a snapshot at one instant we will see that successive wave crests are spaced out at a constant distance from each other, known as the wavelength λ of the wave (measured in metres; λ is the Greek letter 'lamda').

The ripple on the pond surface is the easiest type of wave to visualize, because the disturbance affects the position of a visible feature, the pond surface. The concept of 'waves' recognized by physics has a much wider scope, however, because physical quantities other than position may undergo periodic oscillation. A good example is a sound wave, in which it is the pressure of the air that fluctuates as the wave passes by. These changes in air pressure generate a periodic pressure difference across the eardrum that we sense as sound; the more times the pressure oscillates per second (the greater the frequency),

the higher the 'pitch' of the note that we perceive. In the absence of air there is no pressure to fluctuate, which is why sound cannot propagate through a vacuum.

An electromagnetic wave (e.g. light) can be visualized as a train of 'ripples' in the intensity of electric and magnetic fields. In a place that is remote from magnets and electrostatic charges, the average or equilibrium values of these fields will both be zero, but the passage of a light wave causes them each to oscillate between positive and negative values. Light waves are characterized by wavelength and frequency in the same way as sound waves, although the values are very different (Box 6.3).

The examples of waves so far considered have been *travelling waves* which propagate energy from one place to another (e.g. from the loudspeaker to the ear). A wave confined within an enclosure of some kind, however, behaves in a different way: when it is reflected from the walls of the enclosure, the interaction between forward and reflected waves makes the wave appear to stand still. This *stationary wave* is simplest to see in the one-dimensional example of a guitar string. An important property of a stationary wave is that it has a clearly defined wavelength determined by the dimensions of the enclosure (e.g. the length of the guitar string). This phenomenon is exploited in organ pipes (sound) and lasers (light).

soon reinforced by the experimental discovery that a beam of electrons can be **diffract**ed by a crystal lattice.

Diffraction is a phenomenon peculiar to waves, which arises when any kind of wave encounters a periodic (regularly repeated) structure whose repeat-distance is similar in magnitude to the wavelength of the wave concerned (Box 5.3). In the case of a crystal, this periodic structure is the regular three-dimensional array of its component atoms. The waves scattered from different parts of this structure **interfere** with each other (Figure 5.3.1b and c), generating a characteristic spatial pattern of high and low beam intensities called a *diffraction pattern*, which can be recorded by a roving detector or on a photographic plate. (Mineralogists and crystallographers use diffraction, either of X-rays – electromagnetic waves – or electrons, to investigate the internal structure of minerals and other crystalline materials.) Classical physics offers no explanation for the diffraction of electrons if they behave *solely* as particles, and the phenomenon provides firm evidence that the movements of atomic particles like electrons are determined by an underlying wave-like property.

If an electron is to be considered as a wave phenomenon, how can its position and motion be specified? Figure 5.1 depicts a moving electron as a wave-pulse, travelling in this case along the *x*-axis but frozen for our inspection at some instant *t*, like racehorses at a photo-finish. The electron's position at time *t* cannot be defined precisely, because the wave-pulse extends smoothly over a range of *x*-values. The range Δx represents a fundamental interval of uncertainty, within which we cannot pinpoint exactly where the electron lies, although we know it lurks there somewhere. This uncertainty interval, which exceeds the physical size of the electron, is not the result of any experimental error, but must be seen as a fundamental physical limitation to the concept of 'position' where waves are concerned. Heisenberg, in 1927, was the first to recognize this. He expressed it in a quantitative form called the *Uncertainty Principle*, but the details need not concern us here.

In examining the atom, therefore, we cannot look upon electrons as minute planets, orbiting the nucleus with precisely determined coordinates and motion. The wave nature of the electron rules out this precise classical image, introducing in its place a view of the electron in which position is subject to a degree of uncertainty. Although – as we shall see – electrons occupy well-defined spatial domains around the

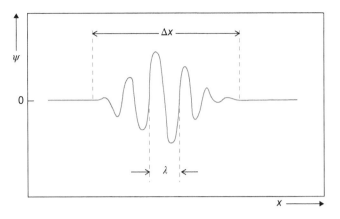

Figure 5.1 A wave train of restricted length, loosely illustrating the wave behaviour of an electron moving in free space. The horizontal axis is the *x*-direction along which the wave is travelling. The vertical axis represents the physical property (as yet unidentified) whose oscillation transmits the wave. It is called the *wave function*, and is symbolized by the Greek letter ψ (psi).

A familiar example of such a short wave 'pulse' is the wave observed when a stone is dropped into a still pond: viewed in profile, the pond wave (somewhat amplified) would look similar to this figure. In this case the wave function would be the vertical displacement of the pond surface from its equilibrium (flat) state.

nucleus, the precise manner in which each electron patrols its own territory is concealed from us by fundamental limits of physical perception, which must be embraced in any model of the atom that we develop. The mathematical certainty enshrined in the laws of Newtonian mechanics thus gives way, on the scale of the atom, to a statistical interpretation of particle mechanics based on probability.

Stationary waves

An electron that belongs to an atom is confined to a small volume of space close to the nucleus. How does a wave behave in these circumstances? To answer this question it is helpful to look at waves of a more familiar kind: those on a vibrating string.

When a guitar string is plucked, the lateral displacement *y* introduced by the player's finger initiates a rapid transverse vibration of the string, which is communicated as waves moving away towards each end of the string. On a string of infinite length, each of these disturbances would continue to propagate outwards indefinitely (like ripples on a pond). The waves carry the energy of the disturbance away from the point of

Box 5.3 Diffraction

Diffraction is a phenomenon that occurs when electromagnetic waves such as light interact with some regularly repeated geometric pattern. A familiar example is provided by the colour fringes seen in light reflected from the surface of a compact disc or DVD. The key requirement for diffraction to occur is that the repeat distance of the pattern should be similar in magnitude to the wavelength(s) of the light being diffracted. The digital signal on a CD is etched in a spiral track with a constant spacing between successive turns; in a manufactured *diffraction grating* the same effect is achieved by engraving straight grooves on a glass plate or mirror (with a typical spacing of 500–1000 nm). In each case, the spacing of track or grooves happens (or is designed) to lie within the wavelength range of visible radiation (400–760 nm). For the same reason X-rays ($\lambda = 10^{-2}$ to 1 nm) are diffracted by the regularly repeated atomic structure of a crystal (typical repeat distance 0.1 to 2 nm), which acts as a 3D diffraction grating.

As a glance at any CD demonstrates, the effect of a diffraction grating on visible radiation is very similar to that of a prism: white light is spread out into a continuum of different colours. The physical process involved, however, is entirely different. Incoming waves are scattered in all directions by individual grooves or tracks. For rays scattered in certain specific directions, waves of a particular wavelength emanating from neighbouring grooves (or atoms in the case of X-ray diffraction) will reinforce each other and give an enhanced intensity of that wavelength or colour (see Figure 5.3.1), whereas in other directions the neighbouring waves are 'out of phase' and will eliminate each other, reducing the intensity of that wavelength. When many wavelengths are present in the incoming light beam, the effect is to disperse one wavelength in one specific angular direction but not in others, so that a spectrum of colours is formed.

Diffraction finds many uses in spectrometry (Chapter 6) for separating visible, ultraviolet or X-ray spectra (Box 6.3) into spectral 'lines' to allow the emissions of individual elements to be measured separately. The wavelength of the line can be calculated from the angle of diffraction at which the wavelength is detected and the spacing of the grooves or atomic planes. This relationship is expressed most simply for X-ray diffraction, through the well-known Bragg equation:

$$n\lambda = 2d \sin\theta \tag{5.3.1}$$

where n is a whole number (1, 2 ...), d is the spacing between adjacent planes of identical atoms in the crystal and θ (Greek theta) is the diffraction angle at which the X-ray wavelength λ will be diffracted with maximum intensity. Using a crystal with known d-spacing, the Bragg

---

plucking, where the string will soon resume a stationary condition. A guitar string, however, has a restricted length, and waves reaching the fixed ends of the string are reflected back on themselves. The string is therefore deflected by waves travelling in opposite directions at the same time, and the overall effect is to set up a *stationary wave*: the string vibrates rapidly up and down within a stationary *envelope* that gradually contracts with time. This envelope on the guitar string is just visible to the naked eye. Its form is shown (much exaggerated) at the top of Figure 5.2. The stationary wave is the characteristic form adopted by any wave that is trapped in a restricted region of space (like a guitar string, or an organ pipe). Standing waves are fundamental to the generation of musical notes, but the phenomenon is not restricted to acoustic waves.

When an electron is captured by an atom, by the electrostatic attraction of the nucleus, the attendant wave becomes trapped within the volume of the atom. It responds to confinement in the same way as the guitar string: it becomes a stationary wave, an oscillating disturbance inside a fixed envelope.

Before developing this analogy further, one must acknowledge two obvious limitations:

(a) On the guitar we see a wave distributed along a one-dimensional string, whereas the electron must be treated as a wave in three-dimensional space.

(b) The vibration of a plucked string decays away quite rapidly because its energy is being dissipated into the surrounding air, through which the sound

equation can be used to investigate the various wavelengths present in a complex X-ray spectrum from a geological specimen (Box 6.4); conversely, if the X-ray spectrum consists of a single known wavelength from an X-ray tube, the equation provides a tool for investigating the atomic structure of an unknown crystalline material by measuring the *d*-spacings of various sets of atomic planes. Since the work of the Braggs in the 1930s, *X-ray diffraction* has been an essential technique for determining the atomic structure of minerals, and for the identification of fine-grained crystalline materials such as clay minerals (Box 8.2).

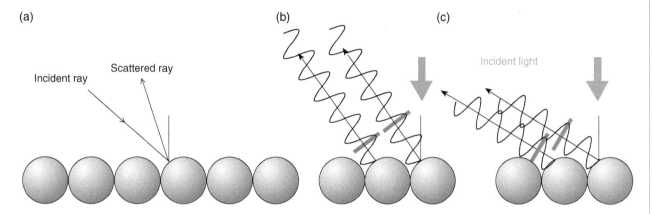

(a)   (b)   (c)

Incident ray   Scattered ray   Incident light

Figure 5.3.1 The principle of the diffraction grating illustrated by a row of atoms. (a) Incident ray, and ray scattered by atom at an arbitrary angle. (b) Waves scattered from neighbouring atoms such that they are *in phase*[1] (constructive interference – rays pull together – enhanced ray diffracted in this angular direction). (c) Waves scattered from neighbouring atoms such that they are exactly *out of phase*[2] (destructive interference – rays cancel each other out – zero intensity observed in this direction). Broad bars illustrate relative positions of equivalent wave-fronts.

---

[1] Meaning the wavefronts of the two rays coincide in space and reinforce each other (Figure 5.3.1b).
[2] Meaning the wavefronts of one ray are displaced relative to those of the other (Figure 5.3.1c) and tend to eliminate each other.

reaches our ears. The electron standing wave does not experience this 'damping' effect, and continues vibrating.

The characteristics of these two types of standing waves differ in detail, but the underlying principles are identical.

Harmonics

The stationary wave shown at the top of Figure 5.2 is the simplest possible mode of vibration, obtained by plucking the string at its centre. This waveform, the *fundamental*, produces the lowest musical note obtainable from the string (frequency v_0). To produce higher notes from the same guitar string one can shorten its length by pressing it down at some intermediate point, but it is more relevant to consider what other notes are obtainable on the open (full-length) string. It is possible to generate a number of higher notes from the open guitar string by making it adopt alternative forms of stationary waves called *harmonics*, as described in the caption to Figure 5.2.

Because the ends of the string are immobilized, only a restricted number of harmonics is possible. These all have wavelengths (λ_2, λ_3, etc.) and frequencies (v_2, v_3, etc.) that are related in simple ways to the fundamental, as shown in Figure 5.2. Whereas in the fundamental mode the string is stationary only at its ends, the harmonics share the property of having intermediate points where the string remains stationary too (that is, where the lateral displacement of the string, y, remains zero). These points are called *nodes*, denoted by the symbol 'N' in Figure 5.2.

The fundamental and the harmonics are members of a restricted series of stationary waveforms that can reside on a string of length L. They can all be represented by one general equation:

$$y = A_n \sin\left(\frac{n}{2} \cdot \frac{x}{L} \cdot 360°\right) \tag{5.1}$$

where y represents the lateral displacement of the string (from its equilibrium position) at distance x from the end of the string. A_n measures the maximum displacement, and is called the *amplitude*. The integer n has a different value for each of the possible harmonics. Entering '$n=1$' into Equation 5.2 yields the equation describing the fundamental; '$n=2$' gives the equation for the *first harmonic* (Figure 5.2), and so on. Note that the value of n defines several important features of a harmonic:

(a) n indicates the number of maxima and minima (the number of 'lobes') of the waveform;
(b) the number of nodes is equal to $n-1$;
(c) the wavelength of the stationary wave is equal to $2L/n$ (Figure 5.2);
(d) the frequency of vibration of each harmonic is proportional to n.

The important conclusion is that a string fixed at both ends can accommodate only a restricted number of stationary waves compatible with its length L, so the vibration frequency can adopt only a limited number of values (the musical pitches of the fundamental and harmonics). A variable that is allowed by the properties of the system to have only certain discrete values is said to be *quantized*. The integer n which enumerates these permitted values is called a **quantum number.** These terms are used chiefly in the context of atomic physics, but they identify characteristics common to all types of stationary waves, regardless of scale.

Electron waves in atoms

The harmonics of a vibrating string can be investigated using simple apparatus, so a thorough mathematical treatment is not called for. Because of its size, the atom does not lend itself to such simple experimentation, and our understanding of how it works comes mainly from theoretical physics. In 1926 the Austrian physicist Erwin Schrödinger formulated an elegant and highly successful mathematical theory of particle mechanics incorporating the wave-like properties of the electron. His method of analysis is known as *wave mechanics*, and it provides the foundation of modern atomic (and nuclear) physics. Although the mathematics is difficult, the underlying physical concepts are straightforward, and have a lot in common with the analysis of the vibrating string.

Schrödinger set up a general *wave equation* (analogous to Equation 5.1) describing the physical circumstances of the electron in an atom: the nature of the electrostatic force attracting it toward the nucleus (the inverse-square law of classical physics) and the (then) newly recognized wave properties of the electron itself. Schrödinger's work suggested that the electron trapped in an atom behaves in much the same way as any stationary wave, including the one on a stringed instrument (Table 5.2).

The Schrödinger wave equation is a **differential equation**, which offers a number of possible mathematical 'solutions' called *stationary states*. Each simply describes a different stationary 'waveform' a trapped electron can adopt, analogous to those shown in Figure 5.2. Each distinct electron waveform with its own specific three-dimensional geometry is called an **orbital**. One speaks of an electron 'occupying' a particular orbital, reminding us that an orbital is the wave-mechanical equivalent of a planetary orbit.

Equation 5.1 expresses the way the displacement y varies along the length of the vibrating string (the x dimension). In the vocabulary of wave theory, the mathematical function y is called the *wave function*. Each solution of the Schrödinger equation expresses how a wave function ψ (the Greek letter 'psi') varies in three-dimensional space (x, y, z) around the nucleus. To understand the physical significance of ψ, we need to square it. It is true of most types of wave that the intensity of the physical sensation is proportional to the square of the wave function. For example, the loudness of the sound emanating from the guitar string is proportional to y^2, not y.

The magnitude of ψ^2 at each point in the atom tells us the probability of finding the electron at that point in space. In keeping with the Uncertainty Principle, this

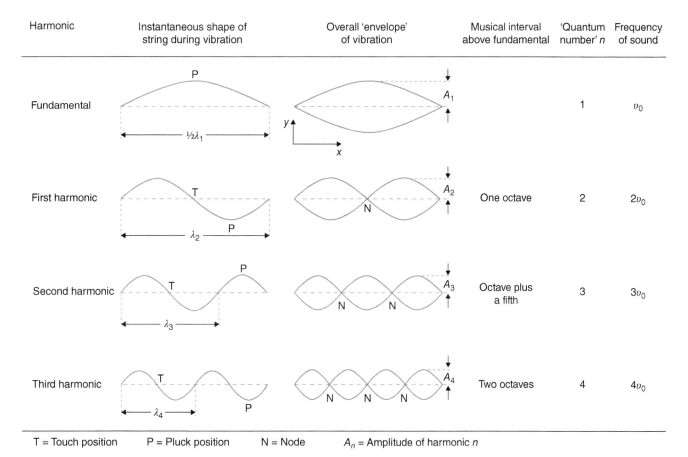

T = Touch position   P = Pluck position   N = Node A_n = Amplitude of harmonic n

Figure 5.2 The harmonics of a plucked string. To obtain such harmonics on the guitar, the player touches the string lightly at the appropriate nodal point (identified as T), whilst plucking it sharply near to the other end (point P). The finger at T is removed immediately after plucking. The idea is to establish a nodal point at T while allowing the string to oscillate along its entire length.

Table 5.2 Parallels between a vibrating string and an atomic orbital

	Stationary wave on a guitar string	*Electron in an atom*			
Nature of wave	Wave in one dimension (along string). Extent of wave is restricted by fixed ends.	Three-dimensional wave (orbital) around nucleus. Spatial extent of electron standing wave is restricted by the electrostatic pull of the nucleus.			
Stationary states	One quantum number (n) is sufficient to define stationary states: $n=1$ fundamental $n=2$ first harmonic $n=3$ second harmonic	Four quantum numbers (n, l, m, s*) are required to define all possible electron stationary states in an atom; n and l are the most important.			
			$l=0$	$l=1$	$l=2$	$l=3$
		$n=1$	1s	–	–	–
		$n=2$	2s	2p	–	–
		$n=3$	3s	3p	3d	–
		$n=4$	4s	4p	4d	4f
Significance of wave function	Wave function $= y$ (lateral displacement of string) Perceived quantity $\propto$ 'loudness'	Wave function $= \psi$ (psi) Physical significance: ψ^2 = probability of finding the electron at (x, y, z) = 'electron density' at (x, y, z)			

*n = principal quantum number
l = angular momentum quantum number
m = magnetic quantum number
s = spin quantum number

probability is less than one at every individual point (that is, there is no one point at which we can locate the electron with certainty), but considering the atom as a whole it must add up to one for each of the electrons present.

Solutions to the Schrödinger equation for the electron in an atom are independent of time (since, like Equation 5.1, they contain no t terms), and so they cannot indicate the precise trajectory or orbit followed by the electron, showing how its x, y and z coordinates vary as a function of time. The electron must be visualized as a constant but diffuse cloud extending throughout the volume of the orbital, the 'electron density' of the cloud varying from point to point according to the magnitude of ψ^2.

For the vibrating string, one quantum number n is sufficient to enumerate the various types of stationary wave observed. Unsurprisingly the electron wave in the three-dimensional atom presents a more complicated picture, and four quantum numbers (Table 5.3) are required to encompass all possible stationary states that the electron wave can adopt. For most purposes, we shall only need to consider two of these four quantum numbers: n, known as the *principal quantum number*, and l, the *angular-momentum quantum number*. (The physical origins of these names need not concern us.) The significance of each quantum number, summarized in Table 5.3, will become clear presently.

The shapes of electron orbitals

Atomic orbitals come in all shapes and sizes, and some acquaintance with their geometry will be helpful in understanding the shapes of molecules and the internal structures of crystals (Chapter 8). The internal structure of the mineral diamond, for instance, is a direct expression of how electron density is arranged within each constituent carbon atom. In a similar way, the disposition of electron orbitals in the oxygen atom is responsible for the bent shape of the water molecule, upon which the unique solvent power of water depends (Box 4.1).

The *symmetry* of an orbital is determined by the value of the quantum number l, in the manner outlined in Table 5.3. Orbitals for which l is zero have simple spherical symmetry, but as l increases we encounter progressively more complex symmetry. We shall begin with the simplest type of orbital symmetry, in which the parallel with the vibrating string is most apparent.

s-orbitals

The simplest solutions to the Schrödinger wave equation, for which l is zero, all possess spherical symmetry and therefore ψ and ψ^2 can be depicted simply in terms of a radial coordinate r, representing the distance from

Table 5.3 Physical significance of quantum numbers

Quantum number		Permitted values		Influence on geometry of orbital
Name	*Symbol*			
Principal quantum number	n	integer 1, 2, 3 ..	(a)	Determines the *size* of the orbital: low n: compact orbital high n: spread-out orbital (Figures 5.3 and 5.4)
			(b)	$(n-1)$ is the number of nodal surfaces, where $\psi^2 = 0$.
Angular momentum quantum number	l	integer 0 to $n-1$		Determines the *shape* of the orbital:
			$l=0$	s-orbital: **s**pherical symmetry
			$l=1$	p-orbital: **p**olar symmetry – electron density forms 2 balloons on opposite sides of the nucleus (Figure 5.4).
			$l=2$	d-orbital: electron density forms 4 balloons (Figure 5.4)
			$l=3$	f-orbital : still more complex
Magnetic quantum number	m	integer $-l$ to $+l$		Determines the *orientation* of the orbital: e.g. indicates whether a p-orbital is aligned along the x, y or z axis.

Note: The spin quantum number, s, only becomes relevant when multi-electron atoms are concerned. It has only 2 permitted values, $-\frac{1}{2}$ and $+\frac{1}{2}$.

the nucleus. Such solutions are called *s-orbitals*. The two simplest cases are shown in Figure 5.3. The upper half shows what we may think of as the atomic counterpart of the fundamental mode on a vibrating string (Figure 5.2). In keeping with this interpretation, the principal quantum number n is equal to 1 and the number of nodes is zero. This orbital is designated '1s'. On the right-hand side of Figure 5.3 is an attempt to show what a cross-section of this orbital would look like. The electron density (represented by the density of dots) is greatest immediately around the nucleus, and decreases smoothly away from it with an exponential-like profile. This diffuse outer fringe is common to all types of orbital, and in this respect wave-mechanical waveforms differ significantly from vibrating string harmonics, which of course terminate abruptly at the end of the string. All atoms and ions, therefore, have diffuse outer margins.

The lower part of Figure 5.3 illustrates a more complex spherical orbital, designated 2s, which resembles a first harmonic: n has the value 2, and there is a node-like feature where the wave function ψ and therefore the electron density ψ^2 are both zero. In three dimensions, this is actually a spherical nodal surface,

separating a core of electron density from an outer fringe. These two parts of the orbital together accommodate the same electron which, somewhat paradoxically, distributes itself statistically between them. Note that the electron density extends significantly further from the nucleus than was the case for the 1s orbital (Figure 5.3).

Each of these orbitals is uniquely defined by the values of the two quantum numbers n (=1 or 2) and l (=0). A series of progressively larger, and more complex, spherical orbitals exists corresponding to values of n up to about 7, and these are distinguished as 3s, 4s and so on, according to the value of n.

p-orbitals

The p-orbitals are a second class of solution (identified by l having the value 1), in which electron density is concentrated into two 'balloons' which stick out from the nucleus in opposite directions (Figure 5.4). The lack of spherical symmetry here requires the introduction of arbitrary x, y and z axes, centred on the nucleus, in order to specify the varying orientations of these balloons.

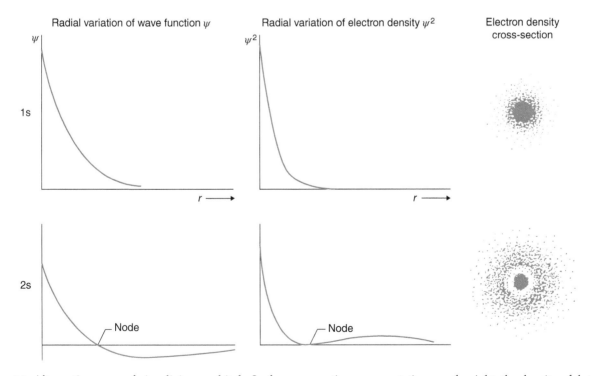

Figure 5.3 Alternative ways of visualizing s-orbitals. In the cross-section representations on the right, the density of dots indicates the electron density.

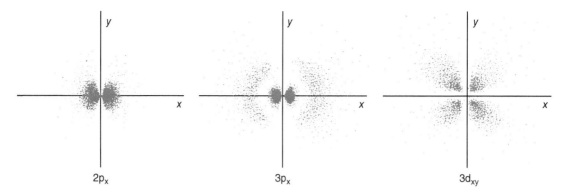

Figure 5.4 Cross-sections of electron density distributions in 2p, 3p and 3d orbitals. Electron density is represented by the density of dots. The $2p_x$ orbital has $(n-1)$=one nodal surface in the y–z plane passing vertically through the origin/nucleus. The $3p_x$ orbital has two: one as for $2p_x$ and a spherical surface separating inner and outer lobes. The $3d_{xy}$ orbital also has two nodal surfaces in the x–z and y–z planes, both passing through the origin/nucleus.

The Schrödinger equation indicates that there are three equivalent, but independent, forms of p-orbital, designated p_x, p_y and p_z, in which these balloons are aligned along the x, y and z axes respectively (Figure 5.5). These variants share the same values of n and l, but are distinguished by having different values of the *magnetic quantum number, m* (Table 5.3), which serves to define the different *orientation* of each of these orbitals.

For reasons we need not discuss, the Schrödinger equation prohibits values of l greater than $n-1$ (Table 5.3). Accordingly there is no p-orbital when $n=1$, and the simplest p-orbitals ($2p_x$, $2p_y$ and $2p_z$) are encountered when $n=2$. (Consistent with this value of n, we observe the existence of a nodal plane separating the two balloons, which – although different in shape – has a similar significance to the nodal surface in the 2 s orbital.) As with s-orbitals, the maximum electron density in each balloon is found close to the nucleus (Figure 5.4). Three such p-orbitals exist for each value of n greater than 1, and these families are designated 2p, 3p, 4p, and so on, according to the value n. (For most purposes, the subscripts x, y and z may be omitted.) As Figure 5.4 illustrates, the size of the orbital and the number of nodal surfaces increases as the value of n increases.

d-orbitals

Setting l equal to 2 generates a third class of orbital called a d-orbital, which is first encountered when $n=3$. Except for one special case, d-orbitals consist of four elongated balloons of electron density, extending out from the nucleus at right angles to each other (Figure 5.4). As stated in Table 5.3, m can now have the values -2, -1, 0, $+1$ and $+2$, and we therefore find five equivalent d-orbitals for each value of n (greater than 2). Their orientation in space is shown schematically in Figure 5.5. Three orbitals, d_{xy}, d_{yz} and d_{zx}, have balloons extending diagonally between the coordinate axes, whereas the other two have their electron density directed broadly along the axes. Our experience of wave theory so far suggests that, because $n=3$, we should encounter two nodal surfaces in each of the 3d orbitals, and the nodal planes or surfaces evident in Figure 5.4 confirm this expectation. A separate family of five d-orbitals exists for every value of n greater than 2 (=3d, 4d, and so on).

As we shall see in Chapter 9, d-orbitals arc responsible for the distinctive chemistry of transition metals such as iron, copper and gold, and play an important role in the aqueous-complex formation upon which their hydrothermal transport and deposition in ore bodies depend.

f-orbitals

The final class of solutions to Schrödinger's equation are those for which $l=3$. Their geometry is too complex to consider here, but we should note that they first occur when $n=4$, and that seven equivalent orbitals having various orientations will exist for each value of n above 3. The f-orbitals only become important chemically in the context of heavy elements such as cerium and uranium.

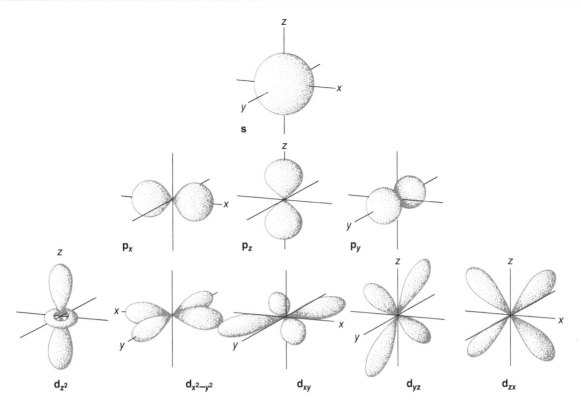

Figure 5.5 Simple 'balloon' diagrams showing where electron density is concentrated in s, p and d orbitals. Each balloon can be considered as a three-dimensional contour of electron density.

Electron energy levels

Each stationary state of the Schrödinger equation possesses a well-defined value of the total electron energy E (the sum of the electron's **potential** and **kinetic energies**). The energy of an electron in an atom is therefore not free to vary continuously, but like the frequency of a guitar string is *quantized* and has to conform to one of these permitted **energy levels.** The Danish physicist Niels Bohr had been the first to suspect this on empirical grounds in 1913, but the theoretical basis remained obscure until Schrödinger's work showed it to be a straightforward consequence of the stationary wave(s) set up by electron(s) trapped in an atom.

The energy levels of the various orbitals are shown in relation to each other (for the simplest example, the hydrogen atom) in Figure 5.6. The zero on the energy scale is defined as the energy of a 'free electron at rest', one that does not belong to any atom (see caption) and possesses no kinetic energy. This convention provides a common baseline that allows us to compare directly

the energy levels of electrons in different types of atom. All of the energy levels have negative electron energies, indicating that an electron enjoys greater stability within an atom than outside it.

Each box in Figure 5.6 represents an **orbital**. Considered together, the orbitals resemble an irregular set of 'pigeonholes' in energy space, offering the electron a variety of alternative accommodation in the atom. The 1 s orbital has by far the lowest energy level, indicating that in this state the electron is most firmly bound to the nucleus. This interpretation is consistent with the very small size of the 1 s orbital (Figure 5.3), which confines the electron very closely to the nucleus, where its electrostatic pull is strongest. This is the most stable stationary state the electron can adopt in the atom, the one in which its energy is minimized (Chapter 1). When the electron resides in this orbital, the hydrogen atom is said to be in its *ground state*.

The 2 s and the three 2p orbitals share the same energy level, some distance up the energy scale. In spite of their different spatial configurations,

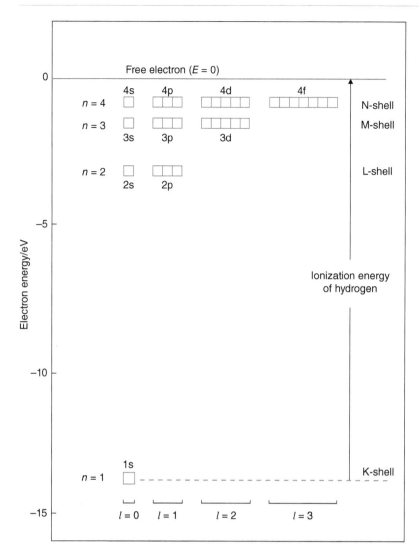

Figure 5.6 A scale diagram of the energy-level structure of orbitals in the hydrogen atom (with only one electron), the electron energy being expressed in **electron-volts** (eV). Note that in hydrogen the 2s and 2p orbitals have the same energy (together they constitute a 'shell', in this case the 'L-shell'), as do the 3s, 3p and 3d orbitals ('M-shell'). Note also that the zero on the electron energy scale is equivalent to the energy of a free electron at rest. A negative electron energy therefore signifies that the electron is trapped in the atom by the nuclear field. The more negative the energy, the more tightly bound the electron has become, and the harder it is to remove from the atom. Positive energy values signify free electrons having appreciable kinetic energy.

therefore, they are energetically equivalent, at least in the hydrogen atom. The same is true of the one 3s orbital, the three 3p orbitals and the five 3d orbitals, all of which share a still higher energy level. Note that there is a direct relationship between an orbital's relative size and its energy level. Evidently an electron must possess quite a high energy before it can overcome the nuclear attraction sufficiently to spread itself into these more far-flung provinces of the atom's territory. In the hydrogen atom these orbitals are normally unoccupied, except temporarily (Chapter 6).

As n increases further, the energy levels get progressively closer together, and so the highest levels have been omitted from Figure 5.6 for clarity.

Multi-electron atoms

In order to describe the chemistry of elements other than hydrogen, the Schrödinger model needs to be extended to explain how more than one electron can be accommodated together in a single atom. Can separate electrons adopt identical waveforms in the same atom, or does wave mechanics organize them into different spatial domains around the nucleus?

The answer to this vital question was provided by another Austrian, Wolfgang Pauli, who in 1925 (actually a year before Schrödinger's paper) formulated the *Exclusion Principle*. Stated in wave-mechanical terms, it says: 'No two electrons in the same atom may possess identical values of all four quantum numbers'.

At this point the fourth quantum number s, representing 'spin' of the electron, enters the discussion. According to the Exclusion Principle, two electrons in the same atom may possess the same values of n, l and m (that is, they may 'occupy the same orbital') *only if* they have different values of s. Wave mechanics allows s only two possible values, $-\frac{1}{2}$ and $+\frac{1}{2}$ ('$\frac{1}{2}$' is the contribution that electron spin makes to the atom's angular momentum, but such details need not concern us.) The important property of the spin is that it can have one of two directions, a positive or negative 'sense'. Although 'spin' in this quantized, wave-mechanical form is rather an abstract concept, it leads to an important practical outcome: each of the orbitals we have been considering can accept *two* electrons, subject only to the proviso that their spins are opposed.

We can now predict the arrangement of electrons in any atom. Each electron entering an atom will of course occupy the orbital that offers accommodation at the lowest available energy level. In the atom of helium, the two electrons present can share the 1s orbital. However in the lithium atom, with three electrons, the third electron cannot – according to the Exclusion Principle – enter the 1s orbital, and must make do with 2s instead, in spite of the much higher energy that entails. One could in principle continue feeding electrons into Figure 5.6, filling orbitals in order of ascending energy, to build an electronic model of any species of atom, but before we can do so accurately we must recognize two features of multi-electron atoms which have been neglected in Figure 5.6.

In the first place, Figure 5.6 has to be modified a little to allow for electrostatic repulsion between electrons, an effect we did not have to consider in the hydrogen atom. Mutual repulsion in multi-electron atoms, when incorporated into the Schrödinger equation, leads to solutions with a slightly modified energy structure, as shown in Figure 5.7. Orbitals that are not spatially equivalent no longer share the same energy level, but have energies which depend on l as well as n. Thus although all of the 3d orbitals still have a common energy level (at least in an isolated atom), their energy now exceeds that of the three 3p orbitals, which in turn is greater than the 3s energy. Figure 5.7 provides a more general framework for comparing the electronic structure of different elements. However, note that the energy axis in Figure 5.7 is non-linear (see caption).

The second point to be remembered in discussing multi-electron atoms is the increased nuclear charge (owing to the greater number of protons, Z, in the nucleus), which causes each electron to experience a stronger electrostatic attraction towards the nucleus. This changes the orbital picture quantitatively but not qualitatively. The shapes of the various orbitals stay the same, but with increasing nuclear charge they all diminish in size as electron density is confined ever more closely to the immediate vicinity of the nucleus. The overall form of Figure 5.7 changes little from element to element, but the negative energy associated with a particular orbital becomes progressively greater (the level becomes 'deeper' in energy space) with increasing nuclear charge, as the energy scales on the left of Figure 5.7 illustrate. In energy terms the two 1s electrons of the uranium atom are held 2000 times more tightly ($\sim -10^5$ eV) than is the 1s electron of a lithium atom (–55 eV).

Electronic configurations

The electronic configuration of an atom is a symbolic code describing the location of its electrons in the various orbitals, something one can readily work out from Figure 5.7. The element boron, for instance, has an atomic number (Z) of 5, so that its nucleus contains five protons and has five positive units of charge (5+). Accordingly the boron atom has to accommodate five electrons, and their distribution can be written:

$$\text{boron} : 1s^2 2s^2 2p^1$$

The requirement to minimize total electron energy is satisfied by putting two electrons into the 1s orbital (this is what the code '$1s^2$' means), a further two into the 2s orbital (hence '$2s^2$'), and the one remaining electron into one of the three 2p orbitals ('$2p^1$'). We do not need to specify – indeed we have no way of knowing – which of the three 2p orbitals receives this single electron.

Another example is the element sodium ($Z = 11$). Its electronic configuration is

$$\text{sodium} : 1s^2 2s^2 2p^6 3s^1$$

In the sodium atom, the three 2p orbitals have accepted their joint quota of six electrons, but one more electron still remains to be accommodated. This goes in the

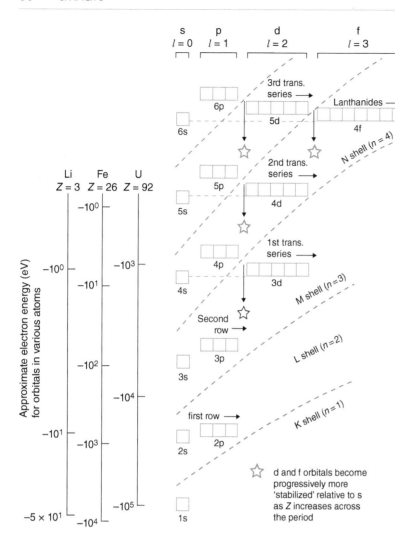

Figure 5.7 A generalized diagram of the energy-level structure in more complex atoms. Note that the vertical axis is non-linear (unlike Figure 5.6), as shown by the scales on the left. It approximates to an inverted logarithmic scale. In contrast to Figure 5.6, orbitals sharing the same value of n no longer share the same energy, but increase with the value of *l* as well (i.e. s<p<d<f).

next-lowest energy level, 3 s. Here, however, it has a conspicuously higher energy than the other electrons in the atom. Because it extends much further from the nucleus in the 3 s orbital, and requires much less energy to remove it from the atom altogether, this solo electron dominates the chemical behaviour of sodium. It is called the **valence electron**, because it can be used in transactions with other atoms, rather like a current account at the bank.

The electrons occupying the 1 s, 2 s and 2p orbitals of the sodium atom, on the other hand, are much more tightly held and they never participate in sodium's chemical reactions. They comprise what is called the **electron core** of the atom, and resemble personal wealth tied up in stocks and shares, too immobile to be used in day-to-day transactions.

Because of the way orbitals are grouped in energy space by their value of n (Figure 5.7), it is sometimes useful to speak of electron *shells*. Electrons in the 1 s state comprise the *K shell*, those in 2 s and 2p comprise the *L shell*, and so on, as shown in Figure 5.6. In hydrogen, all orbitals in the same shell have the same energy, but this is not true of multi-electron atoms (Figure 5.7). One can likewise speak of valence electrons collectively as the *valence shell*. One must, however, be careful in attributing spatial significance to the term 'shell'. It is useful in suggesting that electrons in the M shell project further from the nucleus than those in the L shell, but any impression that electrons are sharply segregated into hollow shells – as some elementary textbooks suggest – is of course false.

Box 5.4 The absolute essentials

Of all the chapters so far in this book, understanding Chapter 5 requires the most physical insight. The reader is strongly encouraged to persevere until the end of the chapter because – armed with the wave-mechanical view of the atom – you will be best equipped to cope with suc-

ceeding chapters of the book. Nevertheless, for the reader to whom the concept of electron 'waveforms' is unacceptably remote from everyday experience, the main conclusions of Chapter 5 are presented below in non-wave terms.

Z = 3, lithium

The essential properties of any atom are determined by the number of **protons** (positively charged nuclear particles) in the nucleus. This is known as the **atomic number** Z of the atom. The value of Z identifies the chemical element to which the atom belongs. The other particles shown in the figure are neutrons (uncharged nuclear particles), the number of which determines the isotope of the element.

The positive charge (Z+) is neutralized by a cloud of Z electrons (negatively charged particles) surrounding the nucleus, held in position by its electrostatic attraction.

The atom forms chemical bonds with other atoms by 'trading' electrons with them. The Z electrons in an atom may be divided into:

(a) **valence electrons** (light shading), constituing the 'liquid assets' with which the atom can trade to form chemical bonds with other atoms, and

(b) **core electrons** (darker shading), the 'capital reserve' of electrons too tightly bound to the nucleus to be involved in bonding.

The number of valence electrons, among other factors, determines the number of bonds that an atom can form (its '**valency**').

Each electron in an atom resides in a specific region of space close to the nucleus called an **orbital,** each orbital accommodating up to two electrons. The various orbitals differ in their symmetry around the nucleus, and this determines the geometry of directional bonds such as those responsible for the structure of diamond.

The size and shape of each orbital is indicated by the values of **quantum numbers** n and l (Table 5.3).

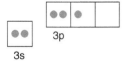

3p

3s

The distribution of electrons between orbitals, and the division into 'core' and 'valence' categories, is determined by the energy-level diagram shown in Figure 5.7. Each box represents an orbital accepting up to two electrons. To predict the chemical properties of an atom, Z electrons are 'fed in' from the bottom (lowest energy, most stable) upward. Valence electrons are those residing in the highest occupied energy levels (least strongly bound electrons, most easily removed).

Review

The important conclusions of this chapter may be summarized in the following terms:

(a) The behaviour of an electron, when trapped in an atom, is dominated by its wave-like nature. The electron distributes itself in space about the nucleus in a manner analogous to a stationary wave on a fixed string.

(b) Each electron in an atom can adopt one of a variety of 'waveforms' having different spatial configurations of electron density. These are called orbitals. Each corresponds to a solution of the Schrödinger wave equation. An orbital's size and shape are defined by the values of various quantum numbers, in a manner reminiscent of the harmonics of a vibrating string.

(c) Electron energy in the atom is quantized, like the frequency of a guitar string (and for the same reason). Each orbital has its own discrete energy level and collectively they give rise to an electron energy structure resembling an irregular set of pigeon-holes (Figure 5.7), which in qualitative terms is common to all types of atom.

(d) Two electrons can share the same orbital provided that they have opposed 'spins'. Each electron normally occupies the lowest-energy orbital in which space is available. Energy considerations thus dictate the geometric distribution of electron density about the nucleus.

As the following chapters will show, these principles, together with the energy level scheme shown in Figure 5.7, form the foundation upon which modern inorganic chemistry is built.

Further reading

Atkins, P., Overton, T., Rourke, J., *et al.* (2010) *Inorganic Chemistry*, 5th edn. Oxford: Oxford University Press.

Barrett, J. (2002) *Atomic Structure and Periodicity*. Cambridge: Royal Society of Chemistry.

Fyfe, W.S. (1964) *Geochemistry of Solids. An Introduction.* Chapter 2. New York: McGraw Hill.

Exercises

5.1 Work out which electron orbitals have the following quantum numbers:

n	l
2	1
3	0
4	3
5	2

5.2 Determine the electron configurations of the chemical elements having the atomic numbers 6, 11, 13, 17, 18, 26.

6 WHAT WE CAN LEARN FROM THE PERIODIC TABLE

A chemical element is identified by its atomic number Z, which defines both the number of protons in the nucleus (and hence the nuclear charge) and the number of electrons in the neutral atom (Box 6.1.). In the present chapter we consider how the atomic number, in conjunction with the electron energy structure developed in Chapter 5 (Figure 5.7), determines the chemical properties of the element concerned. The structure of the energy-level diagram leads to a *periodic repetition* of chemical properties, which is conveniently summarized by tabulating the elements on a grid known as the Periodic Table (see inside rear cover).

Although the architecture of the Periodic Table can be thought of as an outcome of wave-mechanical theory, it was originally worked out from chemical observation. It was first published in its modern form by the Russian chemist Dimitri Mendeleev in 1869, almost 60 years before Schrödinger published his paper on wave mechanics.

Ionization energy

The bonds formed by an atom involve the transfer or sharing of electrons. It therefore makes sense to illustrate the periodicity of chemical properties by looking at a parameter that expresses how easy or difficult it is to remove an electron from an atom. The **ionization energy** of an element is the energy input (expressed in $Jmol^{-1}$) required to detach the loosest electron from atoms of that element (in its ground state). It is the energy difference between the 'free electron at rest' state (the zero on the scale of electron energy levels) and the highest occupied energy level in the atom concerned. What this means in the simplest case, the hydrogen atom, is shown in Figure 5.6. A low ionization energy denotes an easily removed electron, a high value a strongly held one.

We can picture how ionization energy will vary with atomic number by considering the highest occupied energy level in each type of atom (Figure 5.7). In

Chemical Fundamentals of Geology and Environmental Geoscience, Third Edition. Robin Gill.
© 2015 John Wiley & Sons, Ltd. Published 2015 by John Wiley & Sons, Ltd.
Companion Website: www.wiley.com/go/gill/chemicalfundamentals

Box 6.1 Chemical symbols

A few of the one- or two-letter codes for the chemical elements will be familiar to most readers. The majority are abbreviations of the English element names, but a few refer to Latin names such as Na (natrium) for sodium and Ag (argentum) for silver, or to alternative continental names like 'wolfram' for tungsten (W). A list of chemical symbols is given at the end of the book (Appendix C).

Using subscripts and superscripts, chemical symbols can be augmented to specify every detail about a particular atom (see Figure 6.1.1). It is rarely necessary to specify more than one of these numbers in a given context. They should always be written in the positions shown, to avoid ambiguity. For example, to specify a particular isotope (Chapter 10) one writes ^{40}Ar. (Older literature may use the obsolete notation Ar^{40}.)

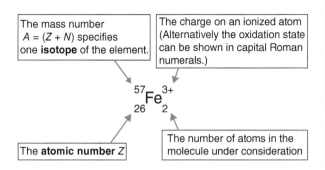

The mass number $A = (Z + N)$ specifies one **isotope** of the element.

The charge on an ionized atom (Alternatively the oxidation state can be shown in capital Roman numerals.)

The **atomic number** Z

The number of atoms in the molecule under consideration

$$^{57}_{26}Fe^{3+}_{2}$$

Figure 6.1.1 How key parameters of an atom are codified in an element's chemical symbol.

The atomic number Z can generally be omitted, its value being implied by the chemical symbol itself (although see Figure 6.2).

---

lithium (Li; $Z = 3$, electronic configuration $= 1s^2 2s^1$) and beryllium (Be; $Z = 4$, $1s^2 2s^2$) it is the 2s level; in boron (B; $Z = 5$, $1s^2 2s^2 2p^1$) it is the 2p level; and so on. If we were to disregard the increasing nuclear charge, we would predict that the energy needed to strip an electron from this 'outermost' level would vary with atomic number as shown in Figure 6.1a. One would expect a general decline in ionization energy with increasing Z, punctuated by sudden drops marking the large energy gaps between one 'shell' and the next one up (Figure 5.6); the downward series of steps in Figure 6.1a thus reflects the occupation of progressively higher energy levels in Figure 5.6. There is no suggestion of periodicity.

However, because nuclear charge – and therefore the strength of the nuclear field – increases with atomic number, we find that each of the level steps anticipated in Figure 6.1a is actually a *ramp* (Figure 6.1b), whose rising profile reflects the increase in nuclear attraction experienced by each electron in the atom in passing from one atomic number to the next. The ionization energy of helium is (nearly) twice that of hydrogen, for example, because its doubly charged nucleus attracts each electron twice as strongly as the singly charged hydrogen nucleus. The ramps are separated by the sudden drops noted in Figure 6.1a, producing a markedly periodic variation of ionization energy.

At the top of each ramp is an element that hangs on tenaciously to all of its electrons. These elements are the **noble** (or **inert**) **gases**, helium (He), neon (Ne) and argon (Ar). (Two others, krypton (Kr) and xenon (Xe), lie beyond the Z-range of the diagram – see Figure 6.1b inset.) Their electronic structures are characterized by completely filled shells, in which all electrons are held so firmly that the exchange of electrons involved in chemical bonding is ruled out. Noble gases therefore exhibit no significant chemical reactivity. Indeed, the electronic structure of the noble gases is so stable that other elements seek to emulate it by losing electrons, or by acquiring additional electrons from other atoms (as happens for example with the element chlorine, Cl). Instead of forming **diatomic** molecules like O_2, N_2 and Cl_2, the noble gases are *monatomic*.

Immediately to the right of each noble gas in Figure 6.1b lies an element with a conspicuously low ionization energy. Lithium (Li), sodium (Na) and potassium (K) are *alkali metals*, whose electronic structures consist of the filled shells of the preceding noble gas, plus one further electron which has to occupy the next shell at a significantly higher energy (Figure 5.7). It projects further from the nucleus than the core electrons, and is **screened** by them from the full attraction of the nuclear charge, making it even easier to remove or involve in bonding. The chemistry of the alkali metals is dominated

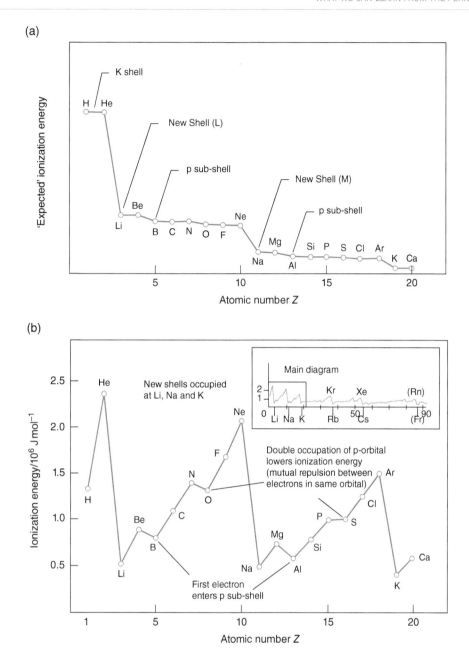

Figure 6.1 (a) A notional plot of ionization energy against atomic number, predicted without regard to the effect of increasing nuclear charge. (b) The variation of *measured* ionization energy with atomic number Z among the first 20 elements. (The whole Z-range is shown in the inset.) The rising profile between each abrupt drop reflects increasing nuclear charge.

by this single **valence electron**. All are **monovalent** metals: removing a second electron would be much more difficult, because to do so would mean breaking into the stable noble-gas core (Chapter 5). Because of their low ionization energies, the alkali metals readily form singly charged M$^+$ cations.

Beryllium (Be; $1s^2 2s^2$), magnesium (Mg; $1s^2 2s^2 2p^6 3s^2$) and calcium (Ca; $1s^2 2s^2 2p^6 3s^2 3p^6 4s^2$) all have two electrons in their valence shells, both of which are fairly easy to remove (although not as easy as the single electron in the valence shell of an alkali metal). These *alkaline earth metals* utilize both of these electrons in their bonding and – with the exception of Be (Box 9.3) – readily form the doubly charged M^{2+} cation.

Boron (B) and aluminium (Al), to the right of Be and Mg in Figure 6.1b, each have three valence electrons to utilize in their chemical reactions; and carbon (C) and silicon (Si) each have four. The increasing nuclear attraction makes these electrons progressively harder to remove, however, and the tendency among these elements to form cations is significant only for Al; the chemistry of B, C and Si is dominated by bonds in which electrons are shared (Chapter 7).

The periodic pattern becomes more complicated at Z values above 20 (inset in Figure 6.1b), owing to the presence of electrons in d-orbitals. The overall periodicity nevertheless persists, with the minimum ionization energies belonging to the alkali metals rubidium (Rb) and caesium (Cs), and the maximum values coinciding with the noble gases krypton (Kr) and xenon (Xe).

The Periodic Table

Having arranged the elements in order of atomic number, they can be divided into several *periods*, each one beginning with an alkali metal – except the first, which begins with hydrogen – and concluding with a noble gas. Writing successive periods underneath each other on a page gives a coherent layout of the elements that highlights their common chemical properties. This is shown for the first three periods (including hydrogen and helium) in Figure 6.2. Notice that each of the groups of similar elements that emerged from Figure 6.1b now forms its own column. The first contains the alkali metals Li, Na and K, to which by tradition we add the element hydrogen, because it resembles the alkali metals in having just one electron in the valence shell. In column 2 we find the alkaline earth metals, Be, Mg and Ca; in column 3, B and Al, and so on.

Numbering the columns in Figure 6.2 from left to right divides the elements into eight groups, identified by Roman numerals, according to the number of electrons in the valence shell. Numbering the periods indicates the n-value (the principal quantum number) of the current valence shell: thus for Al in period 3 the valence electrons are in orbitals that have $n=3$ (the M shell). The noble gases at the end of each period mark the *closure* (filling up) of the p-orbitals of that shell. It is therefore logical to place helium at the head of this eighth column rather than column 2, as its two electrons close the K (1s) shell.

This is the rationale of the *Periodic Table*, which is shown in its complete form on the inside cover. Notice that it has been split between Groups II and III in order to accommodate elements like scandium (Sc), titanium (Ti) and iron (Fe). This is where electrons begin to occupy the lowest-energy d-orbitals (3d). The ten elements from

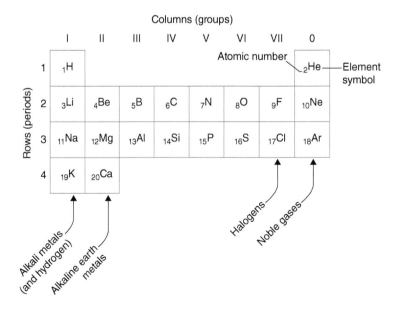

Figure 6.2 A condensed Periodic Table for the first 20 elements. Chemical symbolism is discussed in Box 6.1.

Sc to Zn (zinc) comprise the *First Transition Series*. They include many industrially important metals (Figure 9.7).

The next six elements, from gallium (Ga) to the noble gas krypton (Kr), correspond to the filling of the 4p orbitals, the next subshell in order of energy (Figure 5.7).

Subsequent periods follow the same pattern, except for the added complication of the availability of 4f and 5f orbitals in very heavy atoms. The entry of electrons into 4f orbitals generates a series of geochemically important trace elements called the **lanthanides** (cerium, Ce, to lutetium, Lu), although in geological literature they are more commonly known – together with the preceding element lanthanum, La – as the *rare earth elements* (**REEs**; see Chapter 9).

In geological terms, the most important heavy elements of all are thorium (Th) and uranium (U). Owing to their radioactivity, they make a major contribution to heat generation in the Earth, and are important in geochronology. They belong to a similar series of elements arising from the filling of 5f orbitals, known as the *actinides* (Chapter 9). None have stable nuclei, only Th and U having sufficiently long half-lives to occur in nature today (Chapter 10).

Electronegativity

Ionization energy provides a useful indication of the periodicity of element behaviour, but its chemical applications are limited to elements that lose electrons to form cations. *Electronegativity* is a more versatile concept, summarizing the chemistry of all kinds of elements and the bonds they form. Electronegativity is a number that indicates the *capacity of an atom in a molecule or crystal to attract extra electrons*. In the alkali metals, this capacity is hardly developed at all; elements that tend to give electrons away rather than attracting additional ones are called *electropositive* elements. They appear on the left-hand side of the Periodic Table. Their electronegativity values are low, beginning at 0.8 (for the alkali metals K, Rb and Cs).

The most *electronegative* elements are those having nearly complete valence shells, on the right-hand side of the Periodic Table (Figure 6.3). The nuclear charge effect draws the valence orbitals closer to the nucleus, offering an *incoming* electron a state of low energy, compared for example with the valence shell of an alkali metal where the attraction of the nucleus is only weakly felt.

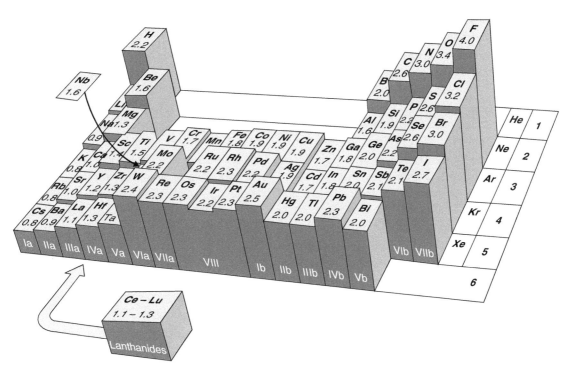

Figure 6.3 Variation of electronegativity (height of each block) across the Periodic Table (stable elements only). Groups and periods are shown in Roman and Arabic numerals respectively. Data from Henderson and Henderson (2009) and other sources. The gap between molybdenum (Mo) and ruthenium (Ru) belongs to the radioactive element technetium (Tc – see Exercise 6.4).

Electronegative atoms thus have the power to attract and retain additional electrons, in spite of the net negative charge that the atom thereby acquires (making it an **anion**). The explanation for this phenomenon is partly wave-mechanical, to do with the special stability of the noble-gas configuration that such an atom can achieve with the aid of borrowed electrons. The most electronegative element is fluorine (F, electronegativity 4.0).

Figure 6.3 shows that electronegativity varies in a fairly regular manner in the Periodic Table. It increases strongly from left to right, and more gently from bottom to top (although the latter trend is reversed in the central area). As a rule, metals have electronegativities less than 2.0, whereas non-metals have values greater than 2.5.

Valency

The number of bonds that an atom can form as part of a compound is expressed by the **valency** of the element. In chemical reactions, elements adjust their electron populations to achieve a noble-gas configuration. Because the single valence electron of alkali metals like Na ($1s^22s^22p^63s^1$) allows them to form only one bond (Chapter 7), they are said to be **monovalent** (valency = 1). Magnesium ($1s^22s^22p^63s^2$) has a valency of two (it is **divalent**), because two of the electrons in the neutral Mg atom are in the valence shell, and can be used to establish bonds. For strongly electropositive elements like these (Figure 6.3), valency is equal to the number of electrons in the valence shell, and can therefore be determined from the column in the Periodic Table in which the element occurs. The valencies of the elements B and Al (**trivalent**), C and Si (quadrivalent) and P (pentavalent) also conform to their position in the Periodic Table.

The electronegative elements like oxygen and chlorine require a complementary definition of valency. These are elements with nearly complete valence shells. Because they can achieve a noble-gas electron structure by *accepting* extra electrons from other atoms (or sharing them), it is the number of *vacancies* in the valence shell that determines the number of bonds that can be formed. Oxygen, with six valence electrons and two vacancies, can establish two bonds with other atoms and is divalent. Chlorine (seven valence electrons, one vacancy) needs only one electron to complete the valence shell and is therefore monovalent.

Many of the elements in the central parts of the Periodic Table behave in a more complicated manner. In Chapter 4 we discussed a few elements that can adopt one of several **oxidation states** depending on the oxidizing or reducing properties of the environment. Each of these oxidation states represents a separate valency of the element concerned. Multiple valency is most characteristic of the transition elements (Figure 9.7). The best-known example is iron, which in addition to the metallic state (valency 0) can exist in the geological world in ferrous (divalent) or ferric (trivalent) compounds. It often happens – as in the case of iron – that none of the oxidation states corresponds to the valency suggested by the element's position in the Periodic Table. Iron occurs in Group VIII (Box 6.2, Figure 6.3), but exhibits valencies of 2 and 3. It is clear that such elements do not utilize all of the electrons that nominally belong to the valence shell in forming bonds, for reasons that will be examined in Chapter 9. The same is true of many heavier elements in the p-block (Box 6.2.). Tin, for instance, occurs in the geological

---

Box 6.2 Sub-groups and blocks

The condensed form of the Periodic Table for the first 20 elements (Figure 6.2) is easy to divide into eight columns, but the introduction of the transition series – an additional ten columns – makes necessary a revision of these column headings. In order that elements like B, C and Al retain group numbers reflecting their valency, groups are divided into 'a' and 'b' *subgroups* as shown in Figure 6.3 and on the inside rear cover, leaving three columns in the middle of the transition series (headed by Fe, Co and Ni) which are lumped together as Group VIII. Formal chemical similarities exist between corresponding a and b subgroups (e.g. valency, as in subgroups IVa and IVb), but they are generally outweighed by the differences in electronegativity (Figure 6.3).

It is useful to divide the Periodic Table into *blocks* reflecting the kind of subshell (s, p, d or f) that is currently incomplete or just filled. The transition series comprises the d-block and the lanthanides and actinides the f-block. It is important to recognize the direct relationship of these blocks, and of the structure of the Periodic Table as a whole, to the energy-level diagram shown in Figure 5.7.

environments as Sn(II) and Sn(IV) compounds. Another example is arsenic which, as Figure 4.2 shows, occurs naturally as both As(III) and As(V).

Atomic spectra

An atom is said to be in its *ground state* when all of its electrons occupy the lowest energy levels allowed to them by the Pauli Principle (Chapter 5). This lowest-energy configuration (Figure 6.4a) is the one normally encountered at room temperature. But atoms can absorb energy from their surroundings, for example when they are heated or exposed to energetic radiation, and this causes one or more electrons to jump from a stable, low energy level into one of the vacant orbitals at higher energy, or perhaps even to be ejected from the atom altogether (Figure 6.4a). The unstable *excited state* of the atom so produced, with a vacancy in a low energy level (Figure 6.4b), soon reverts to the stable ground state by filling the vacancy with an electron from a higher level. The electron

making this downward transition must dispose of an amount of energy ΔE (Figure 6.4) exactly equal to the difference between its initial and final energy levels, and this energy output takes the form of electromagnetic radiation (Box 6.3.). Excited atoms therefore emit a series of sharply defined wavelength peaks (Figure 6.5) that provide detailed information about their electronic energy structure: these wavelengths constitute the *emission spectrum* of the element(s) concerned. Because the electron energy levels in an atom, and therefore the wavelengths it emits, are Z-dependent, the spectrum of one element is readily distinguishable from that of another (Box 6.3.). Atomic spectra thus provide an important practical means of identifying elements present in a sample – be it a rock powder or a solution – when suitably excited, and of determining their relative abundances.

The success in explaining why an atomic emission spectrum consists of a series of sharp lines rather than a continuum is one of the triumphs of wave mechanics.

Just as an atom *emits* radiation at characteristic wavelengths when electrons fall from excited energy levels to lower ones, so the excitation of an electron from the ground state to an excited energy level may be associated with (*i.e.* can be caused by)[1] the *absorption* of radiation at the same distinctive wavelengths. The astronomical application of such atomic *absorption spectra* will be discussed in Chapter 11.

Can an electron leap from any energy level to any other level within an atom? Analysis of the peaks present in an atomic spectrum indicates that the answer must be 'no'. Certain transitions are 'forbidden' because they would violate basic physical principles such as the conservation of angular momentum. Wave mechanics recognizes such restrictions in the form of a number of 'selection rules'. For example, a radiative transition in an atom must satisfy the two conditions:

$$\Delta l = \pm 1$$

$$\Delta n \neq 0$$

Thus element spectra do not include lines that correspond to transitions between 3s and 2s states (for

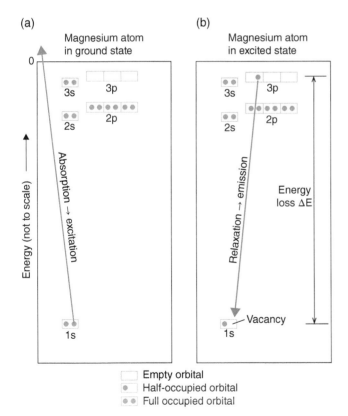

(a) Magnesium atom in ground state
(b) Magnesium atom in excited state

Energy (not to scale)

Absorption → excitation

Relaxation ↑ emission

Energy loss ΔE

Vacancy

☐ Empty orbital
◉ Half-occupied orbital
◉◉ Full occupied orbital

Figure 6.4 Ground states and excited states. The transition shown emits an MgKβ quantum as an X-ray photon.

---

[1] But excitation may be caused by forms of energy other than electromagnetic radiation. For example, in the electron microprobe (Box 6.4) a high-energy electron beam is the agent of excitation.

Box 6.3 Light and other forms of electromagnetic radiation

The light that we see, like other forms of electromagnetic radiation, is an electromagnetic disturbance that propagates energy through space, rather like the radiating ripples on the surface of a pond disturbed by a stone. The source of light excites simultaneous 'ripples' in electric and magnetic field strength, which spread out from the source at the speed of light.

The essential characteristics of any electromagnetic wave are the *frequency* of vibration of the electromagnetic field (v) in hertz (Hz = oscillations per second = s^{-1}) and its *wavelength* (λ) in metres (Box 5.2). These are complementary properties related through the equation:

$$\lambda v = c$$

where c is the speed of light in m s^{-1} ($c = 2.997 \times 10^8$ m s^{-1} in vacuum). The wavelength is the parameter normally used to characterize the quality of visible light that we call colour see Figure 6.3.1. Frequency, however, is the more fundamental property: unlike wavelength and c, it is independent of the refractive index of the medium through which the light is passing.

Light energy is *quantized*: a light beam, though apparently a continuous stream of waves, actually consists of minute packets or 'quanta' of wave energy called **photons,** resembling the wave pulses associated with the electron (Figure 5.1). Planck showed at the turn of the century that each photon has a kinetic energy E_q, related to the frequency of the light of which it forms a part:

$$E_q = hv$$

where h is called Planck's constant and has the value 6.626×10^{-34} J s.

When an electron falls from a high energy level in an atom to a lower one, it emits a quantum of energy in the form of an electromagnetic photon, whose energy is exactly

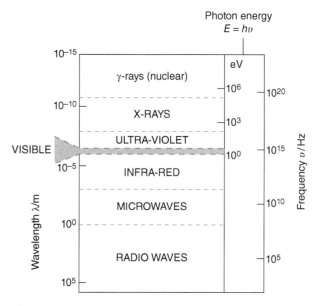

Figure 6.3.1 The electromagnetic spectrum.

equal to the *energy difference* ΔE between the electron's initial and final states. It follows that the light emitted by atoms undergoing this transition has a frequency given by:

$$v = E_q / h = \Delta E / h$$

The corresponding wavelength is $\lambda = \dfrac{hc}{\Delta E}$

Since the energy levels (and ΔEs) in an atom depend on the nuclear charge Z, the wavelengths of atomic spectra vary predictably from one element to the next, and can be used (when separated by a spectrometer into constituent wavelengths) to identify the elements present in a complex sample without separating the elements chemically. The *intensity* of each wavelength 'peak' in the spectrum provides a measure of the concentration in the sample of the element to which it relates (Box 6.4).

---

which Δl would be = 0), between 3d and 2s states (for which $\Delta l = 2$), or between 3p and 3s (for which $\Delta n = 0$).

X-ray spectra

X-rays are electromagnetic waves of very short wavelength (about 10^{-8} to 10^{-11} m) and high frequency (Box 6.3.). They are the most energetic form of radiation that can

be generated in the electron shells of atoms. (γ-rays have higher energies, but are produced in nuclei.) The high energies indicate that X-rays arise from electron transitions involving the deepest, most tightly bound energy levels in the atom, in particular the K and L shells. The energy level structure in these shells is simple (Figure 5.7) owing to the restrictions that apply to the value of quantum number l when n is

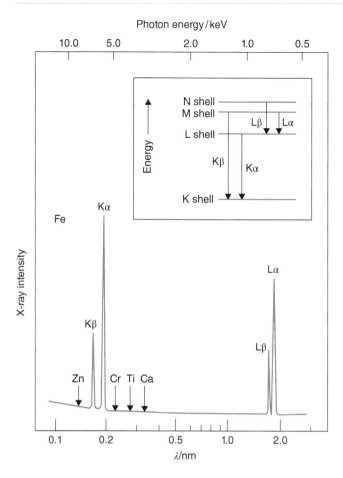

Figure 6.5 The X-ray spectrum of iron as shown by an X-ray spectrometer. The inset shows the electron transitions involved. The arrows in the main diagram show how Kα wavelength shifts from element to element. The width of the peaks has been exaggerated.

small. X-ray spectra therefore consist of relatively few lines, a fact which makes them convenient for the analysis of complex, multi-element samples like rocks and minerals, since superposition of lines is less probable. In common with all atomic spectra, the wavelength of each X-ray peak depends upon the atomic number of the element emitting it (Figure 6.5).

The production of X-rays involves creating a vacancy in the K or L shell by ejecting an electron completely out of the atom (Figure 6.4). Two methods can be used to excite atoms into generating X-ray spectra for analytical purposes:

(a) A very narrow beam of high-energy electrons can be focused on a small area on the surface of the sample (usually a crystal on the surface of a polished thin section). The energy of the electrons

excites the atoms with which they collide in the tiny volume of the sample directly under the area of impact, from which the X-ray spectra of elements present in the sample are emitted. This is the principle of the *electron microprobe* (Box 6.4), widely used for the chemical analysis of mineral crystals *in situ* in geological thin sections. The microprobe can also be used to investigate chemical variation *within* an individual crystal, such as **zoning** (Plate 2) or exsolution (Plate 3).

(b) A powerful beam of X-rays (from an X-ray tube), when directed at a sample in powdered or fused form, will prompt X-ray emission in the sample by **fluorescence**. The photon energy of the incoming ('primary') X-rays, E_q, must be sufficient to eject the relevant electrons from the sample atoms (Figure 6.4a). The sample responds by emitting 'secondary' or 'fluorescent' X-ray spectra characteristic of the elements present in the sample. X-ray fluorescence spectrometry is an important rapid method of whole-rock analysis (of homogeneous samples in powdered or fused form), applicable to major elements and many trace elements.

Z-dependence of X-ray spectra: Moseley's Law

One advantage of using X-ray spectra for rock and mineral analysis is that element wavelengths depend in a very simple way on the atomic number Z. Increasing the nuclear charge 'stretches' the energy level structure downward in energy space, expanding the energy differences ΔE between different levels. This is easily seen from the energy scales in Figure 5.7. It follows that the photon energy for a given transition – the Kα line, for example – increases with atomic number, while the corresponding wavelength decreases (Figure 6.5). This relationship is expressed in a simple equation established empirically by the British physicist H.G.J. Moseley in 1914 and known as *Moseley's Law*:

$$\frac{1}{\lambda} = k(Z - \sigma)^2 \qquad (6.1a)$$

Alternatively this may be written in terms of photon energy E:

$$\frac{E}{hc} = k(Z - \sigma)^2 \qquad (6.1b)$$

Box 6.4 The electron microprobe

The electron microprobe is a scanning electron microscope adapted to analyse the characteristic X-ray spectra emitted when the high-energy electron beam strikes a polished specimen, allowing spatially resolved chemical analyses of the specimen surface to be obtained. The technique revolutionized petrology when it was introduced in the 1950s.

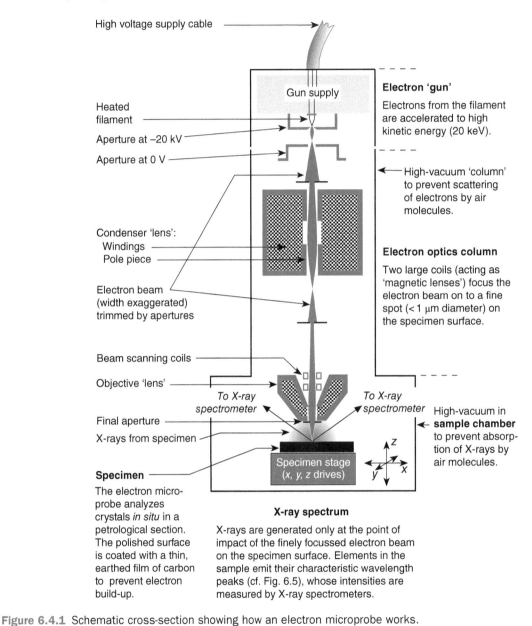

High voltage supply cable

Gun supply

Electron 'gun'

Electrons from the filament are accelerated to high kinetic energy (20 keV).

Heated filament

Aperture at −20 kV

Aperture at 0 V

High-vacuum 'column' to prevent scattering of electrons by air molecules.

Electron optics column

Two large coils (acting as 'magnetic lenses') focus the electron beam on to a fine spot (< 1 μm diameter) on the specimen surface.

Condenser 'lens':
Windings
Pole piece

Electron beam (width exaggerated) trimmed by apertures

Beam scanning coils

Objective 'lens'

To X-ray spectrometer

To X-ray spectrometer

High-vacuum in **sample chamber** to prevent absorption of X-rays by air molecules.

Final aperture

X-rays from specimen

Specimen

The electron microprobe analyzes crystals *in situ* in a petrological section. The polished surface is coated with a thin, earthed film of carbon to prevent electron build-up.

Specimen stage (x, y, z drives)

z, y, x

X-ray spectrum

X-rays are generated only at the point of impact of the finely focussed electron beam on the specimen surface. Elements in the sample emit their characteristic wavelength peaks (cf. Fig. 6.5), whose intensities are measured by X-ray spectrometers.

Figure 6.4.1 Schematic cross-section showing how an electron microprobe works.

The **screening** *constant*, σ, represents the degree to which inter-electron repulsion counteracts the attraction of the positive nuclear charge. $Z - \sigma$ can be regarded as the 'effective nuclear charge' felt by an individual electron. Moseley, who was the first to introduce the concept of atomic number, used this equation to catalogue the chemical elements known in 1914, and to demonstrate from gaps in the sequence

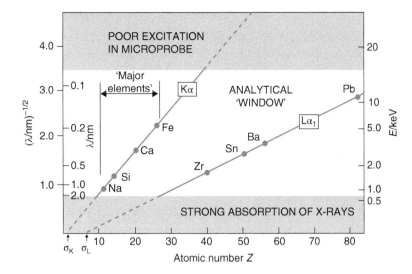

Figure 6.6 Moseley's Law plotted in linear form (Appendix A and Exercise 6.4). Note that the vertical axis is graduated in $\lambda^{-\frac{1}{2}}$ (linear outer scale) and λ (inner scale). The unshaded area shows the range of wavelengths and elements attainable in routine microprobe analysis of minerals and rocks.

of X-ray wavelengths that others (with atomic numbers of 43, 61, 72, 75, 85 and 87) still remained to be discovered.

For plotting the Moseley Equation (6.1) in a graph, it is convenient to express it in terms of the square roots of each side:

$$\left(\frac{1}{\lambda}\right)^{0.5} = k^{0.5}\left(Z - \sigma\right) \qquad (6.2a)$$

or $\left(E_q\right)^{0.5} = \left(hck\right)^{0.5}\left(Z - \sigma\right)$ \qquad (6.2b)

Plotting $(1/\lambda)^{0.5}$ or $E_q^{0.5}$ against Z will produce a straight line with a gradient of $k^{0.5}$ or $(hck)^{0.5}$ respectively (Figure 6.6) and an intercept of σ on the Z axis. Plotting $1/\lambda$ or E_q directly versus Z, on the other hand, would have given a less useful curved line (Appendix A).

The same equation can be used for predicting other X-ray lines such as Kβ and Lα for a particular element, but the constants k and σ will have different values (Figure 6.6).

The analysis of X-ray spectra can be carried out in practice by one of two methods. A crystal of known atomic spacing can be used as an X-ray **diffraction** grating (Box 5.3) to disperse the various wavelength components of the spectrum into a series of peaks according to the Bragg equation (called 'wavelength-dispersive' or 'WD' analysis);

peak intensities can be recorded by a relatively simple X-ray detector driven mechanically to the appropriate angles. Alternatively the incoming X-ray beam can be passed directly into a semiconductor detector that can separate the spectral components according to photon energy ('energy-dispersive' or 'ED' analysis). How such a semiconductor detector works is explained in Chapter 7.

It can be seen from Figure 6.6 that the long-wavelength X-ray spectra from elements of atomic number less than 10 are strongly absorbed, and this limits the effectiveness of X-ray methods when analysing elements having atomic numbers lower than 10, although improved spectrometer design has greatly extended light-element performance in recent years.

Review

(a) The energy level structure in atoms (Figure 5.7), together with the effect of increasing nuclear charge, leads to a periodicity of chemical properties when elements are examined in atomic-number order.

(b) The Periodic Table provides a concise means of summarizing and predicting the variation of chemical properties (such as electronegativity and valency) among the known chemical elements.

(c) Wavelengths of elemental spectra vary systematically with atomic number, providing a powerful means of determining how much of each element is present in a geological sample. X-ray spectra are particularly useful for the chemical analysis of minerals and rocks.

Further reading

Atkins, P., Overton, T., Rourke, J., *et al.* (2010) *Inorganic Chemistry*, 5th edn. Oxford: Oxford University Press.

Barrett, J. (2002) *Atomic Structure and Periodicity*. Cambridge: Royal Society of Chemistry.

Gill, R. (ed.) (1996) *Modern Analytical Geochemistry*. Harlow: Longman.

Scerri, E.R. (2011) *The Periodic Table – a Very Short Introduction*. Oxford: Oxford University Press.

Exercises

6.1 Identify the elements having the atomic numbers listed below. Work out their electronic configurations, distinguishing between core and valence electrons. Establish the block and group to which each element belongs, and work out the valency.

$Z = 3, 5, 8, 9, 14$

6.2 Work out the electronic configurations of the following atoms, representing the electron core by the symbol of the preceding inert gas. To which blocks do they belong?

Ti Ni As U

6.3 Determine the (minimum) values of x and y in the following formulae, consistent with the valencies of the elements concerned:

$$Na_xO_y \ Si_xO_y \ Si_xF_y \ Mg_xCl_y \ Sc_xO_y \ P_xO_y \ B_xN_y$$

6.4 (a) Use the Kα wavelength data below to plot a graph to verify Moseley's Law for the elements Y to Ag (see Appendix B and Figure 6.6). Estimate the values of k and σ.

Element	Z	Wavelength/pm
Y	39	83.0
Zr	40	78.7
Nb	41	74.8
Mo	42	71.1
Ru	44	64.4
Rh	45	61.5
Pd	46	58.7
Ag	47	56.1

(b) The radioactive element technetium (Tc, $Z = 43$; named from the Greek *technetos* = 'artificial' – Figure 6.3) does not occur naturally on Earth, but can be produced artificially. Predict the wavelength of its Kα X-ray line and the corresponding quantum energy (in keV) ($h = 6.626 \times 10^{-34}$ J s $= 4.135 \times 10^{-15}$ eV s; $c = 2.997 \times 10^8$ m s^{-1} in a vacuum.)

7 CHEMICAL BONDING AND THE PROPERTIES OF MINERALS

Few scientists are expected to deal with as wide a range of materials and properties as the geologist. Consider the contrast between red-hot silicate lava and grey Atlantic seawater, between the engineering properties of crystalline granite and those of soft clay or mud, between the electrical and optical properties of quartz and those of gold. The immense physical diversity of geological materials is derived largely from the differences in the chemical bonding that holds them together.

One can distinguish several different mechanisms by which atoms bond together, although the real interaction between two atoms is generally a mixture of more than one bonding type. The extent to which each mechanism contributes to a real bond depends on the difference in electronegativity between the atoms concerned. We begin by examining the type of bond that predominates when the electronegativity contrast is large.

The ionic model of bonding

The **salt** sodium chloride, familiar as common table salt and as the mineral halite, consists of two elements of notably different electronegativity: 3.2 (Cl) and 0.9 (Na) (difference = 2.3). The low ionization energy of the sodium atom (Chapter 6) indicates a readiness to lose an electron, forming a Na^+ cation. The chlorine atom, on the other hand, readily accepts an extra electron, forming a chloride anion Cl^-. When a sodium atom encounters a chlorine atom, one electron may be drawn from the exposed Na 3s orbital into the vacancy in the Cl 3p orbital. The ions resulting from this transfer, having opposite charges, experience a mutual electrostatic attraction that we call *ionic bonding*.

Electrostatic forces operate in all directions, and an ion in an ionic compound like NaCl draws its stability from the attraction of *all* oppositely charged ions nearby. Ionic bonding does not lead to the formation of

Chemical Fundamentals of Geology and Environmental Geoscience, Third Edition. Robin Gill.
© 2015 John Wiley & Sons, Ltd. Published 2015 by John Wiley & Sons, Ltd.
Companion Website: www.wiley.com/go/gill/chemicalfundamentals

discrete molecules like CO_2. Instead ionic compounds exist as solids or liquids (**condensed phases**), which optimize their stability by packing oppositely charged ions closely together in extended structures, or as ionic solutions in which they are stabilized by being surrounded by polar solvent molecules (Box 4.1). Ionic compounds do not exist as gases.

Ionic crystals: stacking of spheres in three dimensions

One can think of most ions as being spherically symmetric. The internal architecture of crystals like NaCl can be understood in terms of stacking spheres of different sizes and different charges into regular three-dimensional arrays. The potential-energy equation for such an array will be the grand sum of (i) negative terms representing the attractive force acting between all pairs of oppositely charged ions in the structure, and (ii) positive terms representing repulsion between all pairs of similarly charged ions. There are three general rules to observe for achieving maximum stabilization (i.e. minimum potential energy):

(a) Ions must obviously combine in proportions leading to an electrically neutral crystal. A halite crystal contains equal numbers of Na^+ and Cl^- ions, whereas in fluorite (calcium fluoride) there must be twice as many F^- ions as Ca^{2+} ions for the charges to cancel.

(b) The spacing (more precisely, the internuclear distance) between neighbouring oppositely charged ions should approximate to an equilibrium bond length r_o for the compound concerned (Box 7.1.), so maximizing the attractive forces holding the structure together. Stretching a bond makes it less stable.

(c) Each cation should be surrounded by as many anions as their relative sizes will allow. This achieves the maximum degree of cation–anion attraction. For the same reason, each anion needs to be closely surrounded by as many cations as possible. The number of oppositely charged nearest neighbours surrounding an ion (in three dimensions) is called its *co-ordination number*, an important parameter in crystal chemistry.

These rules indicate that the atomic arrangement found in a crystal of halite or fluorite is determined primarily by the charge and the size of the constituent ions. The charge can be predicted from the ion's valency, but how can its 'size' be established?

Ionic radius
Because ions have fuzzy outlines (Figures 5.3, 5.4 and 7.1), we face a problem in defining precisely what we mean by the 'radius' of an ion.

Nevertheless, when two oppositely charged ions come into contact they establish a well-defined equilibrium bond length (Box 7.1.). The bond length can be regarded as the sum of two hypothetical 'ionic radii', one for each individual ion, as shown in Figure 7.1. It will be clear that the 'radius' of an ion in a crystal represents not the actual size of the isolated ion (whatever

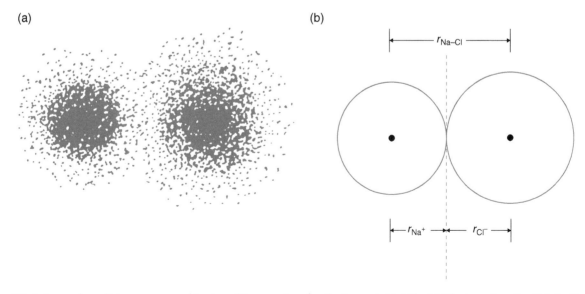

(a)

(b)

Figure 7.1 Inter-nuclear distance $r_{Na–Cl}$ and ionic radii r_{Na^+} and r_{Cl^-} for the ion pair Na^+Cl^-. (a) Electron density distributions of the two ions. (b) Corresponding hard-sphere approximation.

Box 7.1 Equilibrium internuclear distance

Two oppositely charged ions (for example, Na^+ and Cl^-) are attracted towards each other by their opposite net charges ($\pm e$), and the attraction increases as they get closer. The potential energy of this system is:

$$E_p = -\frac{e^2}{r} \qquad (7.1.1)$$

As r – the distance between the two nuclei – gets smaller, the potential energy becomes more negative, indicating greater stability. When the ions get so close to each other that the negative electron clouds begin to intermingle, however, we have to allow for an element of repulsion in the energy equation:

$$E_p = -\underset{\substack{long-range \\ attraction}}{\frac{e^2}{r}} + \underset{\substack{short-range \\ repulsion}}{\frac{be^2}{r^{12}}} \qquad (7.1.2)$$

where b is a constant. The dashed curves in Figure 7.1.1 show the two terms plotted separately. The r^{12} term indicates that repulsion is felt over a much shorter internuclear distance than the attractive force, but rises very steeply as r is reduced below a critical value.

The solid curve in Figure 7.1.1 shows the result of adding the two terms together (Equation 7.1.2). The energy minimum defines an equilibrium internuclear distance or 'bond-length' r_0 at which the isolated ion pair is most stable.

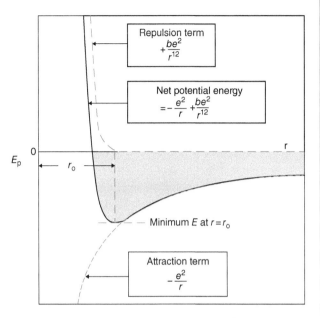

Figure 7.1.1 How the potential energy E_p of two oppositely charged ions varies with internuclear distance r.

(When the cation is associated with more than one anion, as in a crystal, repulsion between the anions makes r_0 somewhat larger.) The rapid rise of the curve to the left of this minimum reflects the observation that the ions strongly resist getting closer together, rather like hard rubber balls.

that may mean), but an empirical 'radius of approach' governing how close to it another ion can be placed.

X-ray diffraction data (Box 5.3) provide accurate equilibrium bond lengths for many binary (two-element) ionic salts like NaCl, NaF and CaF_2. The problem in determining the individual ionic radii of Na^+ and Cl^- lies in deciding how much of the measured Na–Cl bond length to attribute to each ion, the radius of neither being known initially. An early approach to the problem, devised by the American Nobel Prize-winning chemist Linus Pauling in 1927, was to examine compounds like NaF in which the cation and anion happen to have the same number of electrons (10 in this case). Pauling suggested a relationship between ionic radius and nuclear charge in such ion pairs which, together with the measured Na–F bond length, allowed the separate ionic radii to be estimated (see Fyfe, 1964, Chapter 4).

Modern ionic radii are based on more elaborate calculations, although chemists are still not universally agreed on the best values to use: published estimates of the O^{2-} radius, for example, vary between 127 and 140 pm (Henderson, 1982, Chapter 6). Mineralogists and geochemists nonetheless find ionic radii extremely useful in explaining the chemical make-up of crystalline materials. The ionic radii of some geologically important elements are illustrated in Box 7.2.

One can see from Figure 7.2.1 that cation radii vary considerably, from 34 pm (Si^{4+}) to about 170 pm (Rb^+). Note the marked *decrease* in cation radius in proceeding from left to right in the Periodic Table, in response to increasing nuclear charge. Na^+, Mg^{2+}, Al^{3+} and Si^{4+} each possess ten electrons, but they are pulled closer to the silicon nucleus (charge 14+) than to the sodium nucleus (11+). Anion radii are larger, owing to the extra electrons they have acquired and to the mutual repulsion between them. O^{2-}, F^-, S^{2-} and Cl^- are larger than all cations except those of the alkali metals and the heavier alkaline earth elements (Figure 7.2.1). Consequently

Box 7.2 Ionic radii of geologically important elements

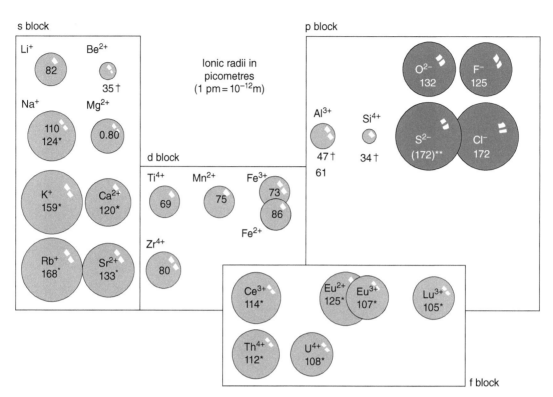

Figure 7.2.1 Ionic radii (in picometres = 10^{-12} m) of geologically important elements.

† Refers to tetrahedral co-ordination
* Refers to 8 fold co-ordination
** The ionic radius of S^{2-} should be used with caution as most metal-sulfide bonds are significantly covalent.

Variation with co-ordination number

The equilibrium length of an ionic bond depends on the co-ordination of the ions (due to repulsion between similarly charged ions – Box 7.1). The Al–O distance in analcime $Na(AlSi_2O_6).H_2O$ (in which Al is tetrahedrally co-ordinated) is 180 pm, whereas in jadeite ($NaAlSi_2O_6$, Al octahedrally co-ordinated) it is more than 190 pm. It follows that individual ionic radii depend on co-ordination too. This dependence is more significant for cations than anions. The data in the chart above refer to octahedral (6-fold) co-ordination unless otherwise specified.

many crystal structures can be visualized as close-packed arrays of large anions, with smaller cations occupying the **interstitial** holes between them.

The radius ratio and its applications

When spheres of the same size are packed together, the most compact arrangement consists of a stack of regular planar layers. Every layer has hexagonal symmetry reminiscent of a honeycomb, each sphere being in contact with six others in the same plane. Similar layers can be stacked on top, each sphere nestling in the depression between three spheres in the layer below (Figure 7.2). The spaces between the touching spheres have two kinds of three-dimensional geometry. One type is bounded by the surfaces of four neighbouring spheres. Joining up their centres generates a regular *tetrahedron* (Figure 7.2a,b, Table 7.1), and in view of their capacity to accommodate small cations (analogous to the ball-bearing shown in Figure 7.2b), such voids are called *tetrahedral sites*.

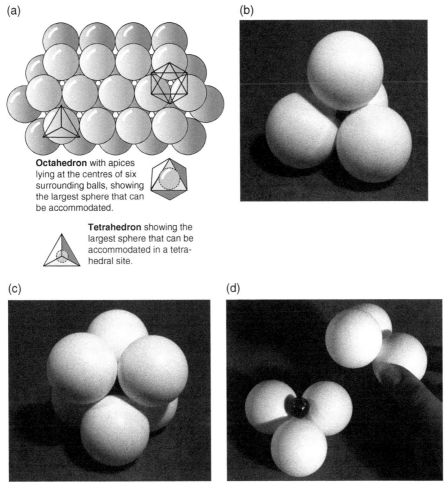

(a)

Octahedron with apices lying at the centres of six surrounding balls, showing the largest sphere that can be accommodated.

Tetrahedron showing the largest sphere that can be accommodated in a tetrahedral site.

(b)

(c)

(d)

Figure 7.2 (a) Two layers in a close-packed array of spheres. The heavy lines show the co-ordination polyhedra of the tetrahedral and octahedral interstitial voids between the spheres. (b) 3D model showing tetrahedral co-ordination of anions around a small cation; the cation (ball bearing) can be seen nestling between the four anions. (c) Octahedral co-ordination around a larger cation. (d) Octahedron with top anion layer removed to show the octahedrally co-ordinated cation within. (Sources: Adapted from McKie & McKie 1974; K. d'Souza.)

The second type of hole in a close-packed array of identical spheres is bounded by six neighbouring spheres, whose centres lie at the six apexes of a regular *octahedron*[1] (Figure 7.2a,c,d). Such holes are called *octahedral sites*. In terms of the size of the largest interstitial sphere that each hole can accommodate (analogous to the glass marble in Figure 7.2d), octahedral sites are 'bigger' than tetrahedral sites. Both are substantially larger than the cavity between three adjacent spheres in the same layer.

In many ionic crystals the anions are assembled in a more or less close-packed array, the cations occupying some of the tetrahedral and/or octahedral sites between them. The kind of site a given cation occupies is determined by the value of the ratio $r_{cation}:r_{anion}$, known as the *radius ratio*. Using three-dimensional trigonometry, it is not difficult to show that, to fit exactly into an octahedral site between six identical 'anion' spheres of radius R, a 'cation' sphere must have a radius of $0.414R$. A cation in this position in a crystal is said to have *octahedral co-ordination*. However, the likelihood of a real ion pair having a radius ratio of exactly 0.414 is negligible, so we must consider the effect on the co-ordination number if the radius ratio deviates from this value.

A radius ratio of exactly 0.414 allows the 'cation' sphere to touch all of the surrounding spheres at once,

---

[1] An octahedron (meaning 'eight sides') has six apexes (points) but eight faces.

Table 7.1 Co-ordination polyhedra

Range of radius ratio r_{cation}/r_{anion}	Co-ordination number N_c	Co-ordination polyhedron
0.225 to 0.414 $\left(1.5^{\frac{1}{2}}-1\right)^{*}$ $\left(2^{\frac{1}{2}}-1\right)^{*}$	4	Tetrahedron
0.414 to 0.732 $\left(2^{\frac{1}{2}}-1\right)$ $\left(3^{\frac{1}{2}}-1\right)$	6	Octahedron
0.732 to 1.000 $\left(3^{\frac{1}{2}}-1\right)$ $\left(4^{\frac{1}{2}}-1\right)$	8	Cube
>1.000	>12	Various

*The upper limit appropriate to co-ordination number N_c is given by $\left(\dfrac{N_c}{2}\right)^{\frac{1}{2}}-1$.

maintaining the optimum bond length (Figure 7.1) with all six anions. This will not be so if the central sphere is smaller than $0.414R$: it will 'rattle in the hole', and the distance to some of the surrounding ions must exceed the optimum bond length, violating the energy-minimizing rule above. This electrostatically unstable situation collapses into a new configuration in which the cation is able to maintain the optimum bond length with fewer surrounding anions. In practice this means that a cation in this size range will occupy a *tetrahedral* site in preference to an octahedral one. A radius ratio of less than 0.414 therefore implies *tetrahedral* (=four-fold) *co-ordination* of the cation (Table 7.1). For this reason we find silicon (Si^{4+}), with a radius ratio to oxide anion

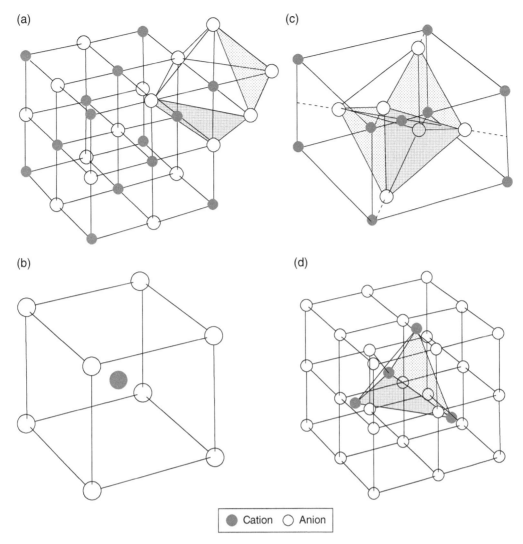

Figure 7.3 Structures of simple binary compounds. Ions are depicted at one-tenth of the appropriate size to allow the three-dimensional disposition to be seen. (a) Sodium chloride structure. (b) Caesium chloride (CsCl) structure. (c) Rutile structure. (d) Fluorite structure.

(O^{2-}) of $34/132 = 0.25$, in tetrahedral co-ordination in all silicate minerals (which are compounds of various metals with silicon and oxygen).

A radius ratio larger than 0.414 does not prevent the cation maintaining the optimum bond length with six equidistant anions. The larger cation prevents the anions remaining in contact with each other (they cease to be strictly close-packed), but in view of their mutual repulsion this will not reduce the stability of the structure. Octahedral co-ordination is therefore consistent with radius ratios above 0.414 (Table 7.1). Na^+ is octahedrally co-ordinated in NaCl (radius ratio $110/172 = 0.64$). Neutrality dictates that Na^+ and Cl^- ions must be equal in number (this is called an 'AB-compound'). It follows that each chloride ion must also be octahedrally co-ordinated by Na^+ ions (Figure 7.3a). This *sodium chloride structure* is shared by the mineral galena (PbS).

In the mineral rutile (one of the **polymorphs** of TiO_2), the Ti^{4+}: O^{2-} radius ratio is $69/132 = 0.52$, and accordingly the Ti^{4+} ion is octahedrally co-ordinated by O^{2-} ions (Figure 7.3c). Because there are twice as many O^{2-} ions (the valency difference makes this an 'AB$_2$'-type compound), each O^{2-} ion is found in three-fold co-ordination, occupying the centre of a triangular grouping of Ti^{4+} ions.

The upper limit for octahedral co-ordination is met at a radius ratio of 0.732, at which the cation is sufficiently large to touch eight equidistant neighbours at the same time (Table 7.1). The requirement of maximum co-ordination leads to a new structure in which the anion nuclei lie at the corners of a cube, with the cation enjoying *eight-fold* co-ordination at the centre (*caesium chloride structure*, see Figure 7.3b). The mineral fluorite CaF_2 (Figure. 7.3d) is an example of an AB_2 compound with the cation in eight-fold co-ordination. Each fluoride F^- ion lies at the centre of a tetrahedral group of Ca^{2+} ions.

If the radius ratio exceeds 1.0, the co-ordination number rises to 12. Eight-fold and larger sites cannot exist as interstitial sites in a close-packed assemblage of anions, and the presence of large ions like K^+ requires the host crystal to have a more open structure, as for example in feldspars.

Analysing ionic crystal structures as if they were 3D assemblages of hard spheres thus explains why the structure of halite, for example, differs from that of rutile, and also provides a basis for understanding the preferences of major elements for specific sites in silicate minerals (Table 8.2). One must recognize, however, that the ionic model is an idealization, and real crystal structures are complicated by other factors. In many minerals the bonding involves a degree of electron-sharing (covalency – see the following section) which undermines the assumption that bonding is non-directional, and where transition metals are concerned (Chapter 9) the presence of d-orbitals introduces still more complications. Predictions of crystal structure based on the radius ratio must therefore be seen only as general guidelines.

The covalent model of bonding

Many substances exhibit chemical bonding between atoms having the same, or very similar, electronegativity. Among them are some of the hardest, most strongly bonded materials known, including diamond, silicon carbide and tungsten carbide. In materials like these, ionic bonding cannot operate and a different bonding mechanism must be at work. Although capable of forming extended crystalline structures like diamond, this bonding is also responsible for small discrete molecules like O_2, CH_4, CO_2 and H_2O, the shape of which often indicates a directional type of bond completely foreign to the electrostatically bonded, close-packed materials so

far considered. These are the characteristics of *covalent bonding,* which operates through the *sharing* of unpaired electrons between neighbouring atoms.

An unpaired electron is one that occupies an orbital on its own. When such singly occupied orbitals in adjacent atoms overlap each other, they coalesce to form a *molecular orbital*, allowing the electrons to pass freely between one atom and the other. In this shared state the electrons have a total energy lower than that in either atom individually (Box 7.3), and consequently the atoms have greater stability attached to each other than they had as separate atoms. The greater the degree of overlap, the stronger the attractive force becomes; however, as with ionic bonding (Box 7.1), the tendency for the atoms to continue approaching each other is restricted by a short-range repulsion between the core electrons of the two atoms, and ultimately between their nuclei. The equilibrium covalent bond length between two atoms can be divided up into the covalent radii of the individual atoms, but the values differ numerically from the corresponding ionic radii.

σ- and π-bonds

The unpaired electron in a hydrogen atom is found in the 1 s orbital, so the coupling in a hydrogen molecule H_2 is due to 1 s overlap. The molecular orbital so formed (Box 7.3) is designated 1sσ, and its occupation by two electrons can be symbolized by the electron configuration $1s\sigma^2$. A *σ-bond* is one with cylindrical symmetry about the line joining the two nuclei (Figure 7.4a). σ-bonds can also form when two p-orbitals overlap end-on, or when a p-orbital in one atom overlaps with an s-orbital in another, as in the water molecule (Figure 7.4b). The participation of p-orbitals, with their elongated shape (Figure 5.5), gives a σ-bond a specific direction, whose orientation in relation to other bonds determines the shape of a multi-atom molecule like H_2O. Note that the two σ-bonds in the water molecule involve separate p-orbitals of the oxygen atom, each of which contributes an unpaired electron; it is not possible for the hydrogen atoms to be attached to opposite lobes of the same p-orbital.

Geometric constraints prevent the formation of more than one σ-bond between the same pair of atoms. There is, however, another way in which p-orbitals can link atoms together. Figure 7.4c shows two atoms that are already joined by a 2pσ bond as described above; the σ molecular orbital is shown by the heavy stipple.

Box 7.3 The mechanism of covalent bonding

When two hydrogen atoms come into contact and their electron orbitals overlap (Figure 7.3.1a), each electron feels the electrostatic attraction of a second nucleus. Both electrons modify their standing waves to extend across the two atoms, forming a *molecular orbital* (Figure 7.3.1b) in which the electrons are identified with the H_2 *molecule* instead of with the separate H *atoms*. The wave function of the molecular orbital can be regarded as the sum of the separate atomic wave functions (Figure 7.3.1c). The *electron density* (ψ^2) in the region between the two nuclei is enhanced (shaded area) at the expense of other parts of the molecule, and **screens** the nuclei from each other. The molecular orbital offers each electron a lower energy than it had in the isolated atom (Figure 7.3.1d). Any two atoms having valence electrons in such *bonding* orbitals have a lower energy together than separately, in spite of the internuclear repulsion. For this reason the hydrogen molecule H_2 is more stable than atomic hydrogen.

A bonding orbital accepts two electrons with opposed spins, accommodating the unpaired electron contributed by each of the bonding atoms. The overlap of atomic orbitals that each contain two electrons (as when two helium atoms touch) does not lead to bond formation. The additional two electrons establish a complementary molecular orbital configuration (an *anti-bonding* orbital, with diminished electron density between the nuclei) whose energy is *higher* than the corresponding atomic orbitals (see Figure 7.3.1d). With electrons occupying bonding and anti-bonding orbitals, there is no net energy advantage in forming a molecule. Thus helium, having no unpaired electrons, cannot form a stable He_2 molecule.

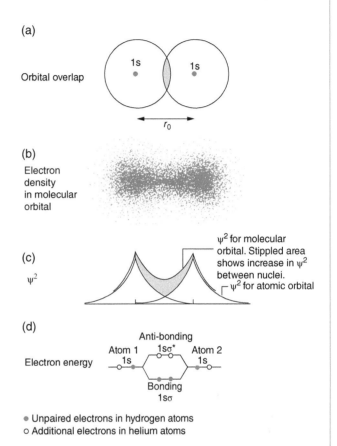

(a) Orbital overlap

(b) Electron density in molecular orbital

(c) ψ^2

ψ^2 for molecular orbital. Stippled area shows increase in ψ^2 between nuclei.
ψ^2 for atomic orbital

(d) Electron energy

Anti-bonding
Atom 1 $1s\sigma^*$ Atom 2
1s   1s
Bonding
$1s\sigma$

● Unpaired electrons in hydrogen atoms
○ Additional electrons in helium atoms

Figure 7.3.1 Ways of looking at a covalent bond. r_0 represents the equilibrium internuclear distance.

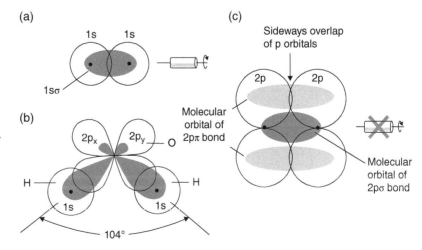

Figure 7.4 Orbital overlap in σ- and π-bonds. The stippled areas show the approximate disposition of the associated bonding molecular orbitals. (a) $1s\sigma$ bond in the hydrogen molecule H_2. Note the cylindrical symmetry. (b) 2p $1s\sigma$ bonds in the water molecule H_2O. (c) $2p\sigma$ and $2p\pi$ bonds in a double-bonded molecule such as O_2.

Suppose each of these atoms has a second unpaired electron in a p-orbital perpendicular to those forming the σ-bond. The two lobes of this orbital lie in the plane of the diagram (large circles), and each can overlap sideways as shown with the corresponding lobe in the other atom, generating two concentrations of electron density (lighter stipple), above and below the σ-bond. Together these constitute the molecular orbital of what is called a *π-bond* between the two atoms. π-bonds form only in conjunction with a σ-bond and are therefore characteristic of molecules containing *double* or *triple bonds*. Such molecules are not cylindrically symmetric. π-bonding occurs in the doubly bonded oxygen molecule (O=O, i.e. O_2), although the configuration is a little more complicated than described above. The triple bond in the nitrogen molecule ($N \equiv N$, i.e. N_2) consists of one σ-bond, with two π-bonds in mutually perpendicular planes.

Covalent crystals

It is only the lightest elements on the right of the Periodic Table (N, O) that form multiple-bonded, diatomic gas molecules. Heavier elements in the same groups (V and VI) do not generally form stable multiple bonds; instead they exist as solids consisting of extended, singly bonded molecular structures. The crystal structure of the element sulfur, for example, consists of buckled rings of six or eight sulfur atoms in which each is bonded to two others. These heavier molecules have too much inertia to exist as gases at room temperature.

In diamond (a form of crystalline carbon) and silicon, each atom is bonded to four others in a continuous three-dimensional network (Figure 7.5d). Each perfect diamond crystal can therefore be regarded in a formal sense as a single 'molecule' of carbon.

Molecular shape: hybridization

Methane (CH_4), the chief constituent of natural gas, consists of molecules with a distinctive tetrahedral shape. Four C–H bonds project from the central C atom, as if towards the four 'corners' of a regular tetrahedron, at angles of 109° to one another (Figure 7.5a). It is hard to reconcile this shape, and the existence of four identical bonds, with the electronic configuration of carbon ($1s^2 2s^2 2p^2$) in which the 2 s electrons are already paired and therefore not available for bonding. Why, in that case, is the valency of carbon 4, not 2?

An unpaired valence electron achieves a lower energy in a molecular orbital (if of the 'bonding' type – Box 7.3) than in its own atomic orbital, which is why molecules can be more stable than separate atoms. In forming methane, carbon can *promote* an electron from the 2 s orbital into the one 2p orbital that is still vacant: the slight energy disadvantage in doing so (Figure 5.7) is outweighed by having *four unpaired electrons* available to establish molecular orbitals with hydrogen atoms, rather than just two (Figure 7.5a).

The Schrödinger wave analysis of the atom, as well as defining the shape of individuals s- and p-orbitals, offers the possibility of *mixing* their wave functions in various proportions to produce *hybrid orbitals* of different geometry. **Hybridization** is another consequence of the wave model of the electron. Just as a guitar string can simultaneously vibrate with two or more harmonics of different frequencies (it is the combination of multiple harmonics that gives the instrument its distinctive musical tone or 'timbre'), so a single atomic electron can adopt more than one waveform. The combined waveform (the 'hybrid' orbital) differs in shape from the individual waveforms from which it is mathematically derived. The result of amalgamating the 2 s and three 2p orbitals of carbon is called an *sp³-hybrid*. It consists of four lobes of electron density, each accommodating an unpaired electron, projecting out from the nucleus with tetrahedral symmetry. Each of these lobes forms a separate σ-bond with a hydrogen atom in the methane molecule.

The geometry of the ammonia molecule (NH_3) also results from the sp³ hybrid. Nitrogen has five electrons in its valence shell (Figure 7.5b). Three of the lobes of an sp³-hybrid are each occupied by an unpaired electron that forms a σ-bond with a hydrogen atom. The fourth lobe accommodates the remaining two electrons, which – being paired – are not available for bonding that involves electron-sharing (covalent bonding). The electron density of this **lone pair** repels each of the N–H molecular orbitals slightly, leading to a distorted hybrid in which the angle between the bonds is only 107°. In the water molecule, *two* of the sp³ lobes on the oxygen atom contain lone pairs, and their combined repulsion closes the angle between the O–H bonds still further (to 104° – Figure 7.5c).

The tetrahedral sp³ hybrid can be recognized in the structure of diamond too (Figure 7.5d), the unique hardness of which reflects the strong bonds that each

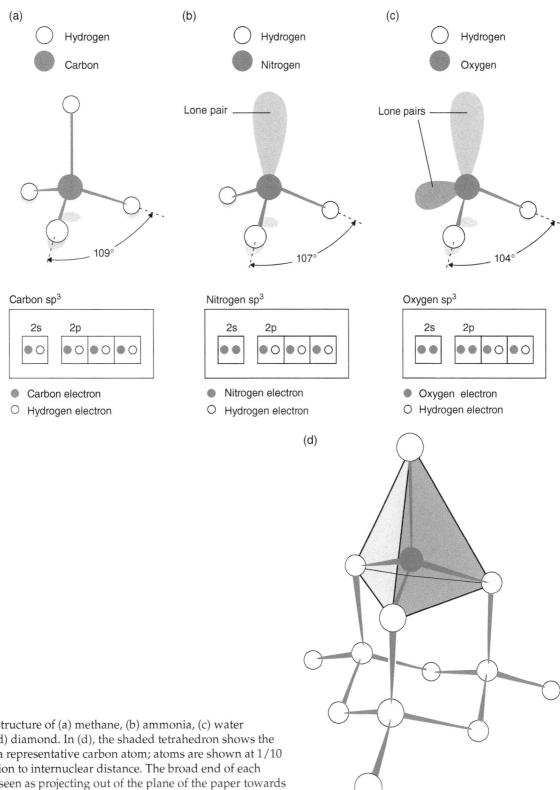

Figure 7.5 The structure of (a) methane, (b) ammonia, (c) water molecules and (d) diamond. In (d), the shaded tetrahedron shows the coordination of a representative carbon atom; atoms are shown at 1/10 true size in relation to internuclear distance. The broad end of each bond should be seen as projecting out of the plane of the paper towards the observer.

Box 7.4 Case study: bonding in graphite and graphene

Graphite is a soft, greasy, black, lustrous mineral, widely used as a lubricant, as an electrical conductor, and of course as pencil 'lead' (the name 'graphite' derives from the Greek verb 'to write'). It is impossible to imagine a greater contrast between these properties and those of the other carbon mineral, diamond. Why are these two forms of carbon so different?

Figure 7.4.1b shows that graphite, unlike diamond, is made up of sheets. Each sheet is a continuous network of interconnected hexagonal rings, in which every carbon atom is bonded to three equidistant neighbours, the bonds sticking out symmetrically at 120° to each other. This 'trigonal planar' bonding geometry (Figure 7.4.1a) is the signature of the sp^2-hybrid, formed by combining the 2s orbital of carbon with *two* of the three 2p orbitals. The three co-planar sp^2 lobes establish σ-bonds with neighbouring atoms.

The C–C bond length within the graphite sheets is 142 pm, a little shorter than the C–C distance in diamond (154 pm). As stronger bonds pull atoms closer, this suggests that the intra-sheet bonding in graphite is actually *stronger* than the bonding in diamond, the hardest substance known. The explanation lies in the fourth unpaired electron of carbon, excluded from the sp^2-hybrid (Figure 7.4.1a). This occupies a p-orbital projecting perpendicular to the sheet, and it allows π-bonds to be formed laterally between neighbouring atoms, reinforcing the intra-sheet σ-bond network.

In classical bonding theory, this π-bond would be added to one of the three σ-bonds, transforming it into a double bond, with the other two bonds remaining single. Figure 7.4.1c shows there are three alternative ways of distributing these double bonds throughout the sheet so that every atom is involved in four bonds. In the wave-mechanical interpretation, however, the real situation is a combination of all three configurations: the p-electron in each atom can be involved in partial π-bonds with all three neighbours at once. Consequently the π-bond 'sausages' of electron density above and below the plane (Figure 7.4c) are not localized between two specific atoms but are spread thinly all round the hexagon and indeed throughout the sheet. The π-electrons are thus **delocalized** in interconnecting molecular orbitals, resembling sheets of chicken-wire, within which the energy structure is like that of a metal (Box 7.6), allowing the π-electrons to migrate across the entire sheet if an electric field is applied. Thus graphite is a good electrical conductor in directions parallel to the sheets. This conducting behaviour is also responsible for the opaque, metallic appearance of graphite.

Intra-layer bonding in graphite ties up all four valence units of carbon. The adhesion of one sheet to another (interlayer bonding) is achieved through a much weaker

carbon atom forms with its four neighbours. The same structure is found in metallic silicon, also a fairly hard material. sp^3 geometry is characteristic of single-bonded carbon compounds, like the **saturated** hydrocarbons (Chapter 9). When carbon forms a double bond, however, it utilizes an alternative *sp^2-hybrid*. This has a different geometry, found in the mineral graphite (Box 7.4) and in the benzene molecule (C_6H_6), the basic unit of aromatic carbon compounds (Chapter 9).

The co-ordinate bond

In some circumstances a covalent-like bond can also be established using a lone pair. This arises in co-ordination complexes, which consist of a central metal atom or ion (commonly a transition metal) surrounded by a group of **ligands**, electronegative ions or small molecules (such as NH_3 or HS^-) that possess lone pairs of valence electrons. By overlapping with an *empty* orbital in the metal atom, each ligand lone pair forms a bonding molecular orbital which can accommodate the two electrons at a lower energy. This *co-ordinate bond*[2] is a variety of covalent bond in which both electrons are supplied by *one* of the participating atoms (the ligand) instead of one electron being supplied by each of them. Such complexes are stable because each electron, enjoying the attraction of two nuclei (cf. Box 7.3), has a lower energy than in the isolated atomic orbital. Current chemical thinking views the formation of a co-ordinate bond as a particular type of acid–base reaction (Box 7.5).

Co-ordination complexes are important in geochemistry because they markedly increase the solubility of

---

[2]   Not to be confused with the **co-ordination polyhedron** of an ionic compound.

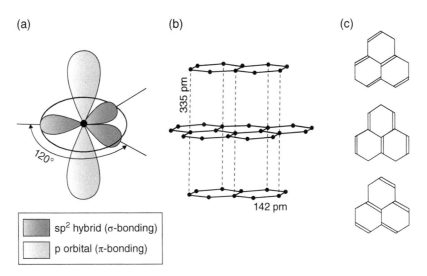

Figure 7.4.1 Different views of bonding in graphite. (a) Disposition of the sp² hybrid and p orbital lobes. (b) Intra-layer and interlayer bond lengths. (c) The three alternative ways of allocating the intra-layer double bonds in classical bonding theory; in the wave-mechanical view, the π-bonds are delocalized throughout the layer.

attraction called the *van der Waals interaction*, described later in the chapter, which is too weak to prevent sheets slipping easily over each other. The extreme softness of graphite and the large interlayer spacing (335 pm) indicate how feeble the interlayer force is. As there is no intercommunication of electron orbitals between sheets, graphite acts as an insulator in directions perpendicular to the sheets.

Noting the dramatic contrast between the extremely strong intra-layer bonding in graphite and the feeble van der Waals interaction binding one layer to the next, it is natural to ask whether individual layers could actually exist on their own. The answer – discovered only in the last decade – is 'yes'. Such monolayers constitute the novel form of carbon now known as *graphene*, one of several newly discovered forms of carbon discussed in Box 9.6.

Box 7.5   Lewis acids and bases

In 1923 the American chemist G.N. Lewis widened the concept of acids and bases to encompass systems in which H^+ ions (upon which the traditional notion of acids and bases is founded – Appendix B) are not available, such as in silicate melts. Lewis' ideas also have particular relevance to co-ordination **complexes**, in which bonding involves the sharing of a 'lone pair' of electrons donated by one of the participating atoms. Lewis defined a base as 'an atom or molecule capable of donating an electron pair to a bond' whereas a 'Lewis acid' is an atom or molecule that can accept a lone pair. The Lewis definition of 'acid' embraces the traditional 'H+ donor' viewpoint (since H^+ can readily attach to a lone pair, forming a covalent bond) but has much wider application.

How can we apply the Lewis concept to co-ordination complexes? As we saw in Chapter 4, the co-ordination complex $Cu(HS)_3^{2-}$ is believed to play an important part in the low-temperature hydrothermal transport of Cu. The electron configuration of the Cu^+ ion is $[Ar]\ 4s^0 3d^{10} 4p^0$. The complex forms because a lone pair of electrons on each HS^- ion overlaps with a vacant Cu orbital (4s or 4p), forming a molecular orbital that allows the electron pair to be associated with two atoms (S and Cu). Here each HS^- ion is acting as a Lewis base (electron-pair donor) and the Cu^+ ion is the Lewis acid.

The Lewis approach provides valuable insights into the chemistry of silicate melts, and helps us to understand why, for example, some metals prefer to be associated with sulfide minerals whereas others have more affinity with silicates (Box 9.8).

many ore metals in saline aqueous solutions such as hydrothermal fluids (Chapter 4).

Metals and semiconductors

In a crystal of pure copper, no electronegativity difference exists between neighbouring atoms, so we expect the Cu–Cu bonding to be covalent. Yet the characteristic properties of metals – lustre, opaqueness, electrical and thermal conductivity, and ductility – make them quite different from the covalent solids considered so far. The explanation of these differences lies in the molecular orbitals by which a metal crystal is held together.

When a hydrogen atom forms a covalent bond, its valence shell becomes in effect fully occupied through electron sharing. The electron configuration in the hydrogen molecule ($1s\sigma^2$), for example, is equivalent from each atom's point of view to the configuration of helium. A sulfur atom participating in two bonds similarly attains a noble-gas configuration (Ar). The picture in a metal, however, differs from these elements in two ways:

(a) Metals have lower electronegativities and ionization energies than hydrogen and sulfur. Electrons in their valence shells are more loosely held.

(b) The formation of covalent bonds does not lead to a noble-gas configuration in a metal atom. Vacant energy levels remain available in the metal's valence shell.

In any covalent crystal, the overlap of valence orbitals between neighbouring atoms leads to a system of molecular orbitals extending throughout the crystal. The electrons occupying them are nominally shared by all of the atoms present. The molecular energy levels available to these electrons are grouped in *bands* (Box 7.6), and the physical properties of a crystal depend on how these bands are arranged in energy space and how they are populated.

Figure 7.6 shows three alternative situations. In diamond (b) – and also in sulfur (not shown) – the valence electrons completely fill the lower band (equivalent to filled valence shells in all of the atoms). Each electron has a molecular wave function that confines it to a particular location in the crystal. In order to migrate it must find a vacancy in another molecular orbital, but

no vacancies exist, except in the upper band at unattainable energy levels. The essential characteristic of such *insulators* is therefore a filled band, separated from vacant higher levels by a large energy gap.

In a metal, valence electrons can fill only the lower part of the band. Upper levels remain vacant and, together with other overlapping bands which are readily accessible (Figure 7.6a), serve as a *conduction band* that enables electrons to migrate to new positions in the crystal. Electrons are not stable in these higher levels, although thermal excitation from the filled levels below is sufficient to ensure that a proportion of such levels is always occupied. A voltage applied across the crystal will produce a net flow of conduction-band electrons that constitutes an electric current. Similar energy bands in the π-bonding network lie behind the electrical conductivity of graphite (Box 7.4).

This leads to the notion of a metal crystal as a regular array of 'cations' (although in this context ionic radii are not applicable) immersed in a fluid of mobile, '*delocalized*' valence electrons. Owing to the mobility of the valence electrons, the bonding in metals is essentially non-directional. Most metals have close-packed atomic structures determined by the stacking rules for spheres of equal size. Twelve-fold co-ordination is therefore the commonest configuration, although a number of metals show eight-fold co-ordination instead.

Hydrogen was cited above as an example of a non-metallic element, as it occurs in the elemental state on Earth solely in the diatomic form H_2. Experiments at extremely high pressure have shown, however, that highly compressed hydrogen can develop the electronic band structure characteristic of a metal (Box 7.6), and it is believed to exist in this metallic form in the deep interiors of Jupiter and Saturn. The strong magnetic fields of Jupiter and Saturn have been attributed to electric currents flowing in such a metallic hydrogen fluid.

Semiconductors

Germanium (Figure 7.6c) illustrates a technologically important intermediate case, the semiconductor. If a *small* gap exists between a filled band and an empty conduction band, the material will behave as an insulator except when activated by an energy impulse from outside. One of the uses of germanium, for example, is as a detector of γ-ray photons (Box 6.3), such as those emitted by certain trace elements in geological materials when irradiated by neutrons in a nuclear reactor

Box 7.6 The energy-level structure of lithium metal

Box 7.3 showed that the 1s orbital overlap between hydrogen atoms generates two molecular orbitals: a bonding orbital ($1s\sigma$), having a lower energy than the atomic orbitals, and an anti-bonding orbital (denoted $1s\sigma^*$), with a higher energy. The same happens when two lithium atoms form a bond.

Because these $2s\sigma$ molecular orbitals extend across the two atoms, the electrons in them are common to both. When three Li atoms are close enough for overlap to occur, they share three electrons and form three molecular orbitals with three distinct energy levels. Four atoms form four orbitals; five atoms form five; and n atoms n orbitals. In a crystal of appreciable size, the value of n is effectively infinite, resulting in a virtually continuous energy *band* (bottom of Figure 7.6.1). As it originates from interaction between s-orbitals, it is called the s-band. The n valence electrons occupy the lower (bonding) half. These lower levels represent fixed electron states, each associated with a particular location in the crystal.

An electron migrates by transferring to a new molecular orbital. The only vacant ones are those with energies in the upper (anti-bonding) part of the band, and promotion of electrons to these accessible upper levels explains the electrical conductivity of metals (and the optical properties derived from it – Chapter 8). Electrons may also have access to higher bands; in Li the hypothetical overlap of empty p-orbitals makes available a 'p-band', partly interleaved with the s-band (Figure 7.6). These vacant levels constitute the *conduction band*.

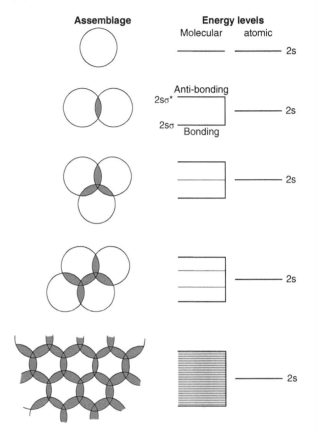

Figure 7.6.1 Illustrating the origin of conduction bands in Li metal.

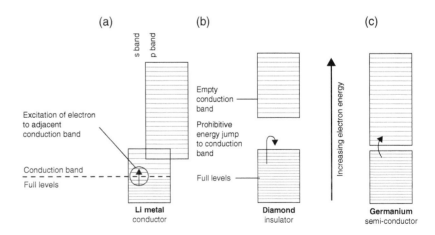

Figure 7.6 Energy bands in (a) a conductor, (b) an insulator, and (c) a semiconductor.

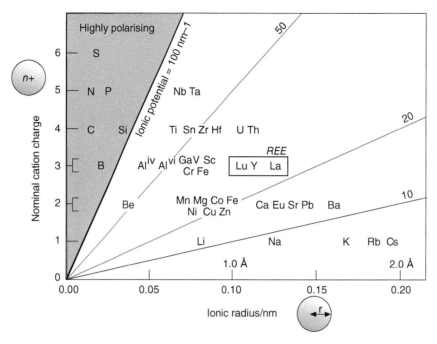

Figure 7.7 Ionic potentials of common 'cations'. 'REE' refers to the rare earth elements La (Z = 57) to Lu (Z = 71) – see Chapter 6. Yttrium (Y) has similar properties.

(an analytical technique called 'neutron activation analysis'). Silicon is used in a similar way as an X-ray detector in the electron microprobe (Box 6.4). A Ge (or Si) crystal, even with several hundred DC volts applied across it, will conduct no appreciable current[3] except when it absorbs a γ-ray (or X-ray) photon. In either case the photon energy is sufficient to promote a number of electrons temporarily into the conduction band, leaving an equivalent number of 'holes' in the valence band. The arrival of the photon is thus marked by a short pulse of electric current through the crystal, after which it relaxes back to its insulating condition. As the number of electrons promoted is proportional to the photon energy, the amplitude of the output pulse will vary in proportion to the quantum energy of the photon detected. Electronically sorting the pulses according to pulse height therefore provides a simple and economical means of isolating the various components of a complex γ-ray (or X-ray) spectrum; this is the basis of energy-dispersive ('ED') spectrometry (Box 6.4). The

intensity of each spectral 'line' is obtained by counting the number of pulses per second that are recorded within the appropriate pulse-height interval.

The energy gap between filled and conduction bands in a semiconductor can be deliberately engineered for specific applications by 'doping' the crystal surface with other elements, notably those of Groups IIIb and Vb; the doping elements introduce additional energy levels into the gap.

Bonding in minerals

We have seen that ionic bonding develops between elements of very different electronegativity, whereas covalent bonds are characteristic of materials, including pure elements, in which the electronegativity contrast is slight. Most of the compounds we meet as minerals fall between these two extremes, however, so we now have to consider what sort of chemical bonding operates between elements that show moderate differences in electronegativity. We shall investigate this intermediate domain by considering how real bonds deviate from the idealized ionic and covalent models.

---

[3]   To prevent thermal excitation of electrons into the conduction band creating a spurious signal, Ge and Si detectors are operated at low temperatures maintained by thermal contact with a liquid nitrogen reservoir ($-196\,°C$).

Anion polarizability: non-ideal ionic bonds

We have postulated an ideal ionic bond in which an electron is completely transferred from a donor atom (which becomes a cation) to a recipient atom (which becomes an anion). Ideally the transferred electron should be associated solely with the anion nucleus, and its spatial distribution should be symmetrical about it. In reality, the net positive charge on each neighbouring cation acts as a competing focus of attraction and to a small extent the electron density of the anion is concentrated in the region between the nuclei. In other words, a degree of electron sharing (partial covalency) occurs in any real ionic bond. The ideal ionic bond does not actually exist in nature.

The degree of polarization of the anion is obviously greater if the cation is highly charged (Mg^{2+} as compared with Na^+, for example). The polarizing field of the cation is also more intense if the cation is small, because it can then be brought closer to the anion. The ratio of the cation's charge to its radius, which is called the **ionic potential** of the cation, is therefore a measure of its power to polarize an anion and thereby recover a fraction of the excess electron density. Large singly charged cations like K^+ possess little polarizing power and form the 'purest' ionic bonds. Small, multiply charged cations like Al^{3+} and Si^{4+} are highly polarizing, and their compounds can be considered partially covalent (Figure 7.7).

A cation's ionic potential – a property correlated with the element's electronegativity – is a useful guide to its behaviour in molten and crystalline silicates (Figure 9.1.1) and in aqueous solution (Figure 9.3.1).

Polarization of a covalent bond: ionicity

In an ideal covalent bond, two valence electrons are shared equally between two atoms. The two atoms must have equal power to attract electron density to keep it positioned symmetrically between them. If they differ even slightly in electronegativity, electron density will be gathered more around the more electronegative atom, giving it a slight negative net charge (and leaving a complementary positive charge on the other atom). Such 'polarization' of the covalent bond is equivalent to transferring a fraction of an electron from one atom to the other, thus introducing a degree of ionic character or 'ionicity' into the bond.

Linus Pauling was the first to point out that a continuous progression in bond type must exist between purely covalent and predominantly ionic bonds. The character of an intermediate type of bond can be quantified in terms of the *proportion of ionic character* present – expressed as a percentage – which Pauling correlated with the difference in electronegativity between the participating atoms, as shown in Figure 7.8a. Although not an exact relationship (and therefore shown as a band in the figure), the correlation offers a valuable insight into the properties of minerals.

Bonding in silicates

One particularly important bond in view of its role in silicate structure (Chapter 8) is the Si–O bond, which appears from Figure 7.8a to possess around 50% ionic character and therefore to combine ionic and covalent bonding in more or less equal amounts. The small nominal ionic radius of the Si^{4+} 'cation' (34 pm Box 7.2) is consistent with its occupation of tetrahedral sites in all silicate minerals. In view of its high charge and small radius, however, the Si^{4+} cation must be highly polarizing and its distortion of the oxygen ion introduces an significant degree of covalency into the Si–O bond. The concept of an Si^{4+} cation is therefore an approximation to be used with caution: the charge residing on a silicon atom in a real silicate structure actually approximates to 2+ rather than 4+. Silicon, like carbon, uses an sp^3-hybrid in forming covalent bonds, and its tetrahedral co-ordination is therefore as much a reflection of covalent bonding between Si and O as of ionic bonding. The relative covalency of the Si–O bond accounts for the structural coherency of the chain, sheet and framework skeletons of many silicate minerals (Chapter 8). Like phosphorus and sulfur, Si shows little tendency to form double or π-bonds. Unlike C, it never forms an sp^2-hybrid.

The other chemical bonds operating in silicates are all more ionic in character than the Si–O bond. Al^{3+} is a small ion with a relatively high charge, but its ionic potential is barely more than half that of Si^{4+} (Figure 7.7) and the Al–O bond is regarded as being nearly 60% ionic (Figure 7.8a). Mg–O is about 65% ionic and Ca–O, Na–O and K–O are all more than 75% ionic. For these elements the ionic model provides the most appropriate predictor of co-ordination and crystal structure. The

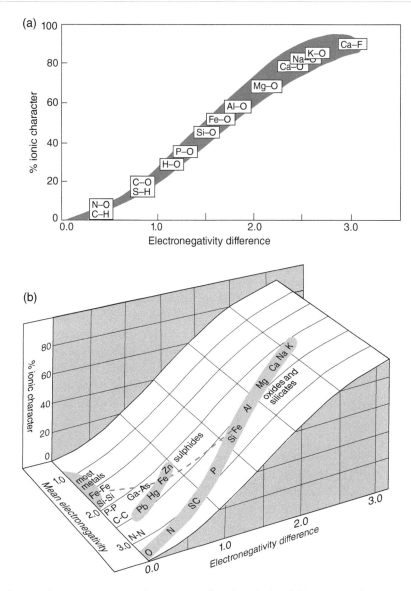

Figure 7.8 (a) The correlation of percentage ionic character in a bond with the difference in electronegativity, after Pauling. The approximate status of geologically relevant bonds is shown. (b) Oxide, sulfide and elemental bond types shown on a 3D figure whose axes are ionicity, mean electronegativity and electronegativity difference. The dashed line links the bonds formed by iron in oxide (silicate), sulfide and metal states.

same, of course, applies to the halides of these elements, such as the minerals halite (NaCl) and fluorite (CaF_2).

Oxy-anions

Oxygen forms bonds with phosphorus, carbon, sulfur and nitrogen which are distinctly more covalent than Si–O (Figure 7.8a). Thus in a carbonate mineral like calcite, for example, we see the different behaviours of two types of chemical bond:

(a) Ca^{2+} and CO_3^{2-} are attracted to each other by ionic bonds which, being electrostatic, break down when calcite is dissolved in a polar solvent like water, releasing separate calcium and carbonate ions stabilized by electrostatic association with surrounding water molecules (hydration – Box 4.1).

(b) The bonds between C and the three O atoms within the carbonate ion are largely covalent, and the anion retains its identity and structure whether in solution or in crystalline form.

Similar behaviour is seen with other **oxy-anion** compounds such as phosphate, nitrate and sulfate. They are related in a formal sense to corresponding oxy-acids (carbonic, phosphoric, nitric and sulfuric acids). Note that OH⁻ is another oxy-anion, important in minerals like brucite [Mg(OH)$_2$] and mica.

Pure elements, alloys and sulfides

Crystals of pure elements, whose atoms have uniform electronegativity, will plot at the origin in Figure 7.8a. This group includes, as we have seen, a wide range of behaviour, from insulators like diamond and crystalline sulfur, through semiconductors, to fully conducting metals. These differences can be appreciated by adding the mean electronegativity as a third dimension to the diagram shown in Figure 7.8a, the curving line of which is transformed into the curved surface shown in Figure 7.8b. Oxide bonds form a diagonal trend on this surface. At the foot of this trend lies the O–O bond, one of a group of non-metal bonds extending along the edge of the surface to phosphorus (P–P). The next element along this edge, silicon, is a semiconductor. Beyond Si lies a (shaded) field in which all of the common metals plot. It extends some distance to the right to include alloys like brass (Cu–Zn).

The metallic appearance of many sulfides, illustrated by the popular reference to pyrite (FeS$_2$) as 'fools' gold', reflects their intermediate position between the oxide and metallic fields in Figure 7.8b. The dashed line shows that the Fe–S bond is intermediate in character between the Fe–O and Fe–Fe bonds. The structural reasons for submetallic behaviour in sulfides are considered in Box 9.8.

Other types of atomic and molecular interaction

We have seen that ionic, covalent and metallic bonding form a unified spectrum of chemical interaction between atoms that possess incomplete valence shells. These bonding mechanisms, considered separately or in combination, account for most of the diversity of appearance, structure and behaviour that we see in minerals (Figure 7.8). There are nonetheless some

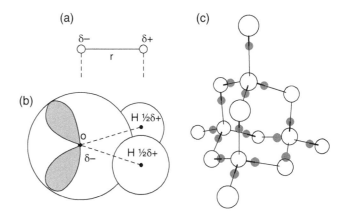

Figure 7.9 Dipole interactions. (a) A dipole consists of equal and opposite charges, a fixed distance apart. (b) The water molecule as a dipole; δ − signifies a partial negative charge (equivalent to a fraction of an electron's charge) arising from the higher electronegativity of the oxygen atom. (c) The structure of ice. Atoms are shown at 1/10 of the appropriate size; heavy lines are covalent bonds, thin lines are hydrogen bonds. Note the difference in O–H and O–H bond lengths.

associations between atoms and molecules that cannot be explained in these terms.

Various types of weak electrostatic attraction operate between all molecules, whether or not they possess overall electric charge. Many molecules, owing to internal electronegativity differences, are slightly polarized; and the electrical field associated with such **dipoles** (Figure 7.9a) makes them exert, and be susceptible to, electrostatic forces. A dipole can be attracted to an ion, or to another dipole. It may, by means of its electric field, even *induce* an unpolarized molecule to become a dipole, and thereby attract it. Such dipole interactions, although much weaker than ionic bonding (an 'ion–ion interaction'), have great mineralogical significance.

Ion–dipole interactions: hydration

The shared electron density in an O–H bond is partially concentrated at the oxygen end, owing to the higher electronegativity of oxygen. This polarization, symbolized by a partial positive charge ½δ+ on each hydrogen atom and a partial negative charge δ− on the oxygen (Figure 7.9b), results in the water molecule as a whole acting as a dipole.

In aqueous NaCl solution, each Na⁺ **cation** is surrounded by a diffuse blanket of water molecules whose negative 'poles' are attracted toward the Na^+ cation, and around each Cl^- **anion** is a similar layer with postive poles aligned towards the negative ion. This phenomenon of **hydration** (Box 4.1) is an *ion–dipole interaction*. The water molecule possesses no *net* charge, but is attracted towards an ion because its oppositely charged (attracted) pole gets closer to the ion than the similarly charged (repelled) pole. The attractive force on the molecule is thus stronger than the repulsive force.

Dipole–dipole interactions: hydrogen bonding

The unique physical and chemical properties of water (Box 4.1) suggest that some kind of attraction exists between the molecules in water and ice. This attraction, known as the *hydrogen bond*, arises from the electrostatic interaction between one water molecule and its neighbours: a *dipole–dipole interaction*. This is seen most clearly in the regular structure of ice (Figure 7.9c), in which each hydrogen atom in the molecule is attracted electrostatically toward the oxygen atom in a neighbouring molecule. The attraction is maximized if the geometry allows the hydrogen to associate with one of the two lone pairs (Figure 7.5c) on the oxygen atom. Owing to the nearly tetrahedral disposition of bonding orbitals and lone pairs in the oxygen atom, this leads not to a close-packed structure but to a three-dimensional network of molecules analogous to the structure of diamond (Figure 7.9c, cf. Figure 7.5d).

This ordered structure breaks down during melting, but about half of the hydrogen bonding is preserved in liquid water, giving rise to its high viscosity, surface tension and boiling point (Box 7.7) compared with other common liquids. Because the molecules are able to pack closer together in the liquid state than in the ordered crystalline framework of ice, water exhibits a slight *increase* in density on melting, and a corresponding increase in volume on freezing (the reason why ice floats), another unique property which we take for granted (Box 2.2).

Hydrogen is the only element capable of 'bridging' between molecules in this way, a capacity it owes to the fact that its nucleus is not shielded by any inner electron shells: it is exposed to the full attraction of an approaching oxygen lone pair. Hydrogen bonding has considerable structural significance not only in water and ice, but in all sorts of compounds in which hydrogen is associated with a strongly electronegative element like oxygen and nitrogen. Hydrogen bonding is largely responsible for the cohesion between layers in minerals like kaolinite, for example. It plays a dominant role in most biological processes, determining for instance the regular arrangement of polypeptide chains in protein molecules (Chapter 9), and cross-linking of the double helix in DNA.

Induced-dipole (van der Waals) interactions

The electrostatic field of a dipole can induce a polarization in an atom that is not itself intrinsically polarized. The induced dipole is aligned in such a way that it is attracted to the original dipole. Curiously this attraction can operate even between two atoms that have no permanent dipoles. To see how, we have to look a little closer at the oscillation of the electron standing wave in the atom (Chapter 5).

We have gained some understanding of the nature of electron orbitals in atoms and molecules by comparing them with standing waves on a vibrating string. Denied by the Uncertainty Principle the opportunity to follow the electron along a precise course in time, we have considered the orbital as a time-independent *envelope* delineating the electron's spatial domain, analogous to the wave envelope just visible in a vigorously plucked guitar string. Just as the string actually oscillates back and forth within its envelope, too fast for us to see, so the electron must also be vibrating within its own orbital. In support of this view one may note that wave mechanics associates an angular momentum with the trapped electron (in orbitals other than s-orbitals), suggesting that the electron, in its own obscure way, does really 'travel round the nucleus'.

The consequence of this oscillation is that if we were able to freeze an atom or molecule at some instant, we would find that electron density in each orbital was not uniformly and symmetrically distributed, but momentarily concentrated in one part or another, making the atom or molecule for that instant a dipole. At each instant the dipole generates an electric field that can polarize neighbouring atoms or molecules in concert with its own oscillation, and attract them closer. Although the electric fields of these synchronous dipoles average out to zero over a period of time, the intermolecular attraction they generate does not.

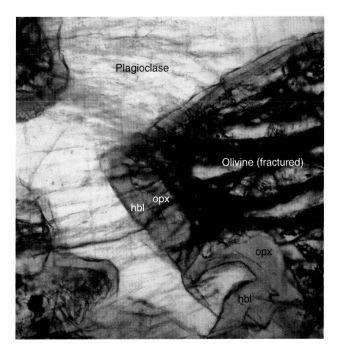

Reaction rim of orthopyroxene (opx) surrounding a fractured olivine crystal in peridotite from the Gemina Gharbia igneous complex, Egypt (Source: Reproduced with permission of H.M. Helmy). The orthopyroxene is the product of reaction between the olivine and late-stage magma (cf. Figure 2.6.1). The orthopyroxene rim is itself overgrown by hornblende (hbl) owing to reaction with later-stage *hydrous* melt and adjacent plagioclase at lower temperatures. Plane-polarized light, field of view 1 mm wide. (Source: Reproduced with permission of H.M. Helmy.)

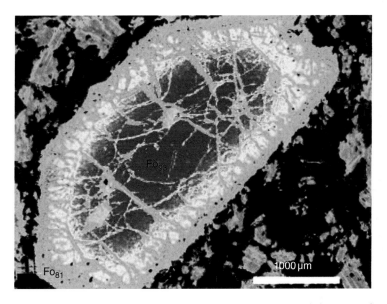

Colour-enhanced X-ray map showing as concentric colour bands the outward decrease of Mg content within a **zoned** olivine crystal, reflecting reaction between peridotite olivine and an infiltrating diorite melt. (Source: Adapted from Qian & Hermann 2010. Reproduced with permission of Oxford University Press.)

Chemical Fundamentals of Geology and Environmental Geoscience, Third Edition. Robin Gill.
© 2015 John Wiley & Sons, Ltd. Published 2015 by John Wiley & Sons, Ltd.
Companion Website: www.wiley.com/go/gill/chemicalfundamentals

Pyroxene crystal (dark) in a Bushveld gabbro hosting two generations of **exsolution** lamellae. The differing interference colours of the host crystal (black), first-generation (vertical, pale turquoise) and second-generation (horizontal, whitish) exsolution lamellae reflect differing crystallographic orientations. Crossed polars, field of view 1.5 mm wide.

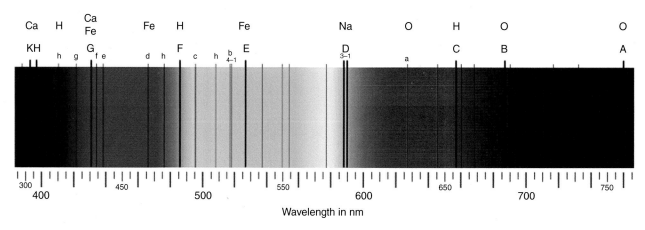

Fraunhofer lines (elemental absorption spectra) superimposed on the visible emission spectrum of the Sun. The chemical symbols at the top identify the elements causing the absorption; the symbols below them show the conventional spectroscopic label for each line. Adapted from an anonymous public-domain on-line source.

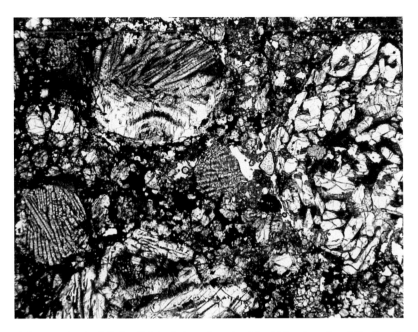

Microscope view (plane-polarized light, field of view 4 mm) of the Etihudna *chondrite* showing chondrules with a variety of internal textures. Chondrule textures in general record a *'complex interplay between the peak temperature, duration of heating, and preservation of nucleation sites, either due to incomplete melting of chondrule precursors or the presence of fine-grained dust in the chondrule-formation region'* (Lauretta *et al.*, 2006). (Source: © The Natural History Museum, London. Reproduced with permission.)

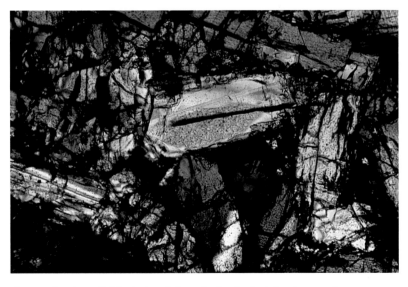

Microscope view (crossed polars, field of view 2.5 mm) of euhedral (i.e. igneous) pyroxene crystals in the Nahkla achondrite meteorite. (Source: © The Natural History Museum, London. Reproduced with permission.)

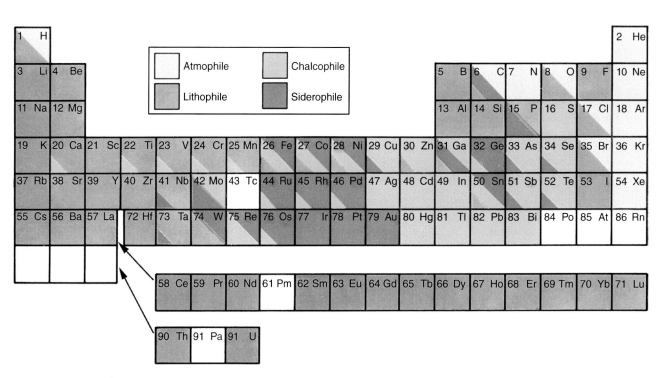

Cosmochemical affinities of the naturally occurring chemical elements (cf. Figure 11.4). White squares with grey writing represent elements with no stable or long-lived isotopes (Chapter 10). The terms atmophile, lithophile, chalcophile and siderophile are defined in Chapter 11.

Box 7.7 Estimating the strength of chemical bonds

The strength of a chemical bond is expressed in terms of the energy required to break it. This is usually quoted in the form of the molar enthalpy change for a specific hypothetical dissociation reaction such as:

$$H_2 \rightarrow H + H$$

This enthalpy change, known as the **dissociation** energy of molecular hydrogen, indicates how much energy (in $kJ\,mol^{-1}$) is required to split every molecule in one mole of hydrogen gas into two separate atoms. Some relevant dissociation energies are given in Table 7.7.1.

Detailed interpretation of these numbers (for example the estimation of the energy of an individual C–C bond) is fraught with pitfalls. We shall simply note the following generalizations:

(i) Covalent and ionic bond energies fall within a similar range. Energies of single covalent bonds are generally between 200 and $500\,kJ\,mol^{-1}$. Note the similarity to the activation energies noted for silicate reactions in Chapter 3 (Figures 3.6 and 3.8).
(ii) Metals have lower values, those for soft metals being lower than those for hard metals (Na < Pb < Ni).
(iii) The energy of the hydrogen bond is roughly a factor of ten less than for covalent or ionic bonds. For an *individual* hydrogen bond the energy is around $24\,kJ\,mol^{-1}$.
(iv) Van der Waals energies are an order of magnitude lower again than hydrogen bonds.

In the absence of quantitative data, one can make a crude estimate of the energy of interaction *between* molecules in a covalent substance from its boiling point. This is the temperature at which the energy of the molecules' thermal vibration (which is roughly $\frac{3}{2}RT$ per mole) exceeds the energy of cohesion holding the molecules together.

Elements and compounds of low molecular weight that are gaseous at room temperature (He, Ar, CO_2, N_2) have weak intermolecular forces, but in a solid or liquid material such as water the forces of cohesion are quite strong.

Table 7.7.1 Energies of various types of chemical interaction

Type of interaction	Defining reaction	Energy/$kJ \cdot mol^{-1}$
Ionic bond	NaCl → Na$^+$ + Cl$^-$ *crystal gas gas*	767
	NaCl → Na + Cl *crystal gas gas*	641
Covalent bond	C → C *diamond gas*	723
	H$_2$O → 2H + O *crystal gas*	466 (x 2)
Metallic bond	Na → Na *crystal gas*	109
	Pb → Pb *crystal gas*	194
	Ni → Ni *crystal gas*	423
Hydration	Na$^+$ → Na$^+$ *aq gas*	405
Hydrogen bond	H$_2$O → H$_2$O *ice gas*	47
Van de Waals interaction	Ar → Ar *crystal gas*	6.3

The induced-dipole or *van der Waals interaction* (named after Dutch physicist and Nobel laureate Johannes van der Waals) is responsible for a weak attraction operating between any pair of atoms or molecules that are sufficiently close (a few tenths of a nanometre), although its effect can be detected only when other inter-atomic forces are absent. Van der Waals forces are responsible for holding the sheets together in crystals of graphite, and no doubt contribute to the cohesion of soft sheet silicates like pyrophyllite and talc. The low hardness of graphite (1–2) and the large interlayer distance (Box 7.4) provide a measure of how feeble the attraction is. The 'bond energy' for the van der Waals interaction is typically two orders of magnitude less than for ionic and covalent bonding (Box 7.7). The interaction explains why all substances, even noble gases, form crystals at sufficiently low temperatures (neon, for instance, melts at 24.6 K and argon at 84 K). These low melting temperatures signify a force so weak that even the very slight thermal vibrations experienced just above these low temperatures are sufficient to overcome it (Box 7.7).

Review

Many properties of minerals and everyday materials (hardness, atomic structure, electrical conductivity, solubility, lustre, etc) are readily explained in terms of the chemical bonds that hold them together on the atomic scale:

- Ionic bonds – involving the donation and acceptance of electrons between atoms – form between elements of *contrasted electronegativity*. Ionic compounds exist in the solid and molten states but not as gases. Their structures are dictated by the packing-together of spherical ions of various sizes, and can be predicted from the radius ratio of cations to anions (Table 7.1).
- Covalent bonds – involving the sharing of electrons between atoms – form between elements of the *same or similar electronegativity*. The structures of simple molecules (e.g. H_2O – Figure 7.5c) and extended structures (e.g. diamond – Figure 7.5d) can be predicted from the geometry of atomic – particularly **hybrid** –orbitals (Figures 7.4, 7.5 and 7.4.1).
- Metallic bonds form between elements of similar electronegativity *when the mean value is low* (Figure 7.8b). Metallic properties derive from the availability of vacant *conduction bands* of electron energy levels that extend throughout the crystal (Figure 7.6). Some sulfides share attributes of metals, such as lustre and measurable electrical conductivity (Figure 7.8b).
- The Si–O bond in silicate minerals is 50:50 ionic:covalent (Figure 7.8). For this reason it plays an important part in determining the architecture of silicate crystals and melts, as described in Chapter 8.

- Ion–dipole and dipole–dipole interactions, although much weaker (Table 7.7.1), are nonetheless important in determining the properties of water and ice (Figure 7.9), the softness and sheet structure of graphite (Figure 7.4.1), and the capacity of even **inert gases** like Ar to form crystals at very low temperatures.

Further reading

Atkins, P., Overton, T., Rourke, J., *et al.* (2010) *Inorganic Chemistry*, 5th edn. Oxford: Oxford University Press.

Barrett, J. (2001) *Structure and Bonding*. Cambridge: Royal Society of Chemistry.

Henderson, P. (1982) *Inorganic Geochemistry*. Oxford: Pergamon.

Exercises

7.1 Predict the co-ordination numbers of the following ions in a silicate crystal (Box 7.2):

$Si^{4+}, Al^{3+}, Ti^{4+}, Fe^{3+}, Fe^{2+}, Mg^{2+}, Ca^{2+}, Na^+, K^+$

7.2 Explain the difference in ionic radius (Box 7.2) between Fe^{2+} and Fe^{3+} and between Eu^{2+} and Eu^{3+}. Calculate the ionic potentials of the ions. What effect does the oxidation state have on the covalency of bonding of these elements?

7.3 Identify the type of bonding between the following pairs of atoms in the solid state. How does it influence the properties of the elements at room temperature? (Use Figure 6.3. Holmium (Ho) is a lanthanide metal.)

He – He   Ho – Ho   Ge – Ge

7.4 Discuss the bonding in the following minerals:

KCl (sylvite), TiO_2 (rutile), MoS_2 (molybdenite), $NiAs$ (niccolite), $CaSO_4$ (anhydrite), $CaSO_4.2H_2O$ (gypsum).

8 SILICATE CRYSTALS AND MELTS

The majority of rock-forming minerals are **silicates**, compounds in which metals are combined with silicon and oxygen. In this chapter we consider how the chemical structure of these compounds, particularly the nature of the bonding, determines the familiar morphological and physical properties of crystalline silicate minerals and the physical properties of silicate melts (Box 8.3). A minimal knowledge of crystallography will be needed.

The relative covalency of the Si–O bond (Figure 7.8) gives silicon a fundamental structural role as the principal **network-forming** element in silicate crystals and melts, establishing the structural skeleton upon which their properties depend. In this respect silicon (with phosphorus and to some extent aluminium) contrasts with the more ionic constituents of silicates like Mg^{2+} and K^+, which are **network-modifying** elements that influence structure only because they restrict the way in which Si–O networks are stacked together.

Silicate polymers

Whether silicates are examined from the ionic or covalent viewpoint, the behaviour of silicon is the same: it lies at the centre of a tetrahedral group of four oxygen atoms or ions. In structural terms, the *SiO_4 tetrahedron* (Figure 8.1) is the basic building brick from which all silicate crystals and melts are constructed. But the silicon atom itself can satisfy only half of the bonding capacity of its four oxygen neighbours (four bonds out of a total of eight). How are the remaining oxygen bonds used?

If, when the silicate crystallizes, there is a high concentration of an electropositive element like magnesium (which forms a *basic oxide* – Box 8.1), relatively ionic bonds are likely to be established between each 'tetrahedral' oxygen and nearby Mg^{2+} ions, the SiO_4 group acquiring in the process an overall negative charge (totalling a nominal 4–). This leads to the chemically

Chemical Fundamentals of Geology and Environmental Geoscience, Third Edition. Robin Gill.
© 2015 John Wiley & Sons, Ltd. Published 2015 by John Wiley & Sons, Ltd.
Companion Website: www.wiley.com/go/gill/chemicalfundamentals

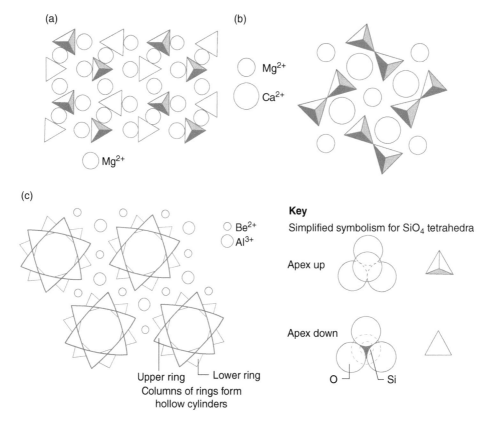

Figure 8.1 Simplified structures of silicate minerals, with cations drawn approximately to scale. SiO_4 groups are shown as bare coordination tetrahedra to clarify the structure. (a) Orthosilicate olivine, viewed down the *a* crystallographic axis; (b) dimer silicate melilite; (c) ring silicate beryl.

simple crystal structure of *olivine* (specifically the Mg **end-member** forsterite, Mg_2SiO_4) in which Mg^{2+} ions alternate with $SiO_4{}^{4-}$ tetrahedra (Figure 8.1a). There is no direct Si–O bonding between one tetrahedron and its neighbour, and the cohesion of the crystal as a whole arises from ionic bonding between Mg^{2+} cations and $SiO_4{}^{4-}$ anions: covalently bonded SiO_4 'bricks' are cemented together by an ionic 'mortar' of Mg^{2+} ions.

If cations Mg^{2+} are scarce, on the other hand, oxygen atoms are more likely to bond directly to two silicon atoms, forming a relatively 'covalent' bridge between them. Taken to the limit, this can lead to every oxygen atom being shared between neighbouring tetrahedra, extending the Si–O bonding to form a three-dimensional network of connected tetrahedra throughout the crystal. This is the situation we find in the mineral *quartz*. Because each oxygen atom is structurally part of two SiO_4 tetrahedra, half as many oxygen atoms are needed to fulfil the co-ordination requirements of silicon. Thus quartz, although in a formal sense built of SiO_4 structural units, has the formula SiO_2.

The formation of extended Si–O networks in silicates is called *polymerization*. Whereas familiar organic **polymers** consist of chains and rings of carbon atoms linked directly to each other (–C–C–C–: Chapter 9), the linkage in silicate polymers is always through oxygen atoms (–Si–O–Si–O–Si–). The degree of Si–O polymerization in a silicate structure is conveniently enumerated by the *number of non-bridging oxygens* (those linked to only one Si atom) per SiO_4 group. This number *p* varies from 4 in olivine (in which SiO_4 tetrahedra are linked only indirectly through other cations) to 0 in quartz. Between these limits, silicate minerals exhibit a wealth of structural diversity, exceeded in complexity only by the chemistry of carbon.

Monomer silicates

Pursuing the analogy with organic **polymers**, olivine (Figure 8.1a) would be called a *monomer*, because the basic unit of the polymer occurs in it uncombined.

Box 8.1 Acidic and basic oxides

The composition of any silicate material can be represented as a combination of various metallic oxides such as MgO and K_2O combined with silicon dioxide (silica). Forsterite, for example, can be regarded as the result of reacting two molecules of MgO with one of SiO_2:

$$2MgO + SiO_2 \rightarrow Mg_2SiO_4 \qquad (8.1.1)$$

The character of a silicate depends upon the proportions in which different oxides combine. Oxides can be subdivided into three types:

Basic oxides

When an *ionic oxide* dissolves in water, the O^{2-} ion released reacts with a water molecule to form two hydroxyl ions:

$$Na_2O + H_2O \rightarrow 2Na^+ + \left(O^{2-} + H_2O\right) \rightarrow 2Na^+ + 2OH^-$$

The production of OH^- ions removes free H^+ ions from the solution, increasing its pH and making it *basic* (Appendix B). For this reason the oxides of the electropositive metals are described as *basic oxides* (Figure 9.4).

Acidic oxides

A *covalent oxide* dissolved in water, instead of producing OH^- ions, will remove them from solution:

$$CO_2 + \left(OH^- + H^+\right) \rightarrow HCO_3^- + H^+$$

(cf. Equations 4.19 and 4.21) resulting in an acidic solution. Similar reactions involving the oxides of sulfur produced by coal-burning power stations are environmentally important in leading to acid rain (see O'Neill, 1998). In the context of silicates, this category of *acidic oxides* includes the oxides of silicon and phosphorus (Figure 9.4).

Amphoteric oxides

Amphoteric oxides and hydroxides share the property of being able to behave as acids (if reacting with a strong base) *and* bases (with a strong acid):

$$\underset{\text{base}}{Al(OH)_3} + \underset{\text{acid}}{3HCl} \rightarrow \underset{\text{salt}}{AlCl_3} + \underset{\text{water}}{3H_2O}$$

$$\underset{\text{acid}}{Al(OH)_3} + \underset{\text{base}}{NaOH} \rightarrow \underset{\text{salt}}{NaAlO_2} + \underset{\text{water}}{2H_2O}$$
$$\underset{\text{(sodium aluminate)}}{}$$

Other elements forming amphoteric oxides are shown in Figure 9.4.

Table 8.1 Silicate polymers

Structural type	p	Z:O	Example mineral	Formula
Orthosilicate (monomer)	4	1 : 4	Forsterite (olivine)	$Mg_2[SiO_4]$
Dimer	3	1 : 3.5	Melilite	$Ca_2Mg[Si_2O_7]$
Ring silicate	2	1 : 3	Beryl	$Be_3Al_2[Si_6O_{18}]$
Chain silicates				
Pyroxene	2	1 : 3	Diopside (pyroxene)	$CaMg[Si_2O_6]$
Amphibole	1.5	1 : 2.75	Tremolite (amphibole)	$Ca_2Mg_5[Si_8O_{22}](OH)_2$
Sheet silicate	1	1 : 2.5	Muscovite (mica)	$KAl_2[AlSi_3O_{10}](OH)_2$
Framework silicate	0	1 : 2	Orthoclase (feldspar)	$K[AlSi_3O_8]$

Silicates built of isolated SiO_4 tetrahedra are commonly known as *orthosilicates*. They include other minerals like garnet (e.g. grossular, $Ca_3Al_2Si_3O_{12}$), zircon ($ZrSiO_4$) and topaz ($Al_2SiO_4F_2$). Notice that in each of these formulae the ratio of Si to O is 1:4, a universal characteristic of orthosilicates. All of the oxygen atoms in these structures are non-bridging, so they share the value $p = 4$ (Table 8.1).

Dimer silicates

The structures of a few minerals involve pairs of SiO_4 tetrahedra linked through a single bridging oxygen ($p=3$). The formula of this dimeric group, consisting of two SiO_4 groups less one oxygen ($=Si_2O_7$), can be seen in melilite ($Ca_2MgSi_2O_7$, Figure 8.1b). The commoner mineral epidote contains both single (SiO_4) and double (Si_2O_7) tetrahedral groups.

Chain silicates

The sharing of two oxygen atoms by each SiO_4 group produces *chains* of tetrahedra of indefinite length (Figures 8.2 and 8.3a) that form the skeleton of one of

Figure 8.2 3D scale model showing the layout of a pyroxene chain, with some upper oxygen anions removed to show the tetrahedrally co-ordinated silicon atoms beneath. (Photo: K. d'Souza).

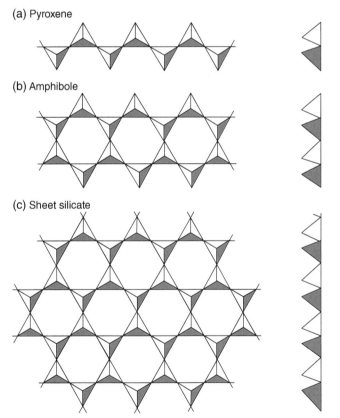

(a) Pyroxene

(b) Amphibole

(c) Sheet silicate

Figure 8.3 Simplified silicate structures in (a) pyroxene, (b) amphibole and (c) sheet silicates, seen in plan and end-on.

the most important mineral groups, the *pyroxenes*. The chains are 'kinked' rather than linear because alternate tetrahedra stick out in opposite directions. Each silicon atom possesses two non-bridging oxygens ($p=2$) and shares two bridging ones, so that the composition of the whole chain can be written $(SiO_3)_n$, where n is the number of tetrahedra in the chain. All pyroxenes therefore have SiO_3 (or Si_2O_6) in their chemical formulae, as for example in diopside $CaMgSi_2O_6$. The chains can be stacked against each other in different ways, allowing pyroxenes the potential to crystallize in both the orthorhombic and monoclinic systems (Box 8.5).

The chains define the crystallographic c-axis in pyroxenes. Parallel to this run several prismatic cleavage planes, such as the perfect {110} cleavage responsible for the characteristic perpendicular cleavages seen in a basal cross-section. These cleavages reflect the stronger cohesion *within* each chain compared with the strength of bonding *between* chains.

Ring silicates

An obvious alternative to forming an infinite linear chain is to link the ends of the chain into a ring. The minerals beryl ($Be_3Al_2Si_6O_{18}$, Figure 8.1c), cordierite ($Al_3Mg_2Si_5AlO_{18}$) and tourmaline are examples of ring silicates, in which the basic structural element is a ring of six SiO_4 tetrahedra. The simpler ring silicates have the p value (2) and Si:O ratio (1:3) of the pyroxenes (Table 8.1).

Double-chain silicates

The *amphibole* structure can be regarded as a pyroxene chain in which alternate tetrahedra share an oxygen atom with one neighbouring chain. This produces a double chain or band (Figure 8.3b), leading to a marked prismatic, sometimes even fibrous, habit. Owing to the wider double chains, the conspicuous prismatic cleavages intersect at about 55°, compared with the 90° characteristic of pyroxenes.

The double chain can be regarded as a row of hexagonal rings, which are not present in pyroxenes. These accommodate additional anions, usually **hydroxyl** (OH^-) or fluoride (F^-). Owing to the presence of these volatile constituents – the main chemical distinction between pyroxenes and amphiboles – the amphiboles are unstable at high temperatures, decomposing into pyroxene and vapour.

Table 8.2 Cation co-ordination in silicates

Cation	Radius ratio	Predicted co-ordination	Occurrence in minerals					
			OLIVINE	GARNET	PYROXENE	AMPHIBOLE	MICA	FELD-SPAR
Si^{4+}	0.26	4-fold (tetrahedral) Z-site						
Al^{3+}	0.36							
	0.46							
Ti^{4+}	0.52	6-fold (octahedral) site						
Fe^{3+}	0.55							
Mn^{2+}	0.56							
Mg^{2+}	0.61							
Fe^{2+}	0.65							
Ca^{2+}	0.91	8-fold site						
Na^+	0.94							
K^+	1.27	≥ 12-fold site						

Of every four tetrahedra in the amphibole structure, two share two oxygens and the other two share three oxygens ($p = 1.5$). In principle this leads to the formula Si_4O_{11} or Si_8O_{22}, as in tremolite ($Ca_2Mg_5Si_8O_{22}(OH)_2$). However, some of the tetrahedral sites may contain Al in place of Si (the ionic radius of aluminium is just small enough to admit it to tetrahedral as well as octahedral sites – Table 8.2). This will of course alter the Si:O ratio in the amphibole formula, as in pargasite ($NaCa_2Mg_4Al$ [$Al_2Si_6O_{22}$] $(OH)_2$). Square brackets are used here to distinguish Al in tetrahedral sites (inside the brackets) from the remaining Al occupying octahedral sites. Notice that both of these examples conform to a more general formula Z_8O_{22}, where 'Z' includes both silicon and 'tetrahedral aluminium' (Box 8.5). Because of its different valency, the substitution of Al for Si requires an adjustment among other cations present (discussed later) in order to maintain electrical neutrality. A similar substitution occurs in some pyroxenes.

Sheet silicates

When every $(Si,Al)O_4$ tetrahedron shares three oxygen atoms with neighbouring tetrahedra ($p = 1$), a continuous covalently bonded sheet structure is formed (Figure 8.3c) which provides the basic framework for the micas – like biotite and muscovite – and a variety of other sheet silicates. All of them are *hydrous*, with OH^- or other anions occupying the rings, and like amphiboles they *dehydrate* at high temperatures (Figure 2.3).

Sheet silicates can be recognized chemically by having Z_4O_{10} in the formula (Table 8.1). The varieties of sheet silicate – micas, chlorite, clay minerals (Box 8.2), serpentine, etc. – can be regarded as multilayer sandwiches, differing from each other in the identity of the ionic 'filling' and the manner in which sandwiches are stacked together. As in many sandwiches, the bread (the silicate sheet) is more coherent than the filling, so sheet silicates have a marked platy cleavage, as developed most obviously in micas like muscovite and biotite.

Framework silicates

Finally we come to a class of silicates in which every oxygen atom is shared between two tetrahedra, extending a semi-covalent network in all directions through the crystal. Because the volume of the crystal is determined wholly by the sp^3-based 'covalent' framework and not by the packing together of ions and separate SiO_4 polymers, such *framework silicates* usually have lower densities than other types of silicate (often less than $2.6\,kg\,dm^{-3}$). The structures of some incorporate

Box 8.2 Clay minerals and ion exchange

Clay minerals are sheet silicate minerals with characteristics that make them of environmental and industrial significance:

- Existing as fine-grained aggregates, they play an important role in determining soil texture, fertility, permeability and moisture retention.
- Some clay minerals (the *smectite* group) can absorb variable amounts of water, swelling as they do so. Conversely they lose water and contract in dry conditions, leading to the polygonal cracks that characterize muddy soils during drought.
- Their surface properties give clay minerals a high cation-exchange capacity.

Clay minerals have numerous technological and environmental applications, from the obvious example of pottery and ceramics (for which natural clay deposits – kaolinite – are the raw material) to uses such as paper fillers and coatings, cat litter, drilling muds, and barrier materials for landfill sites and mine tailings.

Structure and types

Clay mineral crystals are assemblages of three types of layer (Figure 8.2.1):

- layers of SiO_4 *tetrahedra* (with Al substituting for Si in some varieties) – 'T' in Figure 8.2.1;
- layers of Al^{3+}, Mg^{2+} or Fe^{2+} ions, *octahedrally* co-ordinated by OH^- or O^{2-} anions – 'O' in Figure 8.2.1;
- sheets of *large* 'interlayer' cation sites (analogous to 'A'-sites in amphiboles – Box 8.5) that accommodate ions like K^+, Na^+ and Ca^{2+} – 'L' in Figure 8.2.1.

Like varieties of sandwich on a buffet tray, the above layers can be combined in various ways (Figure 8.2.1). The simplest is a stack of tetrahedral sheets, all facing the same way, interspersed on a 1:1 basis with octahedral $Al(OH)_3$ layers (Figure 8.2.1a). This is the structure of the *kaolinite group* of clay minerals, characterized by an interlayer repeat distance (measured by X-ray diffraction – see Box 5.3) of 0.7 nm.

More commonly, the tetrahedral layers combine in inward-facing pairs – like buttered bread slices in a real sandwich – enclosing an octahedral layer as a 'sandwich filling' in between. In this configuration, tetrahedral and octahedral layers are combined in the proportion 2:1. Pairs may be stacked directly against each other as in the mineral *pyrophyllite* (repeat distance 0.9 nm), or successive 'sandwiches' may be separated by a plane of larger interlayer cations such as K^+, as in the clay mineral *illite*, which has a repeat distance of 1.0 nm (Figure 8.2.1b).

All clay minerals can be considered as hydrous owing to the involvement of OH^- anions in the octahedral layer. However, clay minerals of the smectite group, such as *montmorillonite* (Figure 8.2.1c), additionally incorporate discrete sheets of H_2O molecules between adjacent 'sandwiches', closely associated with the large interlayer cations. The presence of H_2O increases the interlayer repeat distance to about 1.4 nm. Water content – and therefore repeat distance – varies with changing temperature and humidity, giving smectites their characteristic capacity to swell and shrink according to conditions. The swelling behaviour poses significant geotechnical challenges when building on smectite-rich ground.

Adsorption

Cations such as K^+, Mg^{2+}, Ca^{2+} and NH_4^+ (ammonium) – all important plant nutrients – are attracted to the negatively charged surfaces[1] of clay mineral particles and attach to them electrostatically, a geochemically important type of surface reaction called **adsorption**. Soils act as storehouses of plant-available nutrients, and the large aggregate surface area provided by fine clay particles regulates the availability of such elements and constitutes – as a result of adsorption – a key part of the soil nutrient reservoir.

Adsorption on mineral surfaces is not restricted to cations. As we saw at the end of Chapter 4, the behaviour of arsenic in groundwater is dictated by the formation of polyanions like arsenate and arsenite, and their capacity to adsorb on to the surfaces of oxide mineral grains. Anion adsorption requires mineral surfaces with positive surface charge, which is more likely to occur in acidic (low-pH) solutions.

---

[1] The causes of electrostatic charge on mineral surfaces are complex and beyond the scope of this book. Summaries are given by Krauskopf and Bird (1995) and White (2013).

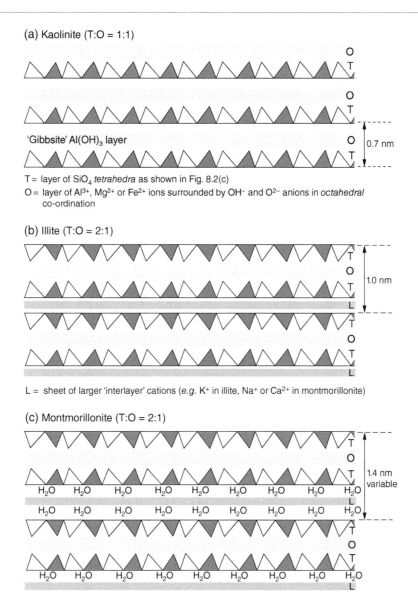

Figure 8.2.1 Simplified representation (not to scale) of the structures of the clay minerals (a) kaolinite, a '1:1 clay mineral', (b) illite, and (c) montmorillonite ('2:1 clay minerals'). Silicate sheets are shown edge-on using the symbolism of Figure 8.3. All three minerals contain octahedral layers (light orange) approximating to the composition of the mineral *gibbsite*, $Al(OH)_3$; other clay minerals may instead contain octahedral layers resembling the mineral *brucite*, $Mg(OH)_2$.

Ion exchange

The population of soluble cations adsorbed on the negatively charged surface of a clay particle interacts chemically with any aqueous fluid in contact with it, exchanging cations in order to arrive at an equilibrium distribution. For example, the reaction

$$Ca_{aq}^{2+} + 2K_{clay}^{+} \rightleftharpoons Ca_{clay}^{2+} + 2K_{aq}^{+} \qquad (8.2.1)$$

illustrates an important control that colloidal clay particles exert on the Ca^{2+} content of river water entering the oceans. Such *ion exchange reactions* play a key part in regulating the composition of natural waters more generally, for example in **buffering** the major element composition of the groundwater in an aquifer. Moreover the interlayer cations within clay mineral particles (the so-called 'exchangeable cations') also participate in

Box 8.2 (Continued)

equilibria like reaction 8.2.1 and contribute to a clay mineral's overall capacity for ion exchange.

Geological materials (clay minerals, sediments, soils) vary widely in their propensity to store and exchange cations, which depends upon composition and mineral structure as well as environmental conditions such as pH. The natural variability among environmental materials is illustrated by typical values of *cation exchange capacity* (CEC) given in Table 8.2.1:

CEC is usually defined and measured in terms of how much NH_4^+ a sample can take up from a 1 M ammonium acetate solution at pH = 7.0. Being dependent upon a number of factors, it is not a precise quantity but it nonetheless provides a useful measure for comparing the ion exchange properties of minerals, soils and sediments.

Table 8.2.1 Cation exchange capacities of geological materials

Soil constituent	Cation exchange capacity/CEC units*
Hydrous oxide minerals	~4
Kaolinite	2–10
Illite	10–40
Montmorillonite	80–150
Vermiculite	120–200
Soil organic matter	150–300

*CEC is usually expressed in arcane units of **meq** per 100 g of mineral or soil.

Note that even the highest CEC values of clays are often exceeded by those of organic colloids in soil.

large cavities and channels through which cations and even molecules can diffuse quite readily.

As there are no non-bridging oxygens in the framework silicate structure ($p = 0$), the Z:O ratio is 1:2. The simplest composition is SiO_2, whose stable form at room temperature is quartz (specific gravity, SG 2.65); at elevated temperatures the more open structures of the polymorphs tridymite (SG 2.26) and cristobalite (SG 2.33) crystallize in its place. The three-dimensional network structure is reflected in the poor or non-existent cleavage of the silica minerals; quartz, for example, has a conchoidal fracture.

Substitution of Al into some of the tetrahedral sites in place of Si makes possible a huge variety of **aluminosilicate** minerals[2] including the feldspars (the most abundant mineral group in the crust), and their silica-deficient cousins the feldspathoids. Replacing Si^{4+} with trivalent Al^{3+} without changing the oxygen content, of course, requires other cations to be introduced to maintain the charge balance. Owing to the openness of the framework structure, these compensating cations can be quite large (Na^+, K^+, Ba^{2+}).

Among the most open of the aluminosilicates are the zeolites. Unlike the feldspars they are hydrous, the water being held loosely in large intercommunicating cavities which can be up to 1 nm across. This framework is so rigid that the zeolites possess the remarkable ability to expel this water continuously and reversibly when heated, without their structure breaking down. Zeolites (natural and artificial) have many uses in chemical engineering, as ion exchangers and as 'molecular sieves' that can separate small molecules according to their size. They perform a vital function as catalysts in the petroleum industry.

The same range of silicate polymers can be found in silicate melts (Box 8.3) too, but with one key difference: whereas each crystalline silicate mineral incorporates *one* (or in rare cases two) types of silicate polymer, *several* polymeric types may coexist within the more disordered structure of a chemically homogeneous silicate melt.

Cation sites in silicates

In most silicates (excluding the framework silicates) the tetrahedra fit together compactly, producing an orderly three-dimensional array of fairly closely-packed

---

[2] Not to be confused with the **aluminium silicate** (Al_2SiO_5) minerals kyanite, andalusite and sillimanite introduced in Chapter 2, in which at least some of the Al^{3+} ions are in *octahedral* co-ordination. This important distinction is often misunderstood.

Box 8.3 Silicate melts

At the atomic level a silicate melt has much in common with silicate crystals: it consists of covalently bonded silicate structural anions such as $(SiO_4)^{4-}$ and $(Si_2O_7)_n^{6n-}$ glued together electrostatically by cations like Mg^{2+} and Fe^{3+}. Over short distances there is the semblance of order (Box 1.2), but a silicate melt, like any liquid, lacks the *long-range order* characteristic of the crystalline state.

Another important difference is that, whereas nearly all silicate crystal structures are each constructed around a single type of silicate skeleton (chains in the case of pyroxenes, for example), the more open structure of a melt allows several types of silicate structural unit to coexist and intermingle. In *basic melts*, those rich in basic oxides and relatively poor in silica, a range of relatively simple polymers predominates (SiO_4^{4-}, $Si_2O_7^{6-}$, $Si_3O_{10}^{8-}$, $(SiO_3)_n^{2n-}$, etc.). This relatively unpolymerized structure favours the crystallization of minerals like olivine and pyroxene, because the appropriate polymers, at least in embryo form, are already present in the melt. An *acidic melt* has a more polymerized structure conducive to the crystallization of minerals like mica (if water is present), feldspars and quartz.

The degree of polymerization of molten silicates is most clearly expressed in their viscosity. Viscosity increases with the degree of polymerization, because large polymeric units get tangled up with each other more than small ones do, and this inhibits the flow of the liquid. Hence rhyolites (acid lavas) are much more viscous than basalts (Figure 3.8), even when allowance is made for temperature differences.

Dissolved water has a dramatic effect on magma viscosity, through its ability to sever polymers and therefore 'depolymerize' the melt as shown in Figure 8.3.1. Magma rich in dissolved water is therefore more fluid and conducive to diffusion than a dry magma of corresponding composition. In this instance water is behaving as a basic oxide.

A high dissolved water content is only stable when confined under high pressure, however. Close to the surface, ascending water-rich melt will become supersaturated and vesiculate to form bubbles of H_2O vapour. Viscosity then rises dramatically as the loss of water allows the magma to polymerize, and the presence of bubbles further enhances viscous behaviour (as stiffly whipped egg white illustrates).

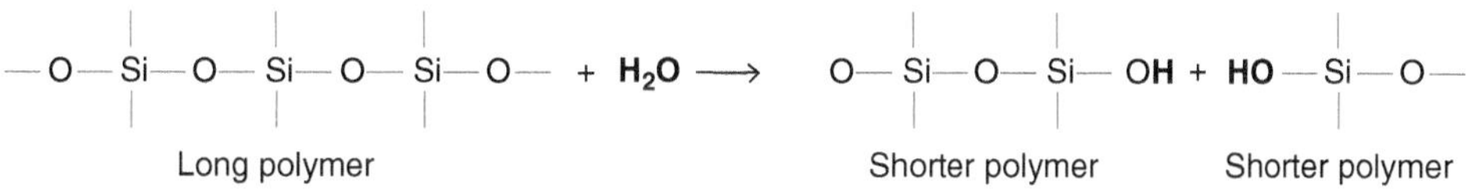

$$-O-Si-O-Si-O-Si-O- \;+\; H_2O \longrightarrow \quad O-Si-O-Si-OH \;+\; HO-Si-O-$$

Long polymer $\qquad\qquad\qquad$ Shorter polymer $\qquad$ Shorter polymer

Figure 8.3.1 Depolymerization of a silicate polymer by a water molecule. The H and OH derived from the H_2O molecule are shown in bold for clarity.

oxygen atoms. Si is known to occupy tetrahedral sites between these atoms. We can work out how the other ions such as Mg^{2+} and K^+ fit in by considering their *radius ratios* (Chapter 7). These are shown in Table 8.2, calculated from the ionic radii given in Box 7.2.

Two radii are given for Al^{3+} in Box 7.2 because, as the radius ratios suggest, its size permits it to fit into both tetrahedral and octahedral sites in silicates. (X-ray crystallography shows there is a slight difference in Al–O bond length in 4- and 6-fold co-ordination, so that a different Al^{3+} radius is associated with each site.) A glance at the formulae in Table 8.1 supports this conclusion. The formula of feldspar only makes sense if

we include Al in Z, and X-ray investigations confirm that Al is tetrahedrally co-ordinated in feldspars. In beryl, on the other hand, it is octahedrally co-ordinated. In many minerals, as in mica, it is found in both types of site.

The radius ratios of titanium (4+), iron (3+ and 2+), magnesium and manganese point to occupation of the octahedral ('Y') sites available in all **ferromagnesian** silicates. Ca^{2+} and Na^+ are somewhat larger, and require 8-fold co-ordination. Such ('X') sites do not exist in olivine, from which these elements are excluded (only traces of Ca are found), but they are available in pyroxene and amphibole (Box 8.5). K^+ clearly requires a still

Box 8.4 Features of a silicate analysis

Table 8.4.1 shows two ways of presenting the chemical analysis of a silicate material (in this case an olivine):

Table 8.4.1 Typical silicate analysis (olivine)

	Percentage of element*		Percentage of oxide†
Si	18.42	SiO_2	39.41
Fe (ferrous)	12.79	FeO	16.46
Mn	0.16	MnO	0.21
Mg	26.10	MgO	43.27
Ca	0.16	CaO	0.23
O	41.95		
Total‡			99.58

*The composition is expressed here in percentages of each *element* (i.e. the number of grams of the element per 100 g of the sample). Such units are commonly referred to as 'weight percent', although 'mass percent' is a more accurate description. Oxygen appears as a separate item, but in fact it is unnecessary to analyse for it: the valency of each element requires that it combines with oxygen in **stoichiometric** proportions, even in complex silicates. It is thus possible to calculate the amount of oxygen present from the percentages of the other elements in the rock and their valencies. Reporting mass % to two decimal places is consistent with the **precision** of analytical methods currently in use for major elements.

†A more convenient format for including oxygen is simply to report the analysis in terms of the percentage of each oxide (g oxide per 100 g sample), as shown in the right-hand column. There is no separate entry for oxygen, as it has all been allocated to individual oxides.

Only five oxides have concentrations significant enough to be listed in this olivine analysis. In other minerals or rocks, a wider range of elements would require analysis. A typical silicate analysis would include the following *major elements* (those making up more than 0.1% of the total and so included in the analysis total): SiO_2, Al_2O_3, Fe_2O_3, FeO, MnO, MgO, CaO, Na_2O, K_2O, TiO_2, P_2O_5, H_2O (see Table 8.5 in Exercise 8.2 at the end of the chapter) and CO_2. One might also look for certain *trace elements* (Ni being the prominent one in the case of olivine) whose concentrations would mostly fall below 0.1% and have no significant effect on the total (see Figure 9.1).

The entries in columns 1 and 2 are related to each other as follows:

$$\text{oxide\%} = \text{metal\%} \times \frac{\textbf{RMM} \text{ of oxide}}{n \times \textbf{RAM} \text{ of element}}$$

where n is the number of metal atoms per molecule of oxide. (It is necessary to specify FeO and Fe_2O_3 separately, or to assume that all of the iron is in one form or the other.)

‡The oxide analysis is followed by a total, which for an accurate analysis should fall between 99.5 and 100.75% (all analytical measurements attract a small statistical error). If the total exceeds 100.75, one would suspect an unacceptable error in analytical procedure or calculation. A low total would suggest an error in the other direction, or neglect of an important constituent.

---

larger site. It occurs in amphiboles and micas in a large ('A') site associated with the silicate rings (Box 8.5), but this site does not exist in the pyroxene structure, which therefore excludes K.

In framework silicates there are no compact Y sites to accommodate Mg^{2+} and Fe^{2+}. Ca, Na and K occupy sites with somewhat irregular geometry, ranging in co-ordination number from 6 to 9. In the more open structures of the zeolites these sites are larger still, and indeed may be substantially larger than the Na^+ and K^+ ions occupying them. This, in combination with the weak charge on these ions, means that they are readily removed and replaced by other cations, a process known as *ion exchange* that zeolites share with clay minerals (Box 8.2).

Calculating site occupancies

The analysis of a silicate mineral (Box 8.4) is easier to understand if it is recalculated into a form directly comparable with the mineral's chemical formula.

Olivine analysis

We have seen that, in the formula of pure forsterite (Mg_2SiO_4), 2 magnesium ions and 1 silicon atom are associated with 4 atoms of oxygen. How many atoms of Mg (and Fe, Mn and Ni) and of Si would on average be associated with 4 oxygen atoms in a general olivine analysis like that given in Box 8.4? To answer this we need to recalculate the analysis in terms of the *relative numbers* of atoms, instead of mass percentages of oxides.

Table 8.3 The mineral formula and site occupancies of an olivine

Oxide	**RMM** of oxide	1 Analysis as mass % oxides*	2 Analysis as moles of oxides[†]	3 Moles of oxygen (as O^{2-})[‡]	4 Cations per 4 oxygens[§]	5 Site totals	
SiO_2	60.09	39.41	0.6558	1.3116	1.0008	Z site sum:	1.001
FeO	71.85	16.46	0.2291	0.2291	0.3496		
MnO	70.94	0.21	0.0030	0.0030	0.0046	Y site sum:	1.998
MgO	40.32	43.27	1.0732	1.0732	1.6378		
CaO	56.08	0.23	0.0041	0.0041	0.0063		
		99.58		2.6210			

$$\times \frac{4}{2.6210} =$$

*See Box 8.4.
[†]Column 1 divided by relative molecular mass.
[‡]Column 2 × number of oxygens per molecule (=2 for SiO_2, 1 for the rest)
[§]Column 2 × 4/2.6210.

Table 8.3 shows how this calculation is carried out. The oxide analysis is written in column 1. The first step is to calculate the number of **moles** of each oxide present. This is achieved by dividing each oxide mass percentage by the **relative molecular mass (RMM)** of the oxide concerned, entering the results in column 2. Because column 1 contains the number of grams of each oxide in 100 g of the olivine (i.e. mass percentage), column 2 contains the *number of moles* per 100 g.[3]

Multiplying each entry in column 2 by the number of oxygen atoms in the corresponding oxide formula – 2 for SiO_2, 1 for the other oxides – gives the number of moles of O^{2-} associated with each oxide (column 3). Adding these up tells us that 100 g of sample contains total of 2.6210 moles of O^{2-}. Our objective, however, is to calculate the numbers of cations associated with 4 moles of O^{2-}. Multiplying each entry in column 2 by 4 ÷ 2.6210 = 1.5261 gives us the number of moles of each oxide which together contain 4 moles of O^{2-} (column 4). As each oxide molecule contains only one atom of metal, the figures in column 4 also indicate the numbers of cations equivalent to a total of 4 oxygens.

The results in column 4 show two notable features. The first entry is a number very close to 1.0000. It represents the number of silicon atoms present for every

four oxygen atoms in the olivine structure, and it indicates that the Z-sites in olivine are filled with silicon alone. Secondly, the remaining entries in column 4 add up to 1.998. These elements collectively represent the average contents of the two Y-sites associated with every group of four oxygens in olivine. These conclusions allow us to write the complete chemical formula for the olivine, showing in what proportions the elements occupy each type of site:

$$\left(Mg_{1.638}Fe_{0.350}Ca_{0.006}Mn_{0.005}\right)Si_{1.001}O_4 \qquad (8.1)$$

The close correspondence of the total site occupancies to the whole numbers of the ideal formula of olivine (Y_2ZO_4) is additional reassurance of an accurate analysis.

Amphibole analysis

Table 8.4 shows the formula calculation for an amphibole analysis.

The sites available in the amphibole structure (Box 8.5) are summarized by the formula:

$$A \quad B_2 \quad C_5 \quad Z_8 \quad O_{22} \left(OH\right)_2$$

although in many amphiboles the large A site is partly or wholly vacant. The following points should be noted:

(a) The formula of an amphibole is normally written with 24 oxygens (including OH), so the analysis is recalculated on this basis.

---

[3] The logic here is analogous to dividing a bag of apples among a group of children: it is more useful to know the *number* of apples in the bag than their weight.

Table 8.4 The mineral formula and site occupancies of an amphibole

Oxide	RMM of oxide	1 Analysis as mass % oxides	2 Analysis as moles of oxides*	2a Moles of metals[†]	3 Moles of oxygen (as O^2)[‡]	4 Cations per 24 oxygens[§]	5 Site occupancy totals	
SiO_2	60.09	57.73	0.9607	0.9607	1.9214	7.786	Z site	8.000
Al_2O_3	101.94	12.04	0.1181	0.2362	0.3543	iv[#] 0.214		
						vi 1.700		
Fe_2O_3	159.70	1.16	0.0073	0.0146	0.0219	0.118		
FeO	71.85	5.41	0.0753	0.0753	0.0753	0.610	C site	5.056
MnO	70.94	0.10	0.0014	0.0014	0.0014	0.011		
MgO	40.32	13.02	0.3229	0.3229	0.3229	2.617		
CaO	56.08	1.04	0.0185	0.0185	0.0185	0.150	B site	1.975
Na_2O	61.98	6.98	0.1126	0.2252	0.1126	1.825		
K_2O	94.20	0.68	0.0072	0.0144	0.0072	0.117	A site[¶]	0.117
H_2O	18.02	2.27	0.1260	0.2520	0.1260	2.042	OH site	2.042
Total		100.43			2.9615			

$$\times \frac{24}{2.9615} =$$

*Column 1 divided by relative molecular mass.

[†]Column 2 × number of cations in oxide molecule. (This column does not appear in Table 8.3 because there all oxide molecules had only one cation.)

[‡]Column 2 × number of oxygens per oxide molecule.

[§] Column 2a × 24/2.9615.

[#]iv represents Al allocated to 4-fold co-ordinated (tetrahedral) Z sites to make up shortfall in Si. vi represents remaining Al allocated to 6-fold (octahedral) C sites.

[¶]The A-site is only partly occupied (i.e. on the atomic scale, some A-sites are filled while others are empty.)

(b) The ions are distributed between a greater variety of structural sites (discussed in Box 8.5) than in an olivine. As a result the agreement between site occupancies and the ideal formula is less close than for olivine.

(c) There is insufficient Si to fill the 8 tetrahedral Z-sites per formula unit. We assume that the remainder are occupied by Al ions (symbolized Al^{iv}), but most of the Al is left over and gets allocated to the octahedral C sites (where it is denoted as Al^{vi})[4] in company with Fe^{3+}, Fe^{2+}, Mg^{2+} and Mn^{2+}.

(d) Ca^{2+} cannot enter the octahedral C sites and must be allocated to the larger B site, which also accommodates Na^+.

(e) K^+ is too large to enter any but the A site. In the example given in Table 8.4, it falls a long way short of filling all of the A-sites available, and in many

amphiboles this site is vacant (a situation indicated in an amphibole formula by '□').

Formula calculations like these have several important applications in mineralogy.

• Firstly, they help to confirm the accuracy of an analysis, as the site totals for a good analysis of a simple mineral like olivine should approximate to whole numbers (Table 8.3).

• Knowing what kind of crystallographic site an element occupies helps in understanding why, for example, pyroxenes contain no potassium, or why the structure of augite is different from hypersthene (Box 8.5).

• Formula calculation also plays a part in classifying mineral groups that involve complicated solid solutions. The nomenclature of the amphiboles, for example, rests heavily on chemical parameters such as the *number of silicon atoms per 8 tetrahedral sites* (item (c) above).

---

4   The 'vi' here signifies the 6-fold co-ordination of octahedral sites (see Table 7.1).

Box 8.5 Cation sites in pyroxenes and amphiboles

Figure 8.5.1a shows a simplified end-on view of the pyroxene chains, indicating how they are stacked together, alternately back-to-back and point-to-point. In addition to the Z-sites (not shown) occupied by Si and a little Al, there are two types of cation site. Between two chains facing each other point-to-point lie two octahedral sites, designated M1, in which Al^{vi} ('octahedral Al'), Fe^{2+}, Fe^{3+}, Mg^{2+}, Mn^{2+}, Cr^{3+} and Ti^{4+} are accommodated. The other type of site

lies between pairs of chains whose tetrahedra face each other base-to-base. They are called M2 sites and their geometry varies according to the ions occupying them. In the absence of Ca or Na, M2 is occupied by Mg^{2+}, Fe^{2+} and Mn^{2+} (for example in enstatite, $Mg_2Si_2O_6$), and has an irregular 6-fold co-ordination. The chains stack together in such a way as to produce an orthorhombic unit cell (orthopyroxene). The substitution of larger ions like Ca^{2+} or Na^+ causes a change of M2 geometry to 8-fold co-ordination. The presence of the larger ion disrupts the stacking, and forces the structure to adopt a lower-symmetry (monoclinic) structure, as in diopside ($CaMgSi_2O_6$).

Owing to the broader bands in the amphibole structure (Figure 8.5.1b), there are three slightly different types of octahedral site (two M1 sites, two M2 sites and one M3 site) in corresponding positions to the pyroxene M1 site. These are the five 'C' sites in the formula given in the main text. A larger site called M4 (or 'B' in the formula) corresponds almost exactly to M2 in the pyroxenes, accommodating small ions like Mg^{2+} in 6-fold co-ordination, and larger ions like Ca^{2+} and Na^+ in 8-fold co-ordination. The occupant of M4 plays the same role as M2 in the pyroxenes in determining the symmetry of stacking: Mg-rich amphiboles are commonly orthorhombic, whereas all other compositions are monoclinic. There are two M4 ('B') sites per formula unit.

The so-called A-site has no equivalent in pyroxenes. It lies sandwiched between pairs of bands whose undersides face each other, associated with the hexagonal rings that account for its large size. It is commonly unoccupied, but may contain Na^+ or K^+. The A-site lies opposite the OH-site, which also has no equivalent in the pyroxene structure. Some sheet silicates exhibit similar architecture (Box 8.2).

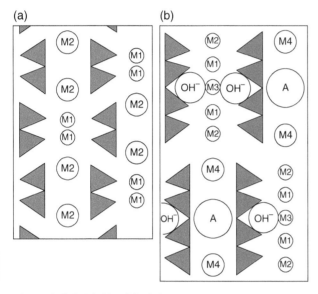

Figure 8.5.1 (a) Simplified view of the crystal structure of a pyroxene. Double triangles represent the silicate chain viewed end-on as in Figure 8.3a; M1 and M2 represent the two types of cation site (see text). (b) Similar representation of the structure of an amphibole. Groups of four triangles represent the silicate double chains end-on as shown in Figure 8.3b. M1, M2 and M3 represent three types of octahedral site; M4 is a larger site, and A (often partially filled) is larger still, nestling in the ring structure of the double chain (Figure 8.3b).

Effects of cation substitution

In the olivine solid solution series extending from the end-member forsterite (Fo = Mg_2SiO_4) to fayalite (Fa = Fe_2SiO_4), Mg ions in the sites are progressively replaced by Fe^{2+} ions. This *substitution* of Fe^{2+} for Mg^{2+} is possible because the ions have the same charge and similar size (Box 7.2). The larger radius of Fe^{2+}

nevertheless causes a slight expansion of the crystal lattice: the *c* cell-dimension, for example, increases from 0.5981 (Fo) to 0.6105 nm (Fa). The substitution of Fe^{2+} for Mg^{2+} gives rise to similar, continuous solid solution series in garnets, pyroxenes, amphiboles and micas.

The effect is different when the substituting ion has a different size. Replacing Mg^{2+} in the pyroxene B-site with the larger Ca^{2+} ion forces the pyroxene to adopt a new structure of different symmetry. Because $Mg_2Si_2O_6$ and $CaMgSi_2O_6$ have intrinsically different structural arrangements, there cannot be a continuous series of compositions between them. There is a *miscibility gap* between these minerals except at high temperature, with the consequence that two pyroxenes of different composition can exist in the same rock in equilibrium with each other. A similar gap operates in the alkali feldspars between albite ($NaAlSi_3O_8$) and orthoclase ($KAlSi_3O_8$). The compositions of co-existing feldspars (or pyroxenes) lie on a **solvus** curve, as illustrated in Figure 2.6.

Coupled substitution

A more complicated form of substitution is seen in plagioclase feldspar. The series of compositions from albite ($NaAlSi_3O_8$) to anorthite ($CaAl_2Si_2O_8$) reflects substitution in two sites at once. Substitution of Al^{3+} for Si^{4+} in tetrahedral sites leaves a charge imbalance, which is cancelled out by the *coupled substitution* of Ca^{2+} for Na^+ in the large cation sites. The overall substitution is of CaAl for NaSi. It follows that the Ca content of plagioclase correlates with the Al content, and the alkali (Na + K) content with Si. In the formula of any accurate plagioclase analysis (calculated to 32 oxygens), Si − (Na + K) will approximate to 8.0 (±0.1), and Al − Ca to 4.0 (±0.1). These requirements provide another measure of the quality of a plagioclase analysis (Exercise 8.3). Similar coupled substitutions occur in pyroxenes: Na^+Al^{3+} for $Ca^{2+}Mg^{2+}$ in diopside to give jadeite ($NaAlSi_2O_6$), for example.

Optical properties of crystals

Crystal optics, a specialized subject for which several excellent textbooks are available, is mostly beyond the scope of this book.[5] It is, however, relevant to take a brief look at how the optical properties of a crystal relate to its chemical bonding.

Refractive index

Light, like any electromagnetic signal, consists of synchronized oscillations of electric and magnetic fields (Boxes 5.2 and 6.3). It is the electric vector that is

relevant to optical phenomena in crystals, because the progress of light in a crystal depends upon the interaction between the electric disturbance and atomic electron clouds. The extent of interaction, which is measured by the refractive index of the crystal, depends on the **polarizability** of the atomic assemblage. The highest refractive indices are found in minerals with metallic bonding, such as sulfides: amongst the highest is galena, with a refractive index of 3.9.

Colour and absorption

Another important optical property is absorption. A mineral is *coloured* in transmitted light because certain wavelength components in the visible spectrum are more strongly absorbed than others, making the crystal appear to have the *complementary colour* (Figure 8.4). The region of absorption is generally a wavelength band rather than specific sharp lines. The green colour of olivine, for example, is due to absorption bands at each end of the visible spectrum, which emphasize the green wavelengths in the middle of the spectrum in the light emerging from the crystal.

The absorption of light energy is related to electron transitions between appropriately spaced energy levels in the crystal. The quantum energy of visible light is much smaller than that of X-rays (Chapter 6), and the transitions concerned occur between valence energy levels, either within an atom or between neighbouring ions. One very important factor in the colour of minerals

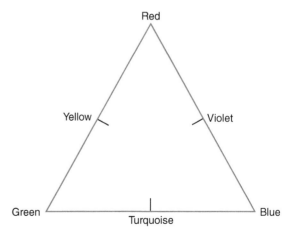

Figure 8.4 Colour triangle showing complementary colours. Absorption of one colour will leave light relatively enhanced in the colour on the opposite side of the figure.

---

[5]   A very simple introduction to crystal optics can be found at http://bit.ly/HMALfx.

is the splitting of d energy levels in transition-metal ions such as iron (Box 9.11). Elements like iron or manganese, if present in a mineral, play a major part in determining its colour. This can be seen in a solid solution series like phlogopite–biotite. The Mg-rich mica phlogopite ($KMg_3\{AlSi_3O_{10}\}[OH]_2$) is colourless or very pale in thin section, whereas the iron end-member biotite is intensely brown or red. Colour attributable to an essential constituent is said to be **idiochromatic.**

Certain other minerals, including some important gemstones, owe their colour to impurity elements. A common agent of such **allochromatic** colours is trivalent chromium, responsible for the deep red colour of ruby (a variety of corundum, Al_2O_3) and the green of emerald (a variety of beryl). The 'chrome diopside' which is conspicuous in many mantle-derived peridotites also owes its brilliant green colour to its Cr_2O_3 content (1% or so).

Colour can also be due to the presence of impurity *phases*. Quartz sometimes takes on a blue colour due to the presence of microscopic needles of rutile (TiO_2) or tourmaline.

Minerals may become coloured from being exposed to natural radiation for long periods. Such colours are due to radiation-induced defects in the crystal structure. The purple colour of fluorite, for example, is attributed to free electrons occupying vacant anion sites called 'colour centres'. An electron trapped in such a vacancy, like an atomic electron, is restricted to a number of quantized energy levels and transitions between these levels generate the observed colours. Heating eliminates most of the vacancies in a crystal and therefore bleaches the colour.

Metals and a number of important minerals – notably graphite and certain sulfides – absorb all light passing through them, so that in thin section they transmit no light at all. The property of **opacity** can be traced to metallic or quasi-metallic bonding in the crystal. Delocalized electrons in a metal behave like a charged fluid (Chapter 7). The alternating polarization of this electron fluid, induced by the incoming light's oscillating electric field, involves work being done in moving electrons around, and this dissipates the energy of the incident light (transforming it into thermal motion) just as the energy of a walker is sapped by walking across unconsolidated dry sand. A seismological analogy is the absorption of shear waves by liquids, which is why such waves do not pass through the Earth's core whereas compressional waves do. Insulating materials like quartz, in which there are no delocalized electrons, are electromagnetically elastic: electrons fixed in atoms cannot be moved around and so absorb practically no energy from the beam, which is therefore transmitted with little loss of intensity like seismic waves through solid rock.

Reflectivity

The striking characteristic of minerals like pyrite and galena is their metal-like *reflectivity*. A reflected-light microscope with a special reflectivity accessory can be used to measure quantitatively the proportion of the perpendicularly incident light that is reflected back at various wavelengths, and this provides a vital diagnostic tool for the ore mineralogist. The reflectivity of a mineral increases with its refractive index and with its opacity.

Anisotropy

Whereas minerals like halite and diamond have optical properties that are uniform in all directions, the majority of minerals are optically **anisotropic**. That is to say, the refractive index, the colour and – for ore minerals – the reflectivity vary according to the direction in which the incoming light (specifically its electric vector) oscillates.

Incoming light entering an anisotropic mineral is split into two rays with mutually perpendicular vibration directions, which experience different refractive indices as they travel through the crystal. The **birefringence** of the crystal is the difference between its maximum and minimum refractive indices. The mineral calcite is among the most birefringent of the common minerals, to the extent that the phenomenon can even be seen without the aid of a microscope. To understand why this is, we need to note that the carbonate anion CO_3^{2-} has a symmetrical planar shape (Figure 8.5), reflecting the geometry of the sp²-hybrid that the carbon atom uses to form σ-bonds with the three oxygen atoms. All the CO_3^{2-} ions in calcite have the same orientation perpendicular to the three-fold symmetry (*z*) axis of the crystal. The remaining carbon p-electron can establish a π-bond with any one of the three oxygens, giving three possible configurations. As with graphite (Box 7.4), the real configuration is a blend of

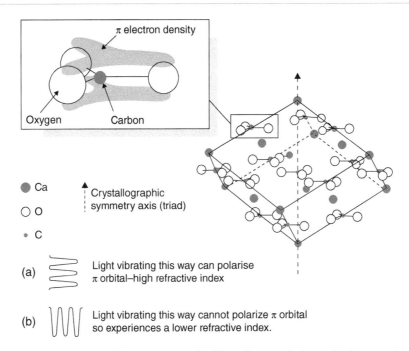

Figure 8.5 Structure of calcite showing (inset) the orientation of sp²-based trigonal planar CO_3^{2-} **oxy-anion**. The highly polarizable concentrations of π electron density perpendicular to the z-axis are shown stippled. Light vibrating in plane (a) encounters a high refractive index. Light vibrating parallel to the z-axis (b) encounters less polarization and therefore a lower refractive index.

all three. Y-shaped π molecular orbitals therefore exist above and below the plane containing the nuclei. To an electric field oscillating perpendicular to the z-axis represented by the dashed arrow (Figure 8.5a), the anion looks highly polarizable, because delocalized electron density is easily shunted from one end of these π-orbitals to the other by the oscillating electric field. Light having this electric vibration direction therefore experiences a relatively high refractive index (1.66). The polarizability parallel to the c-axis is much lower, so light vibrating in this direction encounters a lower refractive index (1.49). Therefore when a mark on a piece of paper is viewed through a clear cleavage rhomb of calcite, two separate images are seen.

The majority of silicate minerals are also optically anisotropic for similar reasons, although the degree of anisotropy (the magnitude of the birefringence) is much lower than for carbonates.

Defects in crystals

The extreme order of the crystalline state conceals the fact that all real crystals incorporate structural defects, which have a profound effect on the growth of crystals and on their mechanical strength.

Crystal growth

The first step in the production of a crystal from a surrounding liquid (melt or solution) is *nucleation*, the formation of the initial embryonic speck of ordered crystalline material upon which the rest of the crystal will be deposited. The free energy of this *nucleus* consists of:

(a) a negative term proportional to its volume, reflecting the cohesive forces between close-packed ions/ atoms in the interior; and
(b) a positive term proportional to the surface area, reflecting the reactivity of unsatisfied bonding potential on the surface.

Thus for a cubic nucleus of edge-length r:

$$\Delta G_L = -r^3 L_v + 6r^2 \sigma_s$$

where ΔG_L is the free energy of the nucleus relative to an equivalent amount of melt. L_v is the free energy of fusion per unit volume, and σ_s is the surface energy per unit surface area (Figure 8.6).

The initial nucleus, having a high surface area:volume ratio, is therefore highly unstable, and it must grow rapidly into a larger, more stable crystal if it is not to be

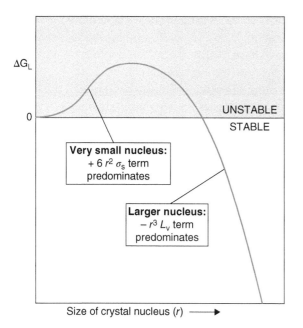

Figure 8.6 Energetics of crystal nucleation. ΔG_L is the free energy of the nucleus relative to the equivalent amount of melt.

redissolved by the liquid. In practice, therefore, melts cool below the liquidus, becoming supersaturated with the crystal components, before the first visible crystals appear, a phenomenon called *supercooling*.

The subsequent growth of a crystal can be seen in terms of adding new material (depicted as rectangular blocks) in layers on an existing crystal surface, as depicted in Figure 8.7a. The blocks here can be interpreted as ions, molecules or entire unit cells. Exposed corners and edges are features of very high surface energy owing to the number of unsatisfied bonds. Adding a block at A is therefore highly unfavourable in energy terms, and would occur only in a strongly **supersaturated** solution. Sites B and C are less hostile, but continued exploitation of such step sites will complete the layer and eliminate the step. In subsequent crystallization on this face, use of sites like A is unavoidable. Calculations suggest that a high threshold of supersaturation or supercooling has to be surmounted before crystallization at sites like A can proceed, and crystallization ought therefore to be an extremely slow process, yet in practice this seems not to be the case.

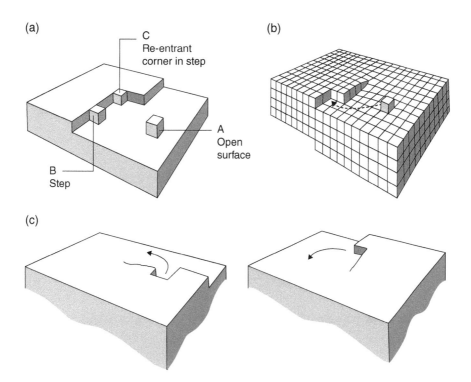

Figure 8.7 (a) Sites of accretion on a crystal face. (b) Screw dislocation, (c) Subsequent positions of the step as crystallization proceeds. (Source: Read 1953. Reproduced with permission of McGraw-Hill.)

The explanation of the discrepancy lies in a lattice imperfection called a *screw dislocation* (Figure 8.7b). Such defects are stacking 'mistakes' incorporated randomly into crystals and perpetuated during growth. A mismatch of layers on one side of the crystal is not seen on the other side, and must disappear at a line (dislocation) extending vertically through the middle of the crystal, where one layer actually twists up into the next one. The crystal resembles the kind of multi-storey car park in which the floors are part of a large spiral leading to the top. The vital feature is that the step affording favourable crystallization sites, like B and C, is perpetuated as crystallization proceeds, spiralling continuously upward as suggested by the arrows and the crystallization 'fronts' in Figure 8.7c. Such dislocations make continued crystallization practicable at modest degrees of supersaturation.

Mechanical strength of crystals

Box 7.7 includes estimates of the strength of bonding in several crystalline materials. From data such as these it is not difficult to calculate values for the shear strength of a number of structurally simple crystalline materials, such as pure metals. However carefully such calculations are carried out, the results are generally 100–1000 times greater than the measured shear strength for the material in question. In resolving this embarrassing discrepancy, we have to recognize the role of another type of crystal imperfection, the *edge dislocation*.

An edge dislocation (Figure 8.8a) marks the edge of an extra half-layer (a–a–a) in a crystal lattice. This may represent a layer overstepped during growth of the crystal, or may be the result of deformation. Note that bonds in the immediate vicinity of the dislocation are either stretched or compressed, and owing to this departure from equilibrium length are not as strong (Box 7.1) as bonds in the undistorted lattice. The high free energy associated with the edge dislocation makes it a particularly susceptible site for initiating chemical reactions: acid etching of freshly broken crystal surfaces leads to the formation of pits at points where dislocations intercept the surface, rendering them visible under an electron microscope.

The mechanical significance of the dislocation becomes clear if we consider the crystal being subjected to *shear stress* as shown in Figure 8.8b. The crystal at first responds with *elastic strain*, in which it is deformed to a minute and reversible extent by the stretching and compression of bonds, without any being broken. Permanent shear deformation (*ductile deformation*) requires bonds to be broken so that the top half of the crystal can be bodily moved to a new position across the lower half.

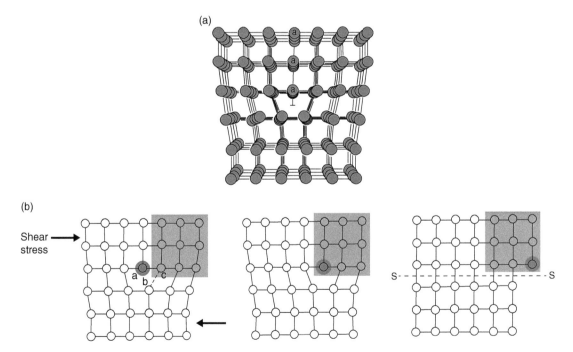

Figure 8.8 (a) Edge dislocation, (b) Migration of edge dislocation in response to shear stress (arrows). The stippled portion of the upper layer *slips* over the lower layer (s–s is the *slip plane*) as the edge dislocation (heavy stipple) jumps from row to row. (Source: Bloss 1971. Reproduced with permission of Holt, Rinehart and Winston, Inc.)

A perfect crystal resists this very effectively, because a whole layer of bonds must be broken simultaneously, requiring a colossal input of energy. The presence of an edge dislocation, however, changes this picture dramatically. When the shear stress is applied, the dashed row of bonds immediately to the right of the dislocation (already stretched in the unstressed crystal) becomes more stretched than other bonds, and will be the first to rupture. This will happen when the distance from row *a* to row *b* becomes less than the stretched bond length *b–c*. These bonds will at this point flip to link rows *a* and *b* instead. Row *a* is now part of a complete layer, and the dislocation has migrated to row *c*. If the stress is maintained, this process will be repeated until the dislocation has migrated to the edge of the crystal. The result is the net movement of the upper half of the crystal over the bottom half (Figure 8.8b), in a manner that requires *only one row of bonds to be broken* at one time. The stress required is orders of magnitude less than that needed to deform a perfect crystal. Edge dislocations therefore explain why the shear strength (and tensile strength) of crystalline materials is less than theoretically expected.

A square centimetre of any crystal intersects 10^8–10^{12} edge dislocations. The exact number depends upon the crystal's deformation history. Deformation – 'working' in the metallurgist's jargon – causes dislocations to multiply and initially makes the crystal more ductile.

Most minerals are considerably more brittle than metals. At high temperatures and pressures, however, *dislocation creep* can become an important mechanism of rock and mineral deformation. Its effect is to make silicate rocks behave in a ductile (plastic) manner if subjected to stress over long periods of time. It thereby becomes possible for solid mantle rocks to *convect* (circulate in response to temperature-induced density gradients), the phenomenon upon which terrestrial heat flow and plate tectonics largely depend. If all crystals were perfect, the Earth would be a very different planet.

Further reading

Cox, K.G., Price, N.B. and Harte B. (1988) *The Practical Study of Crystals, Minerals and Rocks*. London: McGraw-Hill.

Klein, C. and Philpotts, A.R. (2013) *Earth Materials – Introduction to Mineralogy and Petrology*. Cambridge: Canbridge University Press.

Krauskopf, K.B. and Bird, D.K. (1995) *Introduction to Geochemistry*, 3rd edition. New York: McGraw Hill.

White, W.M. (2013) *Geochemistry*. Chichester: Wiley-Blackwell.

Exercises

8.1 Predict the degree of Si–O polymerization of the following minerals:

Edenite	$NaCa_2Mg_5(AlSi_7O_{22})(OH)_2$
Hedenbergite	$CaFeSi_2O_6$
Paragonite	$NaAl_2(AlSi_3O_{10})(OH)_2$
Leucite	$K(AlSi_2O_6)$
Acmite	$NaFeSi_2O_6$

8.2 Express the first analysis in Table 8.5 (Exercise 8.3) below in terms of element percentages.

Table 8.5 Analyses for Exercise 8.3

	Garnet* $X_3Y_2Z_3O_{12}$	Epidote[‡][§] $X_3Y_2Z_3O_{12}(OH)$	Pyroxene[¶] $M_2Z_2O_6$	Feldspar[‖] $X_4Z_8O_{32}$
SiO_2	38.49	39.28	46.92	52.73
Al_2O_3	18.07	31.12	3.49	29.72
TiO_2	0.55[†]	–	1.19	–
Fe_2O_3	5.67	4.15	0.95	0.84
FeO	3.76	0.42	20.31	–
MnO	0.64	0.01	1.13	–
MgO	0.76	0.01	7.30	–
CaO	31.59	23.44	17.35	12.23
Na_2O	–	–	1.24	4.19
K_2O	–	–	–	0.13
H_2O^+	–	1.87[§]	–	–
Total	99.63	100.30	99.84	99.84

*In the garnet structure, all divalent ions reside in the 8-fold ('X') site.

[†]**RMM** of $TiO_2 = 79.9$. (Others given in Table 8.4.)

[‡]Calculate to 13 oxygens (12 + one OH group).

[§]The *total* water content of an analysis consists of:

(a) Internal structural 'water' (actually present as OH but measured as H_2O). This is generally assumed to be released only at temperatures above 110 °C. Denoted H_2O^+.

(b) Adsorbed water (moisture adhering to the surface of the powdered sample grains). Assumed to be released at temperatures below 110 °C. (Denoted 'H_2O-').

[¶]Fill the Z site with sufficient Al to make 2.0000 (cf. Table 8.4). The remaining Al goes in M. In a pyroxene formula it is difficult to distinguish between M1 and M2 (Box 8.5), because Mg, Fe^{2+} and Mn enter both sites. Here they are considered together as two 'M' sites.

[‖]Si, Al and Fe^{3+} reside in the Z-site.

8.3 Calculate the site occupancies of the mineral analyses in Table 8.5, matching your results with the ideal formulae given (cf. Tables 8.3 and 8.4). Carry out all calculations to 4 decimal places.

8.4 Plot the pyroxene composition (Exercise 8.3) in a ternary diagram whose apexes represent Ca, Mg and (Fe^{2+} + Fe^{3+} + Mn). The plotting method was described in Box 2.7. Plot the ideal compositions of diopside, hedenbergite, enstatite and ferrosilite ($Fe_2Si_2O_6$).

9 SOME GEOLOGICALLY IMPORTANT ELEMENTS

Chemical elements are of interest to geoscientists for various reasons. Some, such as silicon (Si) and iron (Fe), are so abundant that their chemical properties govern the behaviour of geological materials. Less abundant elements like rubidium (Rb) and strontium (Sr), though they participate more passively in geological processes, may nonetheless help us to understand how such processes work. Other elements (such as chromium, Cr, neodymium, Nd, or uranium, U) have important commercial uses or environmental impacts (e.g. radon in Box 9.9).

The purpose of this chapter is to introduce the reader to the geochemistry of the more important chemical elements, to show how the Periodic Table can be used to predict element behaviour in the Earth and on its surface as well as in the laboratory, and to illustrate what these elements can tell us about Earth processes. Chapter 10 outlines the additional information that be learned from the study of the **isotopes** of naturally occurring elements.

Major and trace elements

In terms of their abundance in geological materials, elements can be divided into two classes: *major elements* and *trace elements*.

Major elements

These elements, which include Si, Al, Mg and Na, have concentrations in most geological materials in excess of 0.1%. They are essential constituents of rock-forming minerals. Major element concentrations in silicate minerals are usually expressed in terms of *oxide percentages* (Box 8.3; Figure 9.1). The term *minor element* is sometimes applied to the less abundant major elements, such as manganese (Mn) and phosphorus (P), with oxide concentrations below 1%.

Chemical Fundamentals of Geology and Environmental Geoscience, Third Edition. Robin Gill.
© 2015 John Wiley & Sons, Ltd. Published 2015 by John Wiley & Sons, Ltd.
Companion Website: www.wiley.com/go/gill/chemicalfundamentals

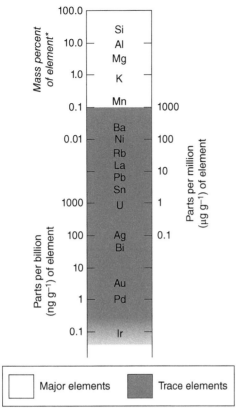

Major elements ☐   Trace elements ■

* Major elements are usually expressed as mass
percent *oxide* (Box 8.3)

Figure 9.1 How units of element concentration compare.
The positions of element symbols illustrate the average
concentration of the selected element in the continental crust.

Trace elements

Trace elements, such as rubidium (Rb) and zinc (Zn),
have concentrations in most geological materials too
low – usually less than 0.1% – for them to influence
which minerals crystallize. They mostly occur as dissolved 'impurities' in major rock-forming minerals,
although some – such as zirconium (Zr) – may form
separate accessory minerals (e.g. zircon, $ZrSiO_4$). Trace
element concentrations in rocks are usually expressed
in **parts per million (ppm** = $1\,\mu g\,g^{-1}$) of the element (not
the oxide) or in *parts per billion* (ppb = $1\,ng\,g^{-1}$) – see
Figure 9.1.

The distinction between major and trace elements
has to be applied in a flexible way, because the same
element can be a major element in one rock type (potassium in granite, for example) and a trace element in
another (potassium in peridotite).

Alkali metals

The alkali metals constitute the first column (Group Ia)
on the left-hand side of the Periodic Table. Their key
chemical features are as follows:

(a) They are strongly electropositive elements (Figure 6.3); their compounds are
characteristically ionic and they form
basic oxides.

(b) The M^+ cations are large (Figure 9.1.1) and
can be accommodated only in relatively
large cation sites, such as the 'A-site' in
amphiboles and micas (Chapter 8) and in
framework silicate minerals. Feldspar is
the main host of these elements in most rocks.

(c) They are very soluble in aqueous fluids, and are
among the first elements to be dissolved during
weathering. Na and K are important constituents
of seawater, and evaporite deposits provide the
main industrial source.

Sodium (Na) and *potassium* (K) are important constituents of feldspar, amphibole and mica and therefore are found as major elements in most rocks of the
continental crust. Potassium is an important plant
nutrient, hence its widespread use in fertilizers.
Rubidium (Rb) and *caesium* (Cs, spelt cesium in North
America), on the other hand, are too scarce to form
their own minerals and occur as trace elements, entering rock-forming silicates only where they can substitute for K^+ ions (e.g. in alkali feldspar).

The large ionic radii of K^+, Rb^+, Cs^+ and to a lesser
extent Na^+ (Figure 9.1.1) lead to their exclusion from
dense ferromagnesian minerals like olivine and pyroxene. Such minerals, with calcic plagioclase, are the first to
crystallize from basic magmas, and because the alkali
metals are excluded they remain in the melt. As the volume of melt decreases with advancing crystallization, the
concentration of dissolved K, Rb and Cs in the remaining
melt increases. A series of lava flows tapping a magma
chamber at successive stages in its crystallization would
therefore exhibit increasing concentrations of these elements in later flows. In plutonic rocks they are enriched
in 'late-stage' granites and pegmatites. Elements like
these, whose exclusion from the main igneous minerals
leads to their concentration in late-stage residual magmas, are called *incompatible elements* (Box 9.1).

Box 9.1 Incompatible elements

Incompatible elements are those whose cations are not easily accommodated in the structures of the principal igneous minerals. Being excluded from these minerals as magma crystallization progresses, they become progressively more enriched in the diminishing amount of residual melt, whose disordered, less compact structure accepts them more readily.

The shaded band in Figure 9.1.1 divides elements whose ions are readily accommodated in relevant igneous minerals (*compatible elements*) from elements that are incompatible. Elements whose symbols overlap the band may fall into either group depending on the minerals that happen to be crystallizing.

The reason why an incompatible element is excluded from a mineral's crystal structure depends on its **ionic potential** (Figure 7.7):

Large-ion lithophile (LIL) elements

(Ionic potential <40 nm^{-1}.) These are elements like Rb and Ba, which most crystals exclude because their ions are simply too large to fit into the cation sites available in the crystal structure (Box 7.2).

High field-strength (HFS) elements

(Ionic potential ≥40 nm^{-1}.) An ion like zirconium (Zr^{4+}) has a radius no bigger than Mg^{2+}, yet its high polarizing power and relatively covalent bonding make it an uncomfortable occupant of a Mg^{2+} cation site in a predominantly ionic crystal.

The behaviour of the **rare earth elements** (**REEs** = La to Lu) and yttrium (Y) is discussed in a later section.

Partition coefficients

A trace element's tendency to behave compatibly or incompatibly during magma evolution is expressed numerically by a number called the *partition coefficient*, represented by the symbol K_i^A, where *i* identifies a specific element and *A* specifies the mineral crystallizing. For example, the partition coefficient describing the equilibrium distribution of nickel between an olivine crystal and a coexisting melt is defined as:

$$K_{Ni}^{olivine} = \frac{\text{concentration of nickel in olivine crystal}}{\text{concentration of nickel in coexisting melt}} \quad (9.1.1)$$

where the concentrations are given in ppm. Nickel is a compatible element in olivine and therefore has a partition coefficient greater than 1.00 (typically about 18).

An incompatible element is characterized by partition coefficient values less than one for all minerals crystallizing from the melt: Rb, for example, has the following partition coefficient values: 0.006 in olivine, 0.04 in clinopyroxene, 0.25 in amphibole, 0.10 in plagioclase.

An element may behave compatibly in relation to one mineral while being incompatible in others. Thus Ni is a compatible element in olivine but incompatible in plagioclase.

Typical partition coefficient values are tabulated in Henderson and Henderson (2009).

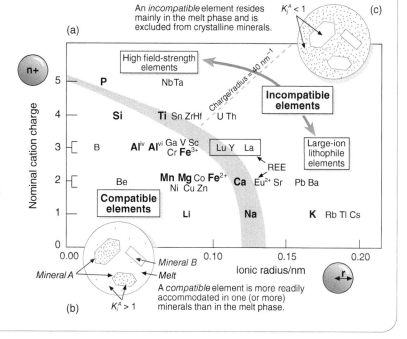

Figure 9.1.1 (a) Compatible and incompatible elements displayed on a graph of cation charge *versus* ionic radius. **Bold** type signifies major elements. (b) A cartoon illustrating the distribution of a trace element *i* that is compatible in mineral A but not mineral B. (c) A cartoon illustrating the distribution of a trace element that is incompatible in minerals A and B. The density of dots in (b) and (c) represents the concentration of element *i*.

Radioactive alkali metals

Two of the alkali metals have radioactive isotopes of great significance to the geoscientist: ^{40}K and ^{87}Rb. Both have long half-lives $(1.25 \times 10^9$ and 48.8×10^9 years respectively), of similar **order of magnitude** to the age of the Earth $(4.55 \times 10^9$ years). The amounts of these isotopes in the Earth are decreasing with time, but too little time has elapsed since the last episode of element-formation (Chapter 11) for them to have decayed away completely. The rates of decay are accurately known from laboratory measurement, and ^{40}K and ^{87}Rb provide isotopic clocks that can be used for dating geological events (Chapter 10).

Radioactive decay generates heat (Box 11.2). Much of the heat currently escaping from the Earth's interior is due to the decay of the radioactive isotopes of potassium (K), thorium (Th) and uranium (U) in the crust and mantle (Box 10.1). ^{40}K is thought to be responsible for about 15% of the heat generated in the crust. Owing to its low abundance and slow rate of decay (long half-life), ^{87}Rb does not make a significant contribution to the Earth's heat flow.

Hydrogen

The position of *hydrogen* (H) in Group I of the Periodic Table suggests that it should behave like the alkali metals, but in fact the similarity is restricted to valency. The ionization energy of hydrogen is much higher than the alkali metals (Figure 6.1b) and, because the valence shell consists of the 1s orbital alone, hydrogen possesses no metallic properties (except under extreme conditions, such as within the interior of Jupiter). The predominant hydrogen compound on Earth is the oxide H_2O, which occurs in the familiar gas (vapour), liquid and solid forms (Figure 2.2.1b). The presence of liquid-water oceans on the Earth's surface is unique in the Solar System, and has been essential to the evolution of life on Earth (Chapter 11). The peculiarities of the ice–water system play an important part in regulating the Earth's climate (Box 4.1) and in driving erosion and the transport of sediment.

Within the Earth, hydrogen occurs in rocks, mainly in the form of OH⁻ions in hydrous minerals such as muscovite $[KAl_3Si_3O_{10}(OH)_2]$ and the clay minerals (Box 8.2). Heating such minerals leads to **dehydration** (Figure 2.3), releasing H_2O-rich fluids that are essential agents in regional metamorphism. The presence of hydrous fluid also lowers the temperature (the solidus) at which a rock begins to melt, and such fluids therefore play an important role in bringing about melting in the mantle and crust above subduction zones, where the hydrous fluid originates by dehydration of hydrous minerals in the altered oceanic crust of the downgoing slab.

Water, as well as accelerating many geochemical reactions (Chapter 3), exerts a powerful influence on the physical properties of melts (Box 8.3). The explosive expansion of steam escaping from ascending magma, as it experiences depressurization close to the surface, provides the energy for the most destructive volcanic eruptions on Earth.

The isotopes of hydrogen are discussed in Chapter 10.

Alkaline earth metals

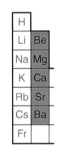

Relations between the divalent alkaline earth elements bear some resemblance to those between the alkali metals. *Beryllium* (Be), like lithium, has rather different chemical properties from the other members of its group (Box 9.2). *Magnesium* (Mg) and *calcium* (Ca) are major elements in most rock types, whereas *strontium* (Sr) and *barium* (Ba) are trace elements showing incompatible behaviour. All are strongly electropositive, reactive metals (Figure 6.3). The alkaline earths form stable, highly refractory, basic oxides (Box 8.1).

The ionic radius of Mg^{2+} is similar to that of Fe^{2+} (Box 7.2). Ferromagnesian minerals such as olivine and pyroxene exhibit complete solid solution between a magnesian end-member (such as forsterite, Mg_2SiO_4) and a ferrous end-member (fayalite, Fe_2SiO_4). Because MgO is a more **refractory** oxide (melting point 2800 °C), the Mg end-member of these minerals has the higher melting temperature (e.g. Box 2.4). The crystallization of ferromagnesian minerals from a magma depletes it in MgO more rapidly than FeO, so the FeO:MgO ratio of the residual melt increases with advancing crystallization and provides a useful indicator of the degree of fractionation of a magma.

Magnesium is an essential constituent of chlorophyll (Box 9.5), and therefore plays a part in photosynthesis.

Box 9.2 Lithium, beryllium and boron

Lithium, beryllium and boron are chemically some-what different from the other members of their respective groups. For example, except for the difference in valency, beryllium has more in common chemically with aluminium than with magnesium or calcium. The reason for this *diagonal relationship* is that beryllium's **ionic potential** is much closer to that of aluminium than to the other alkaline earth metals (see Figure 9.2.1), leading to similar, relatively covalent bonding behaviour and similar crystal chemistry. Both Be and Al form unreactive, very hard oxides that are chemically amphoteric (Box 8.1). Both metals resist acid attack and oxidation – unlike Mg, for example – owing to the formation of a surface film of oxide.

For the same reason boron has much in common with silicon. Both are semiconducting metalloids rather than true metals. B_2O_3 is an acidic oxide like SiO_2. The diagonal relationship between lithium and magnesium, although less pronounced, means that Li is found mainly in the Mg sites in silicate minerals.

Li, Be and B also differ from other elements in their groups in having anomalously low abundance, in the Earth and in the solar system (as explained in Chapter 11). For instance, Li is a trace element in most rocks, whereas Na and K are major elements.

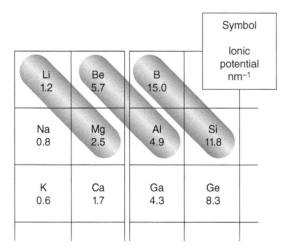

| Li 1.2 | Be 5.7 | B 15.0 | Symbol / Ionic potential nm⁻¹ |

Figure 9.2.1 A condensed version of the Periodic Table, illustrating the diagonal similarity in ionic potential values for first two rows.

After Na^+, Mg^{2+} is the most abundant cation in seawater (Table 4.3) and is recovered from it industrially. Although Ca^{2+} is less abundant in seawater, Ca salts (carbonate and sulfate) are among the first to precipitate from seawater during evaporation; few evaporite sequences include magnesium minerals, which appear only at an advanced stage of evaporation. Most of the calcium in sediments, however, occurs in biogenic limestone.

Aluminium

Aluminium is the commonest metallic element in the crust (Figure 11.7). Its malleability, ductility, electrical conductivity, chemical stability and low density make it an ideal metal for many industrial and domestic uses. The oxide, alumina (Al_2O_3), is **amphoteric**. It forms the extremely hard, refractory mineral *corundum* (hardness = 9), which is widely used as an abrasive (emery) and in ceramics. The gemstones ruby and sapphire are coloured varieties of corundum.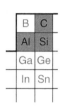

The mobility of aluminium in weathering depends upon the pH of the solution. Al_2O_3 is extremely insoluble in the majority of groundwaters, whose pHs tend to lie in the range 5–6 (Figure 4.2). During the weathering of a granite under such conditions, Na and Ca can be almost entirely removed and K, Mg, Fe^{3+} and even Si partially leached, while Al remains immobile. The minerals left behind are therefore very aluminous, as in the china clay deposits (consisting largely of kaolinite) associated with Cornish granites. The weathering of intermediate volcanic rocks and ash in extreme tropical conditions can proceed even further and remove most of the silica too, leaving behind a mixture of aluminium hydroxides stained by hydrated iron oxides. This material, the chief industrial source of aluminium, is called *bauxite*.

Contact with rotting vegetation or pollution by acid rain can, however, make surface waters sufficiently acid (pH 4 or less) to dissolve alumina. Signs of this can be seen in soil profiles in temperate forests, where an upper light-coloured horizon, from which all components except silica have been leached by acid solutions, passes down into a lower horizon rich in clay minerals, in which Al has been re-precipitated through contact with relatively neutral groundwater at the water table. Precipitation of Al^{3+} (and Fe^{3+}) in this way is an example of *hydrolysis* (Box 9.3).

Box 9.3 Hydrolysis

To a geochemist, the term *hydrolysis* refers to reactions in which either (or both) of the O–H bonds in water is broken. Consider the hydrolysis of atmospheric sulfur dioxide.

$$SO_2 + H_2O \rightarrow H_2SO_3 \rightarrow H^+ + HSO_3^-$$
$$\text{sulfurous} \quad\quad\quad\quad\quad (9.3.1)$$
$$\text{acid}$$

This is one of the reactions contributing to the phenomenon of *acid rain* (O'Neill, 1998). Hydrolysis of an acidic oxide (SO_2, NO_2, CO_2) generally gives an acidic solution.[1] The converse applies to the hydrolysis of a basic oxide:

$$Na_2O + H_2O \rightarrow 2NaOH \rightarrow 2Na^+ + 2OH^-$$
$$\text{alkaline} \quad\quad\quad\quad\quad (9.3.2)$$
$$\text{hydroxide}$$
$$\text{solution}$$

Hydrolysis reactions are important in weathering. Elements of intermediate ionic potential like Al and Fe(III) are soluble only in fairly acidic solutions. If an Al-bearing solution mixes with a less acidic solution, precipitation occurs as a result of hydrolysis:

$$Al^{3+} + 3H_2O \rightarrow Al(OH)_3 + 3H^+$$
$$\text{solution} \quad\quad \text{insoluble} \quad\quad (9.3.3)$$
$$\text{precipitate}$$

Geochemists refer to such elements and their precipitates as *hydrolysates* (see Figure 9.3.1). For certain elements like Fe and Mn, the first step in this process is oxidation. Fe^{2+} and Mn^{2+} are readily dissolved by mildly acid waters during weathering, but once dissolved they are prone to oxidation to Fe^{3+} and Mn^{4+}, whose higher ionic potentials lie in the hydrolysate field and which consequently precipitate as minerals like goethite [FeO(OH)].

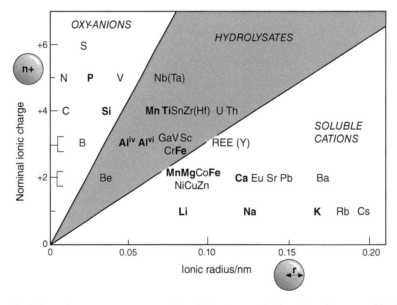

Figure 9.3.1 How metals behave in aqueous systems, in relation to nominal cation charge and ionic radius (cf. Figure 9.1.1). High charge-to-radius elements like P, Si and B exist in solution as the oxy-anions *phosphate* PO_4^{3-}, *metasilicate* SiO_3^{2-} and *borate* BO_3^{3-}. The lowest charge-to-radius elements occur as soluble cations. In between is the hydrolysate field.

---

[1]   The acids involved are sulfurous (H_2SO_3), nitric (HNO_3) and carbonic (H_2CO_3 – see Chapter 4) acids.

Carbon

Carbon is unique among the chemical elements for the huge variety of compounds it can form: the number of known carbon compounds exceeds the number of compounds associated with all of the other elements considered together. So extensive is the chemistry of carbon and so vital to life that – on its own – it constitutes a separate branch of chemical science called *organic chemistry*. Sedimentary rocks commonly contain 0.2 to 2% of organic matter, the highest concentrations occurring in shales (mudrocks).

From the point of view of its occurrence in rocks, it is helpful to consider the chemistry of carbon under two headings: organic carbon and inorganic carbon.

Organic carbon

In a book of this size, one can only scratch the surface of a subject as vast as organic geochemistry. Good summaries written from a geoscientist's viewpoint are given by Krauskopf and Bird (1995) and White (2013); a fuller introduction can be found in the opening chapters of the book by Killops and Killops (2005).

The names of organic compounds used here conform to modern systematic nomenclature (recognized by the *International Union of Pure and Applied Chemistry*). Where names are given in brackets, they are the corresponding traditional or 'trivial' names, which may be more familiar.

Hydrocarbons

Hydrocarbons are compounds of carbon and hydrogen alone. The simplest are the *alkanes* (paraffins), a family of **polymers** related to methane:

$$\begin{array}{ccc}
& & H \\
& & | \\
& H & H—C—H \\
& | & | \\
H & H—C—H & H—C—H \\
| & | & | \\
H—C—H & H—C—H & H—C—H \\
| & | & | \\
H & H & H \\
\text{Methane} & \text{Ethane} & \text{Propane}
\end{array}$$

Many alkanes are chain-like polymers in which each carbon atom (except those at the ends of the chain) is attached to two neighbouring carbon atoms and

two hydrogen atoms – these are called *normal alkanes* (*n-alkanes* for short) – while other alkanes have branched or ring structures. Alkanes have the general formula C_nH_{2n+2} (C_nH_{2n} for ring-shaped *cycloalkanes*).

Natural gas consists mainly of methane. Most of the world's natural gas supply comes from conventional hydrocarbon reservoirs, but recently *shale gas* reserves tapped by hydraulic fracturing ('fracking') have transformed the US gas market. Potential reserves of methane also exist in the form of *methane hydrate* (methane trapped in ice) in marine sediments on continental margins and in onshore permafrost regions.

Methane is a greenhouse gas with a higher global warming potential than carbon dioxide (Table 9.7.1): each molecule is over 20 times more powerful as a greenhouse gas than a molecule of CO_2. Major anthropogenic sources of atmospheric methane include the oil and gas industry, farming (ruminant livestock, rice paddies), biomass burning and landfills. The main process removing methane from the atmosphere is reaction with hydroxyl (∞OH) **free radical**s in the troposphere as summarized in Equation 3.18. This reaction also occurs in the upper atmosphere where it is the chief source of stratospheric water.

Alkanes are examples of **saturated** organic compounds, consisting of molecules constructed entirely with single bonds. Although not shown in the simplified diagrams above, such molecules have a zigzag shape in three dimensions, owing to the tetrahedral (sp³-based) disposition of bonds around each carbon atom (Box 9.4). An example is hexane:

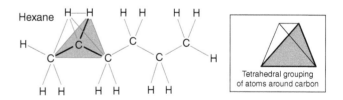

Unsaturated compounds are those containing one or more double C=C bonds or triple C≡C bonds. Examples are ethene (ethylene), C_2H_4, belonging to a family called *alkenes*:

$$\begin{array}{ccc}
H & & H \\
\diagdown & & \diagup \\
& C = C & \\
\diagup & & \diagdown \\
H & & H
\end{array}$$

Box 9.4 Notation for organic molecules

The complexity of most organic molecules necessitates a shorthand notation. Hydrocarbon molecules are often abbreviated to 'matchstick' structures showing just the bonds between carbon atoms: at each angle in the symbol the appropriate CH_x group is implied.

 becomes

The cyclic molecule of benzene is written as a hexagon incorporating three double bonds. This can be written in two alternative ways:

The molecule actually adopts both configurations at once: as with graphite (Box 7.4), the π-electron density is delocalized around the entire ring. This is often symbolized as:

and ethyne (acetylene), C_2H_2:

$$H - C \equiv C - H$$

Hydrocarbons like ethyne that incorporate a triple $C \equiv C$ bond are known collectively as *alkynes*.

When forming double bonds (as we discovered with graphite – Box 7.4), carbon uses the *sp^2-hybrid*, which has 120° 'planar trigonal' geometry. All of the atoms in the ethene molecule therefore lie in the same plane.[2]

---

2   In the triple-bonded C_2H_2 molecule, on the other hand, carbon uses the linear *sp^1-hybrid*.

The same is true of a large and important class of unsaturated organic compounds based on the **benzene** molecule C_6H_6:

Condensed notation (Box 9.4)

Compounds based on the benzene ring, like *styrene* below (from which polystyrene is manufactured), are called *aromatic* compounds, as many possess distinctive odours.

Styrene   Condensed notation (Box 9.4)

Open-chain (acyclic) and aromatic hydrocarbons are the main, and most useful, constituents of **petroleum** (Killops and Killops, 2005).

Carbohydrates

Carbohydrates are a class of organic compounds in which C, H and O combine in the proportion $C_x(H_2O)_y$, where x and y are integers. *Sugars* such as *glucose* ($C_6H_{12}O_6$):

are the simplest carbohydrates. (Like alkanes, these are zigzag molecules in 3D.) Sugars consist of chains of a few of these C_6 units: sucrose, for example, is $C_{12}H_{22}O_{11}$. *Starches* consist of longer chains. The longest-chain carbohydrates are *celluloses* (with $x \sim 10,000$). Cellulose is familiar as the main constituent of paper. Wood consists of about 70% cellulose, the remainder being an aromatic constituent called *lignin*, which gives wood its toughness and structural utility.

Acids, amino acids and proteins

Acidic behaviour in organic molecules is associated with the presence of a *carboxyl* group, as illustrated by *ethanoic* (acetic) acid, familiar as the sharp-tasting constituent of vinegar:

The acidity arises from dissociation of the OH present in the carboxyl group, releasing an H^+ ion (proton).

Groups of atoms like the carboxyl group, which involve the organic molecule to which they belong in certain specific types of reaction, are called **functional groups**. Another example is the *amino group* (NH_2), whose presence in the same molecule as the carboxyl group is the characteristic of particularly important class of organic acids, the *amino acids*. About 20 amino acids occur in the living world. The simplest, glycine, can be recognized as a derivative of ethanoic acid:

In solution, amino acids behave as weak acids: like ethanoic acid, they dissociate by releasing a proton from the carboxyl group:

$$NH_2CH_2COOH \rightarrow H^+ + NH_2CH_2COO^- \tag{9.1}$$

At the same time, however, the amino group has the capacity to accept a proton, thereby behaving as a base:

$$H^+ + NH_2CH_2COO^- \rightarrow {}^+NH_3CH_2COO^- \tag{9.2}$$

Amino acids thus have the remarkable property of forming dipolar ions or **zwitterions** (a German term meaning 'hybrid ion') with opposite charges at each

end. No doubt this contributes to their high solubility in polar solvents like water or ethanol, each end being solvated by a sheath of solvent molecules (Box 4.1).

The capacity of the amino acids to react with acids or bases according to circumstances means that molecules can also react with each other. Reaction between the amino group of one molecule and the carboxyl group of another joins the two molecules together:

Nature uses this *peptide linkage* reaction to assemble amino acid units into huge *protein* molecules, whose relative molecular masses run into thousands. Proteins are the essential constituents of the living cell. Nitrogen plays a fundamental role in other key molecules of living things, such as chlorophyll (Box 9.5)

Proteins in decaying organic matter break down rapidly by *hydrolysis* (Box 9.3), the reverse of the reaction above, leading to the production of simpler proteins and amino acids.

Inorganic carbon

The element carbon exists in the Earth primarily in two crystalline forms, graphite and diamond (whose structures and properties are compared in Chapter 7). Diamond crystallizes naturally only at very high pressure, equivalent to depths within the Earth greater

Box 9.5 Chlorophyll

We have seen the key role that nitrogen plays in the synthesis of proteins. Another compound essential to life, in which nitrogen is a vital constituent, is *chlorophyll*, the green pigment in plants upon which photosynthesis primarily depends. Chlorophyll is actually a class of closely related compounds that share the following characteristics:

(a) The fundamental architecture comprises four molecules of pyrrole (C_4H_5N –Figure 9.5.1), linked together to form a ring structure known as a porphyrin (Figure 9.5.2).

(b) At the centre of the porphyrin ring in chlorophyll (Figure 9.5.3) lies a magnesium atom, forming a co-ordination complex with the surrounding pyrrole nitrogens.

Figure 9.5.1 Structure of pyrrole, using the condensed notation of Box 9.4. Remember that a hydrogen atom is implicitly attached to each carbon atom (*i.e.* at each apex of the polygon) and to the nitrogen atom.

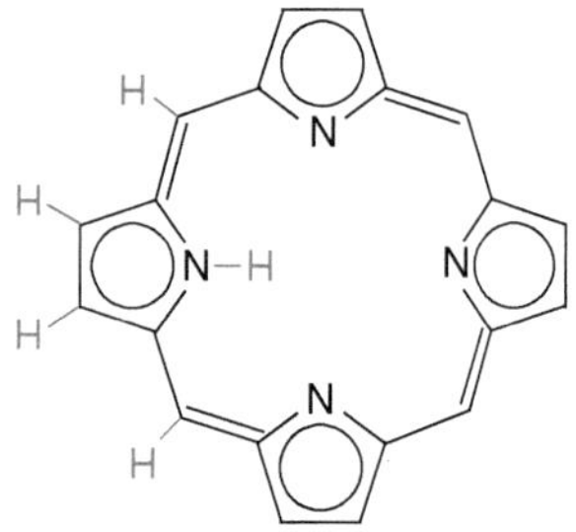

Figure 9.5.2 Structure of porphyrin. The complement of H atoms associated with one of the pyrrole groups is shown in orange as a reminder. The porphyrin ring structure is found in other biologically essential compounds such as haemoglobin and vitamin B_{12}.

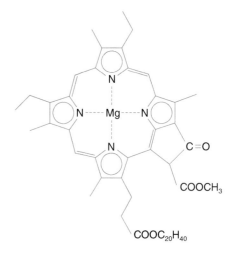

Figure 9.5.3 Structure of chlorophyll *a*.

(c) Various alkyl and carboxyl groups are attached to the porphyrin ring, whose identities differ with the variant of chlorophyll under discussion.

The alternation of double and single bonds in the porphyrin ring resembles that in graphite (Box 7.4) and benzene, and in a similar way the bonds coalesce to form interconnecting molecular orbitals above and below the ring. Chlorophyll absorbs light strongly at the blue and red ends of the visible spectrum (hence its green colour – Figure 8.3), photochemically releasing electrons from these orbitals and setting in train a complex set of biochemical reactions by which CO_2 is ultimately reduced to carbohydrate (reaction 9.3).

than 120 km (Box 2.2). The derivation of diamond from deep in the mantle, and geochemical evidence for CO_2-rich fluids in peridotite nodules brought up from similar depths, indicate that carbon is an important minor constituent of the mantle. Carbonaceous material is a major component of many meteorites (Box 11.1).

In recent decades several novel structural forms of elemental carbon have been discovered (Box 9.6) that hold the promise of major technological innovations.

Carbon occurs in the Earth's crust chiefly as carbonate, most of which is biogenic. Limestones make up about 25% of the total mass of Phanerozoic sedimentary rocks.

Box 9.6 Novel forms of carbon

In 2004, Russian-born physicists André Geim and Kostya Novoselov published the remarkable discovery that *single layers* can be peeled from a crystal of graphite simply by repeated application of sticky tape! The resulting graphite monolayers – just one atom thick and so dubbed '2D crystals' – have been found to have astonishing nanotechnological properties, so distinct from ordinary graphite as to merit a new chemical name – *graphene*. As well as opening an entirely new field of physics, graphene (Figure 9.6.1a) promises a host of novel technological applications, from ultrafast electronic devices to highly efficient photovoltaic collectors (Geim and Kim, 2008).

But graphene is just the most recently discovered of a number of new structural forms of carbon that have come to light in the laboratory – and (in some cases) in nature too – since 1985. That was when British chemist Harry Kroto and teams from Rice (Texas) and Sussex (UK) universities discovered C_{60} molecules with a distinctive spherical geometry that assembled themselves spontaneously from hot ionized carbon vapour. These molecules were made up of alternating 5- and 6-membered rings reminiscent of a soccer ball. The rings have delocalized π-electrons similar to graphene, and indeed the molecule can be thought of as a spherical derivative of graphene (Figure 9.6.1b). Owing to its geometric similarity

to the iconic 'Geodesic Dome' designed for the US exhibit at Expo 67 in Montreal by architect Buckminster Fuller, the C_{60} molecule was named *buckminsterfullerene* (Figure 9.6.1b),[3] although the shorter name 'buckyballs' is more commonly used today! Infrared spectra show that C_{60} and C_{70} exist in young planetary nebulae in interstellar space, but they are also found in ordinary terrestrial soot.

Other members of the fullerene structural family take hollow cylindrical form with diameters in the nanometre range, visualized as 'rolled up' versions of a graphene sheet (Figure 9.6.1c). Such *carbon nanotubes*[4] (which may be single- or multiwalled) can be engineered to have a huge variety of properties of value in nanotechnology. Carbon nanotubes are the strongest and stiffest materials yet discovered.

The π-electrons in graphene make it too conducting for some potential applications. It can be transformed into insulating sheets by attaching one hydrogen atom to each carbon atom in the sheet, forming a new class of compounds, devoid of π orbitals, called *graphanes*.

---

[3] One of a family of similar molecules (including C_{70}) called fullerenes.

[4] Known in the Russian literature since the 1950s, but rediscovered more recently in the West.

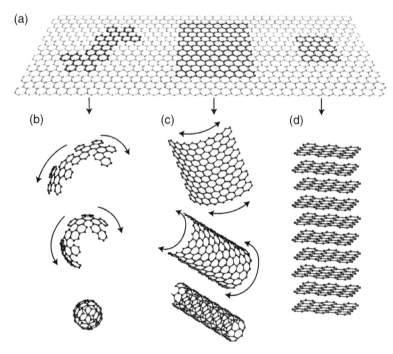

(a)

(b)   (c)   (d)

Figure 9.6.1 (a) A single sheet of graphene, showing how its basic sheet structure relates to (b) buckminsterfullerene (C_{60} 'buckyballs'), (c) carbon nanotubes, and (d) graphite. (Source: Geim and Lovoselov (2007). Reprinted by permission of Macmillan Publishers Ltd).

Carbon dioxide

In the carbon dioxide molecule, each oxygen atom is doubly bonded to a carbon atom. The carbon atom allocates two valence electrons to an *sp¹-hybrid*, whose lobes stick out in opposite directions, and the σ-bonds they form give the molecule a linear O–C–O shape. The two carbon 2p orbitals not involved in the hybrid form π-bonds with the oxygen atoms on either side.

CO_2 is an acidic oxide (Figure 9.4). On dissolving in water it forms a slightly acidic solution (Chapter 4). Rain water in equilibrium with atmospheric CO_2 has a pH of 5.7 (see Exercise 4.3), which partly accounts for its effectiveness as a weathering agent. In the northern hemisphere, however, this acidity has been reinforced by oxides of sulfur and nitrogen introduced into the atmosphere by the burning of fossil fuels.

CO_2 is believed to have been a major constituent of the Earth's primordial atmosphere, as it is for Venus today (CO_2 accounts for 96% of the Venusian atmosphere). Throughout most of the Earth's history, however, photosynthesis has 'drawn down' carbon from the atmosphere into the biosphere. Photosynthesis is a reaction by which plants and algae use chlorophyll (Box 9.5) to generate the reduced carbon they need from atmospheric CO_2, or from CO_2 dissolved in the oceans:

$$xH_2O + xCO_2 \xrightarrow[\substack{absorbed\,by \\ chlorophyll}]{hv\,(solar\,energy)} \left(-CH_2O-\right)_x + xO_2 \quad (9.3)$$
carbohydrate
(simplified)

Such *oxygenic* photosynthesis – releasing the oxygen we rely on today – has transformed the Earth's atmosphere through the course of geological time, as described in Chapters 10 and 11. It is the most complex, and most recently evolved, form of photosynthesis, which relies on water to release the electrons needed to reduce the carbon in CO_2. Earlier *anoxygenic* forms of photosynthesis relied on H_2S or Fe as electron donors instead.

Aquatic biota (zooplankton and higher organisms) have also fixed carbon in carbonate shells which have accumulated as limestone. By progressively sequestering carbon from the atmosphere into the crust, these processes have together reduced the CO_2 content of the Earth's atmosphere today to a few hundred parts per million.

The pre-industrial atmospheric concentration of CO_2 was about 280 ppmv. This represents a biologically mediated balance between oxygenic photosynthesis and respiration (which converts O_2 to CO_2). CO_2 is a

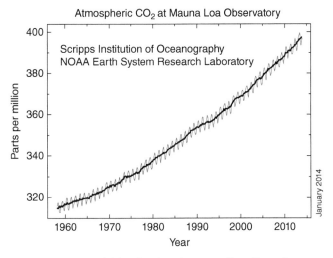

Figure 9.2 Annual (thin line) and seasonally adjusted (heavy line) variation in atmospheric CO_2 content measured at the Mauna Loa observatory in Hawai'i (chosen for its remoteness from centres of industrialization) from March 1958 to January 2014, as depicted on the Earth Systems Research Laboratory website of the US National Oceanographic and Atmospheric Administration (NOAA).[5] (Source: US National Oceanographic and Atmospheric Adminstration (NOAA)).

'greenhouse' gas (Box 9.7). For the past two centuries the atmospheric concentration of CO_2 has been rising due to fossil-fuel burning and deforestation, and it exceeded 400 ppmv for the first time in 2013 (Figure 9.2), higher than at any time in the last 4.5 Ma. This rise in atmospheric CO_2 is the main cause of global warming. The hemispheric balance between combustion and photosynthesis shifts with the seasons, to the left in winter and to the right in summer (Figure 9.2). Current estimates suggest that the *anthropogenic greenhouse effect* (Box 9.7) is warming the Earth by 0.2 °C per decade, which will have serious climatic and social consequences if humanity fails to stem the rise very soon.

Carbon also forms a *monoxide* (CO) which occurs in volcanic gases and in the atmosphere, but at much lower concentrations than CO_2.

Carbon isotopes

Carbon consists of two stable isotopes (^{12}C and ^{13}C) and a short-lived radioactive isotope ^{14}C, which are discussed in Chapter 10.

[5] www.esrl.noaa.gov/gmd/webdata/ccgg/trends/co2_data_mlo.pdf

Box 9.7 Molecular spectra: greenhouse gases

We saw in Chapter 6 that *atoms* emit and absorb light at characteristic wavelengths as a result of electrons jumping between bound (i.e. quantized) energy levels within the atom. *Molecules* too can absorb and emit their own characteristic electromagnetic spectra, although the mechanisms and wavelength range involved are entirely different.

At room temperature a gas molecule like CO_2 is not a rigid body but is constantly stretching and flexing. The precise molecular geometries implied by Figures 7.4 and 7.5 merely indicate the *mean* positions of atoms, about which they constantly oscillate in a variety of ways (and at various frequencies $\omega_1, \omega_2, \omega_3 \ldots$) as illustrated in Figure 9.7.1. Wave-mechanical analysis of such internal molecular vibrations shows they too are quantized, giving rise to a series of discrete *vibrational energy levels*. A molecule can switch between modes of vibration (energy levels) by absorbing or emitting a photon whose energy is equivalent to the difference in energy ΔE between the molecule's initial and final vibrational states.

A pure gas – whether **diatomic** like O_2 or triatomic like H_2O or CO_2 – will therefore emit or absorb radiant electromagnetic energy at a series of discrete wavelengths (or photon energies) characteristic of that molecule. A gas mixture like air will combine the spectra of all of the constituent gases. The photon energies involved are, however, much lower than those associated with electronic transitions in atoms, and correspond to wavelengths in the infrared (IR) range. Molecular *rotation* is likewise quantized, leading to further characteristic absorption and emission lines at IR wavelengths.

The Earth's climate results from a balance between (a) incoming radiant energy received from the Sun and (b) thermal energy radiated into space by the Earth and its atmosphere. The Earth's mean surface temperature of $288\,K = 15°C$ confines it to radiating energy in the infrared (IR) wavelength range. As outgoing radiant energy passes through the atmosphere, some wavelengths are absorbed by the vibrational and rotational transitions in atmospheric gases,[6] inhibiting the Earth's ability to radiate heat into space and upsetting the balance between incoming (solar) and outgoing energy fluxes. This atmospheric spectral 'blanket' – the so-called 'greenhouse effect' – significantly warms the Earth's climate, and the most strongly absorbing species like CO_2, N_2O, H_2O and CH_4 are thus known as 'greenhouse gases'.

Were it not for the greenhouse gases naturally present in the atmosphere, the Earth's mean surface temperature would be colder, about $255\,K = -18°C$. Life on Earth thus benefits from a *natural greenhouse effect* of around $+33°C$. Current concern about climate change relates to the additional *anthropogenic* (or 'enhanced') *greenhouse effect* caused by mankind's accumulated emissions into the atmosphere of CO_2 (from fossil fuel burning and the destruction of tropical rainforests) as well as of N_2O, CH_4 and **CFC**s.

Greenhouse gases vary considerably in their atmospheric lifetimes and their impact on global warming (Table 9.7.1).

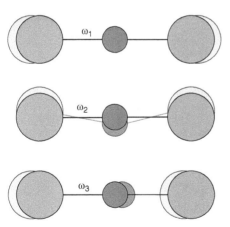

Figure 9.7.1 Symmetric and asymmetric modes of vibration of a linear triatomic molecule (e.g. CO_2). The darker atoms represent mean atomic positions while the fainter ones indicate one direction of oscillation. The three fundamental frequencies of vibration ω_1, ω_2 and ω_3 are shown. An animation of the modes of vibration of a non-linear triatomic molecule such as H_2O can be seen at www1.lsbu.ac.uk/water/vibrat.html.

Table 9.7.1 Global warming potentials of greenhouse gases. H_2O is not shown in the table because – although it is the most abundant greenhouse gas – its atmospheric concentration varies on a short timescale, is poorly mixed, and is little affected by human activity

Name	Formula	Atmospheric lifetime/years	Global warming potential*	
			Over 20 years	Over 100 years
Carbon dioxide	CO_2	Centuries[§]	1	1
Methane	CH_4	12	84	28
Nitrous oxide	N_2O	121	264	265
CFC-12	CCl_2F_2	100	10,800	10,200
HCFC-22	$CHClF_2$	12	5280	1760

*GWP is a dimensionless measure of how much heat a given mass of a greenhouse gas traps in the atmosphere *relative to the same mass of carbon dioxide*. Values from IPCC (2013) Tables 8.7 and 8.A.1.
[§]Defining the atmospheric lifetime of CO_2 is a complex question. See Archer *et al.* (2009).

---

[6] The energy absorbed either raises atmospheric temperature or is radiated in other directions (including downwards).

Silicon

Silicon (Si) is a hard metalloid of intermediate electronegativity (1.9), with a structure identical to that of diamond (Figure 7.5). Like the next element in Group IV, germanium (Ge), it has become very important as a semiconductor (Chapter 7). For this use it must be extremely pure (impurities less than 1 part in 10^8). High-purity silicon metal is also used in 90% of the world's photovoltaic (electricity-generating) solar panels, both in crystalline and amorphous forms.

Silicon the element should not be confused (as sometimes happens in the media) with *silicone*, a class of synthetic *organo-silicon* polymers, in which groups such as CH_3 are attached to $-Si-O-Si-O-Si-$ chains and networks. Such compounds are widely used as lubricants and insulators, having greater thermal stability than equivalent organic polymers.

Silicon is the most abundant of the electropositive elements in the Earth's crust. It invariably occurs in the oxidized state (valency 4), as SiO_2 or silicate polymers (see Table 8.1).

SiO_2 occurs in a variety of structural forms, both crystalline and amorphous. As well as forming megascopic crystals, quartz occurs commonly in the **crypto-crystalline** form chalcedony, familiar as varieties like agate, jasper, chert and flint. The only truly amorphous form of silica is opal. Molten high-purity silica can be drawn into glass fibres a fraction of a millimetre in diameter, widely used for fibre-optic communications.

Quartz has a low but significant solubility in water of about 6 ppm at room temperature. SiO_2 solubility increases markedly with temperature, providing a geothermometer that can be used to estimate deep temperatures in hot springs.

Dissolved silica exists in the hydrated form $Si(OH)_4 = H_4SiO_4$, known as silicic acid (analogous to carbonic acid but weaker). Except near to ocean-floor hot springs and lava eruptions, seawater is undersaturated with silica. Diatoms and radiolaria are nevertheless able, by extracting SiO_2 from seawater, to secrete shells of opaline silica. They do so mainly in the uppermost photic zone of the oceans, where sunlight promotes a high biological productivity. Biogenic precipitation of silica dramatically reduces the level of dissolved silica in the surface layer, to the extent that the available silica actually controls the populations of these organisms; Si is thus an example of a **biolimiting** element. The siliceous hard parts that these organisms secrete dissolve relatively slowly and therefore accumulate on the ocean floor, eventually to be lithified into flinty rock called *chert*, which may often be seen under the electron microscope to consist of radiolarian debris. (Some cherts, however, are abiogenic in origin.)

Nitrogen and phosphorus

Nitrogen is familiar as the unreactive diatomic gas N_2 making up the major part of the atmosphere (Figure 9.3). The electronegative nitrogen atom has three vacancies in the valence shell, allowing three covalent bonds (one σ-bond and two π-bonds) to be established between the two atoms in the molecule.

B	C	N
Al	Si	P
Ga	Ge	As
In	Sn	Sb

Nitrogen adopts a range of valency states. It forms three stable gaseous oxides: nitrous oxide (N_2O), nitric oxide (NO) and nitrogen dioxide (NO_2); 'NO_x' is a convenient abbreviation embracing all three. Significant amounts of NO and NO_2 are produced during combustion of fossil fuels, notably by cars; they contribute to the formation of the photochemical smog that degrades air quality in many large cities on hot summer days (O'Neill, 1998).

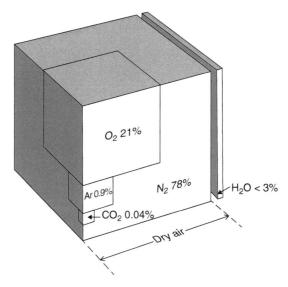

Figure 9.3 The composition (volume percentages) of the atmosphere at sea-level.

Nitrogen is an important constituent of all living matter: the importance of the amino group (–NH$_2$) and amino acids has been discussed under carbon. In decaying organic matter, such compounds are decomposed bacterially to the gas ammonia (NH$_3$), most of which is oxidized by soil bacteria to nitrate (NO$_3^-$), the form of nitrogen most readily utilized by plants. A large amount of nitrate is used as fertilizer which, owing to its high solubility, is washed into streams, rivers, lakes and groundwater, where it may compromise water quality.

Phosphorus, unlike nitrogen, does not occur in the elemental state in nature. In silicate analyses (Box 8.3), it is reported as the oxide P$_2$O$_5$. It exists geologically as the phosphate oxy-anion (PO$_4^{3-}$), most commonly in the accessory mineral *apatite* (Ca$_5$(PO$_4$)$_3$OH). In basic magmas, phosphorus behaves as a high field-strength **incompatible element** (Box 9.1), but as crystallization proceeds such melts eventually become saturated with apatite, whose crystallization then leads to the depletion of phosphorus in subsequent melt fractions.

Nitrogen and phosphorus are both important **biolimiting** inorganic nutrients. When confined bodies of water (lakes, estuaries, coastal waters) receive influxes of nitrate or phosphate – for example, from sewage outflow or the application of excess fertilizer – sunlight may promote rapid blooming of bright green algae at the water surface, preventing normal oxygenation of the water below (hypoxia) that may result in fish-kill. The phenomenon is known as *eutrophication* (from the Greek for 'well nourished').

Oxygen

Viewed from a terrestrial perspective, oxygen is the most important of all chemical elements, being the most abundant element in the crust and mantle, and a major life-supporting constituent of the oceans and atmosphere (Figure 9.3).

B	C	N	O
Al	Si	P	S
Ga	Ge	As	Se
In	Sn	Sb	Te

Oxygen (1s^{2}2s^{2}2p^4) is the second most electronegative element (3.4; Figure 6.3). It is divalent, having two vacancies in the valence shell. Diatomic oxygen O$_2$ makes up 21% of dry air at sea-level. At altitudes between 10 and 60km, in the stratosphere, another type of oxygen molecule plays a significant role: this is the triatomic molecule **ozone**, O$_3$. Although the O$_3$

concentration rarely exceeds 10ppm, it absorbs solar ultraviolet radiation strongly, and this 'ozone layer' protects terrestrial life from the damaging effects of the Sun's UV rays. At sea-level, however, ozone is an undesirable pollutant, contributing to photochemical smog and acid-rain damage (O'Neill, 1998).

Oxygen forms an oxide (oxidation state –II) with almost every other element. Oxides may have basic or acidic properties (Box 8.1), or they may be **amphoteric**, exhibiting both aspects. Because these characteristics depend on the electronegativity of the element bonded to oxygen, they correlate with its position in the Periodic Table (Figure 9.4).

The availability of oxygen determines the stability of many minerals, particularly those containing elements like iron that have multiple oxidation states. The availability of oxygen in low-temperature, aqueous environments is expressed by the **oxidation potential** *Eh* (Figure 4.1b); environments with free access to atmospheric oxygen have high *Eh* values, whereas anaerobic conditions are characterized by low *Eh* values.

In high-temperature systems it is more convenient to express the availability of oxygen in terms of its **partial pressure** (Chapter 4), or the related parameter called the *oxygen fugacity*, f_{O_2}. Partial pressure is appropriate as a measure of concentration only in low-pressure gas mixtures, in which molecules are so dispersed that they behave independently of each other (except when colliding). This state is called an 'ideal gas'. In high-pressure mixtures or when a gas is dissolved in another phase such as an igneous melt, gas molecules interact more strongly with neighbouring molecules, rather like ions in non-ideal aqueous solutions (Chapter 4).

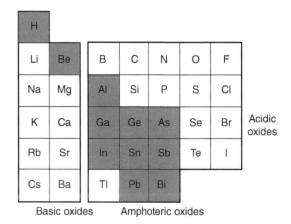

Figure 9.4 Acidic, basic and amphoteric oxides of the non-transition elements.

Oxygen fugacity fo_2 is analogous to *activity* in describing the 'effective concentration' of oxygen in these non-ideal conditions.

Consider the reaction between iron-rich olivine crystallizing from a magma and oxygen dissolved in the melt:

$$3\,\underset{\substack{\text{fayalite}\\\text{(olivine)}}}{Fe_2SiO_4} + \underset{\text{melt}}{O_2} \rightleftharpoons 2\,\underset{\text{magnetite}}{\left(FeO.Fe_2O_3\right)} + 3\,\underset{\text{quartz}}{SiO_2} \qquad (9.4)$$

Fayalite is a ferrous (Fe^{2+}) compound, whose crystal structure tolerates only a minute amount of Fe^{3+}. If sufficient oxygen is present in the system to cause significant oxidation of Fe^{2+} in olivine to Fe^{3+}, the olivine breaks down to a mixture of magnetite ($Fe_3O_4 = FeO.Fe_2O_3$), which contains both Fe^{2+} and Fe^{3+}, and quartz.

The reaction between fayalite and oxygen occurs at specific fo_2–T conditions, as shown in Figure 9.5 (a kind of phase diagram). The lower curve is a univariant equilibrium boundary separating fields where fayalite + oxygen (below) and magnetite + quartz (above) are the stable assemblages. The reaction can proceed in either direction, and the coexistence of all four phases – as recorded by coexisting phenocrysts of fayalite, magnetite and quartz in a fine-grained volcanic rock, for instance – suggests that crystallization occurred under conditions lying somewhere on the equilibrium

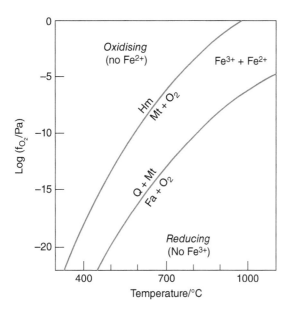

Figure 9.5 f_{O_2}–T equilibrium diagram showing the experimentally determined reaction boundaries for the magnetite (Mt)–hematite (Hm) and fayalite (Fa)–magnetite–quartz (Q) reactions.

boundary. Because this is a univariant equilibrium (cf. Figure 2.2), one would need to estimate the temperature of crystallization of the phenocrysts by some other means before a numerical value of fo_2 could be worked out. Nonetheless, the univariant boundary is a useful reference line, and the assemblage quartz-fayalite-magnetite provides a means of regulating (or **buffering**) redox conditions in high-temperature phase equilibrium experiments; it is often referred to as the 'QFM buffer'.

Under more oxidizing conditions, as shown by the upper equilibrium boundary in Figure 9.5, the Fe^{2+} in magnetite becomes oxidized, causing it to recrystallize as the mineral hematite (Fe_2O_3), in which the iron is entirely ferric.

The isotope geochemistry of oxygen is discussed in Chapter 10.

Sulfur

Native sulfur (oxidation state 0) forms yellow encrustations around volcanic vents and fumaroles, where it crystallizes as a **sublimate**. It can also be deposited from hot springs rich in H_2S or SO_2 and can occur in sedimentary rocks as a result of bacterial reduction of sulfate.

Sulfur can form compounds either with elements less electronegative than itself (hydrogen and the metals) or with oxygen, which is more electronegative. It is useful to distinguish these two tendencies as 'reduced sulfur' and 'oxidized sulfur' respectively.

Reduced sulfur compounds

Sulfur is an essential nutrient element for living things, and organosulfur compounds therefore have considerable biochemical importance. They produce the distinctively pungent flavour and odour of onions and garlic. Hydrogen sulfide (H_2S), familiar as the smell of rotten eggs, is produced by decay of organic matter in anaerobic conditions, for example in stagnant water. Significant amounts of organosulfur compounds are present in oil, natural gas and coal; oxidation of these compounds to SO_2 during fuel combustion is the main anthropogenic cause of acid rain.

H_2S is also a significant constituent of volcanic gases. This suggests that *sulfide* (oxidation state –II) is the predominant form of sulfur in the Earth's interior.

A great many metals of economic importance are deposited from hydrothermal fluids in the form of sulfide minerals (Box 9.8). The other Group VI elements, *selenium* (Se) and *tellurium* (Te), may take the place of sulfur in such minerals. A number of selenide and telluride minerals are known.

Oxidized sulfur compounds

Sulfides are not stable in contact with atmospheric oxygen. H_2S is rapidly oxidized, in water to the sulfate anion SO_4^{2-} (if sufficient dissolved oxygen is present) and in air to gaseous sulfur dioxide (SO_2, oxidation

---

Box 9.8 Sulfide minerals and 'soft' metals

Not all metals exhibit the capacity to form sulfide minerals. The elements that do, known amongst geochemists as **chalcophile** elements (Figure 11.4, Plate 7), mostly lie in the right-hand side of the d-block or the adjacent portion of the p-block (Figure 9.8.1).

Why is chalcophile behaviour restricted to such a specific region of the Periodic Table? Chemists divide metal ions into 'hard' and 'soft' Lewis acids (Box 7.5). 'Hardness' in this sense is a characteristic of the strongly electropositive metals on the left-hand side of the Periodic Table (e.g. the alkali and alkaline earth metals), which form very ionic bonds, particularly with strongly electronegative elements like oxygen. Chalcophile metal ions, on the other hand, behave as 'soft' acids: they are well endowed with d-electrons and – for metals – have relatively high electronegativities (1.7–2.5, Figure 6.3); their ions are easily polarized and therefore, as metals go, they form relatively covalent bonds. Such bonds can be established most effectively with ligands of low electronegativity such as sulfide (which is classed as a 'soft base'). The tendency for hard acids (Na^+, K^+, Mg^{2+}, Ca^{2+}) to combine with hard bases (O^{2-}), whereas soft acids (e.g. Cu^{2+}, Ag^+, Hg^{2+}) associate with soft bases (S^{2-}), is a fundamental distinction in modern inorganic chemistry, corresponding closely to the geochemical division of elements into lithophile and chalcophile groups (Figure 11.4).

We have seen that bonding in sulfides has affinities with bonding in metals (Figure 7.8b). So it is not surprising to find that many sulfide minerals exhibit metal-like features, such as metalloid lustre, opacity and relatively high thermal and electrical conductivity. These arise from the low electronegativity contrast between the soft acid and soft base, the mean electronegativity of PbS (2.45), for example, being little different from Pb metal itself (2.3). In some Fe, Co and Ni sulfides the metallic character is further enhanced by direct metal–metal bonding, an intervening conduction band being formed by interaction between d-orbitals of neighbouring metal atoms.

Zn has the lowest electronegativity of the truly chalcophile elements, and sphalerite (ZnS) shows little metallic character; iron-free samples are translucent.

Sulfur commonly combines with metals in non-integer proportions. An example of such a 'non-**stoichiometric**' sulfide is pyrrhotite, whose composition is best represented by the formula $Fe_{1-x}S$, where x can lie between 0.0 and 0.15.

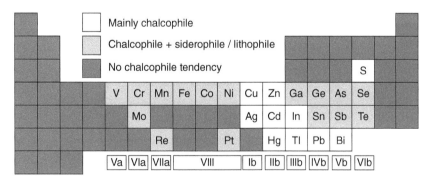

Figure 9.8.1 Distribution of chalcophile elements in the Periodic Table (see also Plate 7). Those in the p-block, even rare ones, have become increasingly valuable owing to many new applications in semiconductor (GaInAs), photovoltaic (CuInSe, CdTe) and other high-tech applications.

state IV). About 10^8 tonnes per year of SO_2 are released into the atmosphere by fossil-fuel burning and other industrial processes. On dissolving in water droplets (Box 9.3), SO_2 oxidizes to form an **aerosol** of sulfuric acid (H_2SO_4, oxidation state VI). This was the main contaminant in *acid rain*, which caused severe and widespread damage to lakes, rivers and forests in the northern hemisphere.

Sulfide minerals are also susceptible to atmospheric oxidation. The weathering of near-surface sulfide ore bodies by the downward percolation of oxygenated water may lead to further enrichment of the ore. Just below the surface (Figure 9.6), sulfides are oxidized to sulfates, which migrate downward in solution. Commonly a number of carbonate, sulfate and oxide ores are precipitated just above the water table, whereas reducing conditions below the water table lead to deposition of *secondary* sulfide minerals. This is called *supergene* enrichment.

Sulfate minerals occur in two other environments. Barite ($BaSO_4$) occurs in low-temperature hydrothermal veins (as in the English Pennine orefield – Chapter 4), commonly in association with sulfides.

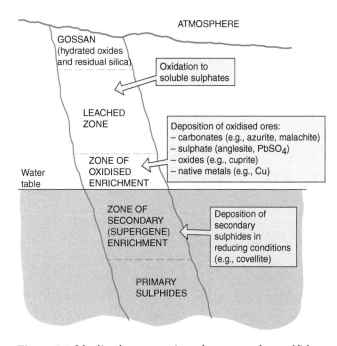

Figure 9.6 Idealized cross-section of a near-surface sulfide ore body, showing the zonation due to percolating oxygen-bearing solutions. The mineral assemblages produced are illustrated by typical copper minerals (stability relations are indicated in Figure 4.1a), but many other minerals occur in such environments.

Minerals like anhydrite ($CaSO_4$) and gypsum ($CaSO_4.2H_2O$) are characteristic of evaporites.

Fluorine

Fluorine has the highest electronegativity of all elements (Figure 6.3), and is the most reactive. It forms strongly ionic compounds. The commonest fluoride mineral (and the chief industrial source of fluorine) is *fluorite*, CaF_2 (Figure 7.3d), which occurs most commonly in hydrothermal veins. The ionic radius of the fluoride anion F^- (1.25 pm, Box 7.2) is similar to those of O^{2-} and OH^-, and fluorine is a common substituent for OH^- in hydrous minerals like amphiboles, micas and apatite.

Being more reactive and electronegative, fluorine can displace oxygen from most silicates. Dissolving hydrogen fluoride (HF, a gas) in water produces *hydrofluoric acid*, which is widely used in analytical geochemistry as it is the only acid capable of attacking silicate rocks (in powdered form) to bring them into solution for analysis. It is a dangerous reagent that requires special training and handling: unlike the more familiar hydrochloric acid, it causes no burning sensation on contact with the skin, but penetrates into deeper tissue and may cause intense pain after a few hours. Because HF (as solution or gas) attacks glass, it must be used only in platinum or plastic containers, in specially designed fume cupboards.

Chlorine and other halogens

Chlorine is the third most electronegative element (Figure 6.3). In silicate rocks it is a trace element, being about four times less abundant in the crust than F. The chloride anion (Cl^-) is, however, the most abundant dissolved species in seawater, and it is the dominant **ligand** in brines and hydrothermal fluids. Chlorine gas (Cl_2) has many industrial uses.

Organochlorine compounds are widely used in industry as solvents, propellants and refrigerants owing to their chemical inertness. Among the most inert are *chlorofluorocarbons* (**CFCs**), hydrocarbon derivatives in which every hydrogen has been replaced by chlorine or fluorine atoms. Whereas elemental chlorine is highly reactive in the atmosphere and is therefore rapidly washed out by rain, volatile CFCs may persist for up to a hundred years (Table 9.7.1). They are very powerful 'greenhouse gases' (Box 9.7), but their persistence once posed a more immediate threat through their capacity to deplete stratospheric ozone: their atmospheric residence time is long enough to allow them to penetrate the stratosphere, where they undergo photochemical decomposition and release Cl radicals that destroy ozone. Following implementation of the Montreal Protocol in 1989, however, CFCs have been replaced for most industrial uses by less persistent *hydrochlorofluorocarbons* (HCFCs); these pose less threat to stratospheric ozone, although their global warming potential is only slightly less than CFCs (Table 9.7.1).

The rarer halogens, bromine (Br, a liquid at room temperature) and iodine (I, a solid), react in a similar way to chlorine.

Noble gases

The elements helium (He), neon (Ne), argon (Ar), krypton (Kr), xenon (Xe) and radon (Rn – see Box 9.9), found in column 18 at the right-hand side of the Periodic Table, have completely filled valence shells. This electronic structure is so stable that these elements display negligible chemical reactivity, and exist (except at extremely low temperatures, Box 7.7) as monatomic gases. They are known as the **noble**, **inert** or **rare gases**.

Although helium is the second most abundant element in the universe after hydrogen (Figure 11.2), it comprises only 0.00052% of the Earth's atmosphere. Unlike other gaseous elements like N_2, helium has not accumulated in the atmosphere because its relative atomic mass is too low for it to be retained by the Earth's gravitational field: the escape velocity of helium atoms (and H_2 molecules) is well below the actual average thermal velocity of such atoms at normal temperatures.

Helium consists of two stable isotopes, ^{3}He and ^{4}He. In natural helium escaping through the Earth's surface (recovered from oilfield brines or hot springs, for example), ^{3}He is a relic of the He originally incorporated during the accretion of the Earth (Chapter 10), whereas ^{4}He – about a million times more abundant – is almost entirely the accumulated product of the radioactive decay of thorium (Th) and uranium (U) throughout the Earth's history.

The geochemist's principal interest in Ar is for radiometric dating, using the K–Ar method (Chapter 10) or the more recently developed ^{39}Ar–^{40}Ar method (Box 10.2). As to environmental impact, the radioactive inert gas radon (Rn) poses significant concern (Box 9.9).

Transition metals

Most metals that we use in the home, in the office and in industry are *transition metals*, making up the d-block of the Periodic Table. The essential feature of a transition metal is the presence in the atom or ion of a *partially filled d subshell* (Chapter 6). The d-block can be divided into first, second and third *transition series* (Figure 9.7), according to whether the 3d, 4d or 5d subshell is the partially filled subshell. The elements of Group IIb (zinc, cadmium and mercury), as they have full d subshells, are not formally classified as transition metals.

The transition metals share a number of important chemical characteristics.

(a) Most transition metals and their alloys are tough, chemically stable metals that have innumerable industrial and domestic uses (Figure 9.7).

(b) Most can utilize more than one **oxidation state** (valency) in geological environments. The most familiar example is iron (Box 4.7). Figure 9.8 shows the range of oxidation states among the first transition series. In the elements up to manganese, all d electrons are able to participate in bonding, and high oxidation states can therefore be attained. In elements like iron, cobalt (Co) and nickel (Ni), the d subshell behaves more like the electron core (Chapter 5): only the 4s electrons and perhaps one of the 3d electrons have energies high enough to be used in bonding, and only low oxidation states occur.

Box 9.9 Radon

The heaviest of the noble gases, radon (Rn, $Z=86$), comprises four naturally occurring isotopes, all of which are highly **radioactive** with half-lives from fractions of a second to 3.8 days. ^{218}Rn, ^{219}Rn, ^{220}Rn and ^{222}Rn are all stages in the decay chains leading from ^{232}Th, ^{235}U and ^{238}U to isotopes of lead (exemplified in Figure 3.3.1).

Uniquely among the many radioactive decay products of U and Th, radon's gaseous state allows it to escape readily from U- and Th-bearing minerals. Radon thus constitutes a potential public health concern in areas underlain by U- and Th-rich rocks, such as granite and shale (and some limestones and sandstones). Radon gas seeping up from bedrock in such areas (particularly ^{222}Rn whose relatively long half-life of 3.8 days facilitates migration) may accumulate in poorly ventilated household basements and under-floor cavities, where it constitutes a radiological hazard to occupants who may breath it in.

Radon isotopes undergo alpha-decay to short-lived isotopes of polonium (Po, $Z=84$), a solid element that may lodge in lung tissue and decay to a succession of α-active daughter nuclides. The α-particles emitted cause intense tissue damage on account of their high mass and charge. In the UK, radon is recognized as the second greatest cause of lung cancer after tobacco-smoking, and it accounts for about half the annual average human radiation exposure.

The effects of radon can be mitigated by ensuring that under-floor spaces are adequately ventilated, sometimes with the aid of a pump venting to the outside air (where Rn is dispersed and diluted to safe levels). In some countries, radon risk maps are available (e.g. www.ukradon.org/article.php?key=indicativemap).

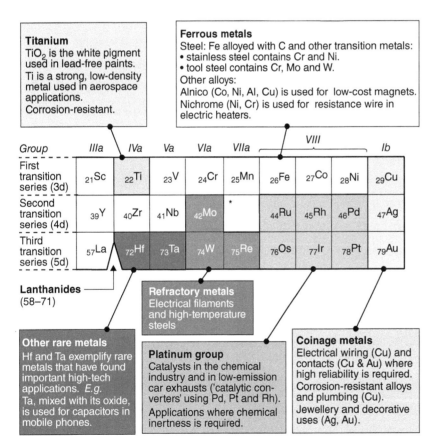

*Technetium (Tc, $Z = 43$) has no stable isotope and is not found in Nature – see Exercise 6.4(b).

Figure 9.7 Transition metals and their uses.

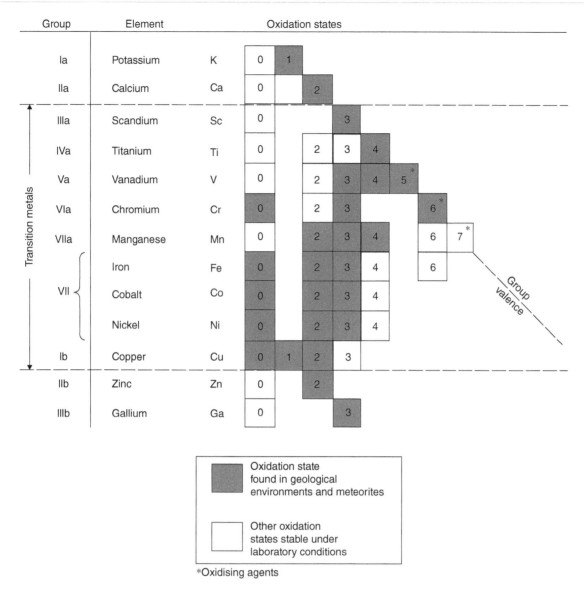

Figure 9.8 Oxidation states of the first transition series.

(c) Transition metals form a wide range of co-ordination **complexes,** some of which play an important part in stabilizing the metals in solution and promoting their transport in hydrothermal fluids (e.g. equations 4.31 and 4.32; Box 7.5).

(d) Transition-metal compounds are often strongly *coloured* (Box 9.10). Many minerals owe their distinctive colours to the presence of a transition metal.

(e) Transition metals are responsible for the *magnetism* of minerals and rocks. This property, most prominent in the later members of the first transition series (Fe, Co, Ni), is due to the presence of *unpaired electrons* in the d subshell. Unlike valence electrons, these 3d electrons may remain unpaired when the metal atom combines in a compound. Paired electrons in an orbital generate equal and opposite magnetic fields which cancel out, but an unpaired electron causes a net magnetic field, which in the case of a few minerals like magnetite and pyrrhotite gives rise to permanent (remanent) magnetism.

Box 9.10 Transition metals and the colour of minerals

d-orbitals project a long way from the nucleus and are highly directional (Figure 5.5). d energy levels are therefore sensitive to the positions of surrounding **ligands**. Figure 9.10.1 a depicts a transition metal in a regular octahedral site in a crystal, surrounded by six equidistant anions, which one can imagine positioned on the reference axes used for describing orbital geometry (Chapter 5). The co-ordination structure is shown cut in half, to clarify the geometry, as is the **co-ordination polyhedron** sketched in the cartoon (b). Because of the potential repulsion between these anions and electron density in the transition metal d orbitals, the most stable d electrons will be those in orbitals that interfere least with the octahedrally positioned ligand anions.

Placing a transition metal in octahedral co-ordination brings two changes in d energy levels. The mean energy increases, due to the overall repulsion by the anion field. Secondly, the energy levels are split. Orbitals like the d_{yz} example shown, whose lobes point at the edges of the co-ordination polyhedron (between the ligands), have a lower energy than orbitals pointing directly at the anions, which experience maximum repulsion.

The split in d energy levels (Δ_{oct}) varies with the identity of the cation and the crystal site. For many transition metals in minerals, the energy difference between the d levels corresponds to the photon energy of visible light. Such ions are therefore capable of absorbing strongly certain wavelengths in the visible spectrum, by promoting electrons from the lower level to a vacancy in the upper level (cf. Figure 6.4). This *crystal field splitting* is the cause of the strong colours of minerals like malachite and azurite (Cu – see front cover) or olivine (Fe).

The presence of d-electrons effectively causes a transition metal ion to deviate significantly from the spherical shape assumed in Chapter 7. This affects the ease with which such an ion can be accommodated in a crystal site. Consider the case of nickel, an important trace element in basalts: Ni^{2+} in a basalt melt crystallizing olivine exhibits an unexpectedly strong preference for the octahedral Mg site in olivine, to such an extent that crystallization of olivine rapidly depletes the Ni content of the melt. The reason is because the olivine site more readily accommodates the d-orbital geometry of Ni^{2+} than do the available sites in the melt.

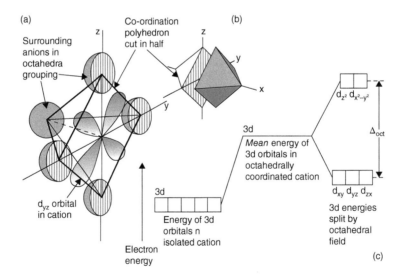

Figure 9.10.1 Crystal-field splitting of d-orbital energy levels in a transition metal cation in octahedral co-ordination.

Rare earth elements

Following the element lanthanum, La (the first member of the third transition series – see Figure 9.7), electrons begin to occupy the seven 4f orbitals, forming the 14 metals from cerium (Ce) to lutetium (Lu) known as the **lanthanides** or **rare earth elements (REEs)**. The distinction between individual rare earths lies in the number of 4f electrons. These are mostly not involved in bonding, and the chemical properties of all 14 elements, together with lanthanum, are therefore remarkably similar. All have stable trivalent states (Figure 9.9).

Owing to the increase in nuclear charge, there is a steady *decrease* of ionic radius in progressing from lanthanum La^{3+} to lutetium Lu^{3+} (the *lanthanide contraction*–Figure 9.9). The 'light rare earth elements' ('LREEs', La–Sm) are incompatible elements. Owing to their smaller ionic radii, however, the 'heavy rare earths' ('HREE', Gd–Lu) are more easily accommodated in the crystal structures of a few rock-forming minerals, particularly garnet and amphibole. The rare earth elements therefore provide the geochemist with an array of trace elements which, though virtually identical in other chemical properties, range continuously in behaviour from incompatible to selectively compatible.

For reasons discussed in Chapter 11, an element of even atomic number tends to be about ten times more abundant in the cosmos than its neighbours with odd Z-values, giving the rare earths in particular (Figure 11.2 inset) a characteristic zigzag abundance pattern that is found in all Solar System matter. When examining the rare-earth pattern of a terrestrial or lunar rock, therefore, geochemists eliminate this sawtooth effect by dividing each REE concentration in the rock (ppm) by the average concentration of the same element in chondrite meteorites (also in ppm). Chondrites (Box 11.1) serve as a sensible reference value here because they provide a reasonable estimate of the primordial composition of the Earth's mantle, the 'starting point' from which all igneous rocks have ultimately been derived. The result is a smooth 'chondrite-normalized' REE pattern showing how much each rare earth has been 'enriched' in the sample of interest relative to a model mantle source.

Figure 9.10 shows a chondrite-normalized REE pattern for two lunar basalts. The lower pattern represents lava formed directly by melting of the lunar mantle; the pattern is relatively flat, suggesting that the lunar

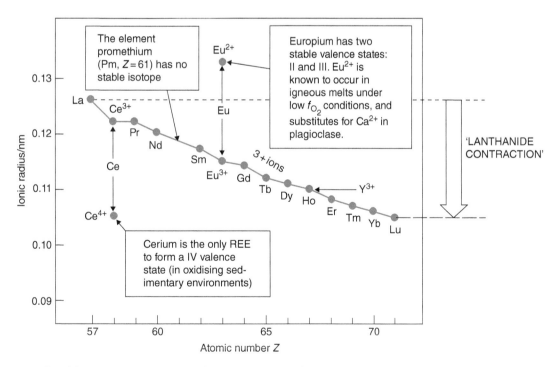

Figure 9.9 Ionic radii of the rare earth elements. Y^{3+} represents the related element yttrium (Figure 9.7).

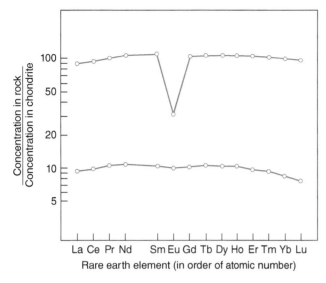

Figure 9.10 Chondrite-normalized rare earth element diagram for two lunar basalts. Note that the vertical (enrichment) scale is logarithmic. REE patterns are prepared by plotting the ratio for each element versus element position in the REE series, then joining up the array of points with straight lines. (Source: Data from Taylor (1982)).

source region differed little from chondrite, at least in terms of REE abundances. The upper pattern is also flat in overall terms but differs in two important ways: REE concentrations are generally about ten times higher, but there is a pronounced deficiency in the abundance of the element europium (Eu), a feature known as a 'negative Eu anomaly'. This basalt evidently represents magma that has undergone partial crystallization. Most REEs are incompatible in basalt minerals, and crystallization of olivine, clinopyroxene, plagioclase, and so on, must therefore concentrate REE in the decreasing amount of remaining melt. The exception is europium which, unique among the REEs, has a 2+ oxidation state (Figure 9.9). Eu^{2+} has a radius similar to Ca^{2+} (Box 9.1), and can substitute for Ca in plagioclase (but not in clinopyroxene, whose more compact structure excludes it). When plagioclase is crystallizing under relatively low f_{O_2} (reducing conditions), therefore, Eu behaves as a *compatible* element and becomes *depleted* in the melt as plagioclase crystallization removes it. Thus a negative Eu anomaly is a valuable geochemical tracer of plagioclase crystallization during magma fractionation.

REEs are useful for detecting the influence of other minerals on igneous processes. For example, the presence of garnet in a basalt source region during melting

causes HREEs to be retained in the residual garnet-bearing solid, leaving the melt formed relatively depleted in HREEs relative to LREEs. The rare earth pattern of such a basalt would show a steep negative slope.

The lanthanides have acquired great technological importance in recent decades, for example in lasers and in the high-field permanent magnets used in wind-turbine generators and electric vehicles. Global demand for REEs is therefore rising faster than supply, and REEs are at the top of the world's 'supply risk list' of technologically vital elements whose future availability is uncertain. Since the 1990s China has become the world's main supplier of REEs.

The trace element yttrium (Y) also forms a 3+ ion. The ionic radius of Y^{3+} is the same as Ho^{3+}, and in geological materials yttrium is always closely associated with the HREEs.

Actinides

The second row of the f-block consists solely of radioactive elements, known collectively as the *actinides* after the element actinium (Ac) that precedes them in the Periodic Table (cf. the lanthanides). Thorium (Th) and uranium (U) have long-lived isotopes ^{232}Th, ^{235}U and ^{238}U (see Table 10.1), so these elements occur as trace elements in nature. All three long-lived isotopes are α-emitters, whose decay generates chains of shorter-lived α- and β-emitting radionuclides that lead eventually to different stable isotopes of lead (see Figure 3.3.1 and Table 10.1). For the Earth scientist, there are two important consequences. Firstly, collision of these multiple α-particles with surrounding nuclei captures their kinetic energy within the host rock[7] and raises its temperature. Decay of Th and U (and to a lesser extent ^{40}K) within the Earth thus makes an important contribution to the Earth's heat budget and the *heat flow* we observe at the surface. Secondly, as the decay constants for these three naturally occurring radioisotopes are accurately known (Table 10.1), they provide an important means of dating the formation of

---

[7] The high charge (2+) of α-particles causes them to interact strongly with surrounding atoms, limiting the distance to which they can penetrate matter: a sheet of paper is sufficient to shield against α-radiation.

rocks and minerals. Various aspects of isotopic dating are explained in the following chapter.

Another important feature of uranium is that the ^{235}U isotope is **fissile**, giving it the key role in nuclear energy generation. However, ^{235}U accounts for only 0.71% of natural uranium. Reactors have in the past operated successfully with natural uranium fuel, but today's more efficient reactors require the ^{235}U to be artificially *enriched* to 3–5% ^{235}U.

Further reading

Archer, D. (2011) *Global Warming–Understanding the Forecast.* 2nd edition. Chichester: John Wiley and Sons, Ltd.

Barrett, J. (2002) *Atomic Structure and Periodicity.* Cambridge: Royal Society of Chemistry.

Geim, A.K. and Kim, P. (2008) Carbon wonderland. *Scientific American*, April 2008, 90–97.

Killops, S.D. and Killops, V.J. (2005) *An Introduction to Organic Geochemistry*, 2nd edition. Chichester: John Wiley and Sons, Ltd.

Krauskopf, K.B. and Bird, D.K. (1995) *Introduction to Geochemistry*, 3rd edition. New York: McGraw-Hill.

Meislich, H., Nechamkin, H. and Sharefkin, J. (2011) *Organic Chemistry*. 2nd edition. New York: McGraw-Hill.

Ninan, A. (2009) *Organic Chemistry - the Nature of Organic Chemistry Explained*. Studymates.

O'Neill, P. (1998) *Environmental Chemistry*, 3rd edition. London: Blackie (Thomson Science).

Scerri, E.R. (2011) *The Periodic Table – a Very Short Introduction.* Oxford: Oxford University Press.

Exercise

9.1 The reaction:

$$Fe_2(SO_4)_3 + 6H_2O \rightleftharpoons 2Fe(OH)_3 + 3H_2SO_4$$

is a balanced chemical equation: summing the total amount of any constituent on the left-hand side should equal that represented on the right-hand side: 2 atoms of Fe, 3 atoms of S, and so on. Below is a series of similar reactions (relating to super-gene enrichment) which have not been balanced. Work out (by trial and error) the integers required in front of each molecule to produce a balanced equation.

(a) $Fe_2(SO_4)_3 + FeS_2 \rightarrow FeSO_4 + S$

(b) $CuFeS_2 + Fe_2(SO_4)_3 \rightarrow CuSO_4 + FeSO_4 + S$

(c) $FeSO_4 + O_2 + H_2O \rightarrow Fe_2(SO_4)_3 + Fe(OH)_3$

(d) $MnO_2 + H^+ + 2Fe^{2+} \rightarrow Mn^{2+} + H_2O + Fe^{3+}$

(e) $ZnS + Fe_2(SO_4)_3 + H_2O \rightarrow ZnSO_4 + FeSO_4 + H_2SO_4$

10 WHAT CAN WE LEARN FROM ISOTOPES?

Nearly all of the elements discussed in Chapter 9 consist of more than one **isotope**. To say that an element consists of several isotopes means that – for the unique value of the atomic number Z that identifies the element (Box 6.1) – nuclei exist in nature with two or more *alternative* values of the neutron number N. This is illustrated for the trace elements Rb (rubidium) and Sr (strontium) in Figure 10.1a: each square represents a naturally occurring isotope of the relevant element. Natural Rb comprises two isotopes, a stable isotope ^{85}Rb ($Z = 37$, $N = 48$, $A = 85$) and a *radioactive isotope* ^{87}Rb ($Z = 37$, $N = 50$, $A = 87$). Of these, ^{85}Rb is the more abundant in nature (Figure 10.1b). ^{87}Rb decays slowly to ^{87}Sr (Figure 10.1b), although its half-life is long enough (Table 10.1) for some ^{87}Rb still to be present in the Earth, the remnant of an episode of heavy-element production that preceded the formation of the Solar System (see Chapter 11).

Natural Sr, on the other hand, consists of four isotopes, ^{84}Sr, ^{86}Sr, ^{87}Sr and ^{88}Sr, all of which are stable, and of these ^{88}Sr is the most abundant (Figure 10.1b). Although ^{87}Sr is a stable isotope (it doesn't itself undergo radioactive decay), its abundance in the Earth increases with time owing to decay of radioactive ^{87}Rb.

The isotopes of an element share the same electron configuration (Chapter 5) and thus nominally have identical chemical properties, yet their relative abundance in Earth materials varies to a small yet measurable extent. As we shall see, studying how the isotopic compositions of particular elements vary in the natural world – the subject of the present chapter – has proved an immensely fertile source of information on how geological and environmental processes operate.

Chemical Fundamentals of Geology and Environmental Geoscience, Third Edition. Robin Gill.
© 2015 John Wiley & Sons, Ltd. Published 2015 by John Wiley & Sons, Ltd.
Companion Website: www.wiley.com/go/gill/chemicalfundamentals

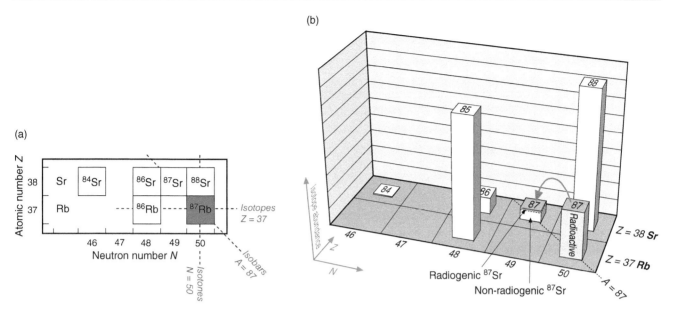

Figure 10.1 (a) The naturally occurring istopes of the elements rubidium (Rb) and strontium (Sr) plotted by atomic number Z and neutron number N. The shading of the ^{87}Rb box signifies its radioactivity. The meanings of 'isotope', 'isotone' and 'isobar' are illustrated. (b) Relative abundance of the naturally occurring isotopes of Rb and Sr. The base of the solid figure is the Z-N plot shown in (a). The height of each column represents the relative abundance of the isotope (shown here for equal amounts of the elements Rb and Sr); the number on the top of each column is the mass number $A = Z + N$ that serves to identify each isotope. Note that the ^{87}Sr column comprises radiogenic and non-radiogenic parts.

Some nuclear terminology

Figure 10.1a illustrates the jargon used when discussing relationships between isotopes. Each individual square in Figure 10.1a – having its own unique combination of Z *and* N – represents a specific **nuclide**. Three terms are used to describe groupings of nuclides that share a common attribute (given here with mnemonic hints to help in memorizing):

Isoto**p**es: nuclides that share the same **Z** value (*i.e.* have the same number of **p**rotons, forming a *horizontal* row in the nuclide chart).

Isoto**n**es: nuclides that share the same **N** value (same number of **n**eutrons – a *vertical* column of squares in Figure 10.1.1, *e.g.* ^{40}Ca, ^{39}K and ^{38}Ar are isotones).

Iso**b**ars: nuclides that share the same **A** value (a *diagonal* array in Figure 10.1.1, *e.g.* ^{40}Ca is isobaric with ^{40}K and ^{40}Ar).

Isotope systems

Measuring the abundance of a single isotope (e.g. ^{87}Sr) in a geological sample reveals little about the sample's origins. Useful information – such as

age – only emerges when the isotope's abundance is considered *in relation to other relevant isotopes* and elements. In Rb–Sr geochronology, for instance, the age of a suite of rocks is determined by plotting the ^{87}Sr/^{86}Sr isotope *ratios* of several samples against the corresponding ^{87}Rb/^{86}Sr ratios (Figures 10.4 and 10.5b). The interrelated isotope measurements and constants (e.g. decay constant) that allow an age to be calculated – or other aspects of sample origin to be inferred – together constitute the *Rb-Sr isotope system*.

Isotope systems of current interest in the Earth sciences fall into three categories:

Radiogenic isotope systems

^{87}Sr is described as the *daughter* isotope or daughter nuclide that results (at least in part) from the **radioactive** decay of the *parent* isotope ^{87}Rb (Figure 10.1b). The Rb–Sr isotope system, in which every radioactive ^{87}Rb nucleus that decays is replaced by a new **radiogenic** nucleus of ^{87}Sr, is one of several so-called *radiogenic isotope systems* (Table 10.1), in which relative isotopic proportions change progressively with the passage of time. The ^{87}Rb content of a geological material decreases with time in relation to ^{85}Rb,

Box 10.1 The nuclide chart

The clearest way to visualize the nuclides, both stable and radioactive, is to plot their Z values against N values as shown in Figure 10.1.1 (of which Figure 10.1a forms a small part).

The *stable* nuclides (solid circles) lie in a narrow, slightly curved ribbon. The most abundant light nuclides have $N \approx Z$, but the stable isotopes of heavier elements become increasingly more neutron-rich ($N \leq 1.5Z$). On each side of the stable-nuclide ribbon lies a band of radioactive nuclides, most of which are too short-lived to be found in nature (so are not shown individually in Figure 10.1.1). Neutron-rich (high-N) radionuclides lying to the right of the stable-nuclide ribbon decay by ejecting a high-energy electron (for historical reasons electrons emitted by nuclei are called β-**particles**, and being negatively charged are symbolized β^-):

$$\underset{\substack{\text{parent} \\ \text{nuclide}}}{^{87}\text{Rb}} \rightarrow \underset{\substack{\text{daughter} \\ \text{nuclide}}}{^{87}\text{Sr}} + \underset{\substack{\text{beta} \\ \text{particle}}}{\beta^-} + \underset{\text{antineutrino}}{\bar{\upsilon}} \qquad (10.1.1)$$

Loss of the β-particle transforms one neutron in the nucleus into a proton, increasing Z by one at the expense of N. Such decays therefore appear in Figure 10.1.1 as short 'up-to-the-left' transverse arrows, as illustrated by ^{87}Rb (see inset).

Proton-rich radionuclides on the left of the stable-nuclide ribbon decay by ejecting a **positron**:

$$\underset{\text{parent}}{^{40}\text{K}} \rightarrow \underset{\text{daughter}}{^{40}\text{Ar}} + \underset{\text{positron}}{\beta^+} + \underset{\text{neutrino}}{\upsilon} \qquad (10.1.2)$$

A positron (β^+) is the **antimatter** particle equivalent to the electron, but with a positive charge. Its emission signals a reaction in the potassium nucleus by which a proton is transformed into a neutron ($Z \rightarrow Z - 1, N \rightarrow N + 1$). The same transformation can also be accomplished by *capturing* an orbital electron. In either case the decay appears in the nuclide chart as a short 'down-to-the-right' transverse arrow (see decay of ^{40}K in the inset in Figure 10.1.1).

Both β^--decay and β^+-decay are **isobaric** reactions, leaving A unchanged. **Isobars** – nuclides sharing the same mass number or A-value – lie on diagonal lines in Z–N space – see the dashed line in the Sm–Nd inset in Figure 10.1.1.

The third category of nuclear decay shown in Figure 10.1.1 is the ejection of an alpha-particle α^{2+} (consisting of 2 protons + 2 neutrons = ^{4}He nucleus):

$$^{147}\text{Sm} \rightarrow {}^{143}\text{Nd} + \alpha^{2+}$$

Such α-decay reactions reduce both Z and N by 2 and A by 4, so appear in Figure 10.1.1 as a trend-parallel arrow pointing down to the left (see Sm–Nd inset). The complicated decays of uranium (^{238}U and ^{235}U) and thorium (^{232}Th) to different isotopes of lead (^{206}Pb, ^{207}Pb, ^{208}Pb) involve multiple α and β^--decay steps (Box 3.3).

^{87}Rb, ^{40}K, ^{147}Sm, ^{232}Th, ^{235}U and ^{238}U are all long-lived, naturally occurring radionuclides which are used extensively in geochronology.

---

while ^{87}Sr increases with time in relation to other Sr isotopes (Figure 10.1b). Radiogenic isotope systems not only lie at the heart of geochronology, but also shed important light on the provenance of igneous magmas, metamorphic rocks and sediments.

Take care not to confuse **radiogenic** with **radioactive**.

Stable isotope systems

The three isotopes of oxygen (^{16}O, ^{17}O and ^{18}O) are neither radioactive nor radiogenic and so their relative proportions *do not vary with time*. Nonetheless, because the difference in mass number A between ^{16}O and ^{18}O ($18 - 16 = 2$) is relatively large in comparison to their mean mass number (17) – with consequent slight differences in quantitative chemical parameters – the ^{18}O/^{16}O ratio is **fractionated** to a measurable extent by geological processes such as crystallization and hydrothermal alteration. The minute natural variations that are seen in this ratio between minerals and fluids therefore serve as a useful **tracer** for detecting and quantifying such processes.

Since their isotope ratios (e.g. ^{18}O/^{16}O and ^{34}S/^{32}S) do not change with time, such systems are designated *stable isotope systems*.

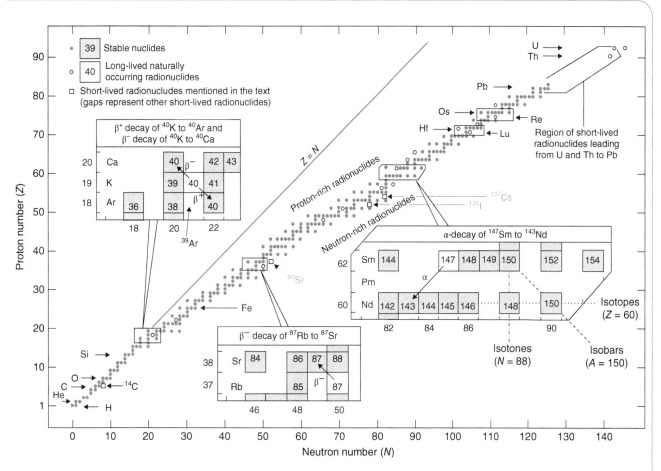

Figure 10.1.1 The complete 'nuclide chart' of atomic number Z versus neutron number N showing the stable nuclides as filled circles, and naturally occurring long-lived ($t_{1/2} > 10^8$ years) radionuclides as open circles. Also shown for illustration (as small open squares) are one **cosmogenic** nuclide (^{14}C) and three short-lived anthropogenic fission-products of environmental concern (^{90}Sr, ^{131}I and ^{137}Cs). Enlarged insets show the decay reactions from ^{40}K to ^{40}Ar and ^{40}Ca, from ^{87}Rb to ^{87}Sr, and from ^{147}Sm to ^{143}Nd. Other rectangles identify the Lu–Hf and Re–Os radiogenic isotope systems (Table 11.1).

Cosmogenic radioisotope systems

The naturally occurring radioisotopes upon which radiogenic isotope systems rely are long-lived relics of an episode of heavy-element formation (Chapter 11) *pre-dating* the formation of our Solar System 4.6 Ga ago. Relatively few radionuclides possess such long half-lives, which is why the number of radiogenic isotope systems available in the geoscientist's toolkit is small (Table 10.1). Radionuclides are, however, being formed today by the action of high-energy cosmic rays on atmospheric gases, and these shorter-lived **cosmogenic radionuclides**, such as the ^{14}C used in radiocarbon dating (formed mainly by cosmic ray bombardment

of ^{14}N nuclei in atmospheric nitrogen), help us to understand recent geological processes.

Radiogenic isotope systems

Table 10.1 and Figure 10.1.1 (Box 10.1) summarize the principal radiogenic isotope systems currently used by geoscientists.

K–Ar geochronology

The potassium isotope ^{40}K is radioactive, having a **half-life** of 1.25 Ga. The proportion of ^{40}K in natural

Table 10.1 Radiogenic isotope systems

Name	Reaction	Decay constant/y^{-1}	Half-life/y	Applications[†]
K–Ar	$^{40}K \rightarrow {}^{40}Ar + \beta^{+} + \upsilon$ $^{40}K \rightarrow {}^{40}Ca + \beta^{-} + \bar{\upsilon}$	$\lambda_{Ar} = 0.581 \times 10^{-10[\S]}$ $\lambda_{Ca} = 4.962 \times 10^{-10[\S]}$	$1.250 \times 10^{9[\S]}$	Geochronology of K-bearing minerals
Rb–Sr	$^{87}Rb \rightarrow {}^{87}Sr + \beta^{-} + \bar{\upsilon}$	1.42×10^{-11}	4.88×10^{10}	Geochronology, seawater evolution, sediment correlation, magma genesis
Sm–Nd	$^{147}Sm \rightarrow {}^{143}Nd + \alpha^{2+}$	6.54×10^{-12}	1.060×10^{11}	Precambrian geochronology, sediment provenance, crustal and mantle evolution, stony meteorite and lunar studies, magma genesis
Lu–Hf	$^{176}Lu \rightarrow {}^{176}Hf + \beta^{-} + \bar{\upsilon}$	1.94×10^{-11}	3.57×10^{10}	Geochronology, mantle evolution, crustal growth models
Re–Os	$^{187}Re \rightarrow {}^{187}Os + \beta^{-} + \bar{\upsilon}$	1.666×10^{-11}	4.16×10^{10}	Geochronology including iron meteorites, mantle and lithosphere evolution
U–Th–Pb	$^{232}Th \rightarrow {}^{208}Pb + 6\alpha^{2+} + 4\beta^{-} + 4\bar{\upsilon}$ $^{235}U \rightarrow {}^{207}Pb + 7\alpha^{2+} + 4\beta^{-} + 4\bar{\upsilon}$ $^{238}U \rightarrow {}^{206}Pb + 8\alpha^{2+} + 6\beta^{-} + 6\bar{\upsilon}*$	4.9475×10^{-11} 9.8485×10^{-10} 1.55125×10^{-10}	14.010×10^{9} 0.7038×10^{9} 4.468×10^{9}	Geochronology, crustal evolution, meteorite studies, magma genesis

[†] After Henderson and Henderson (2009).
[§] The combined rate constant λ is the sum of the two individual rate constants $= 5.543 \times 10^{-10}$ yr^{-1}. The concept of half-life is applicable only to the combined decay of ^{40}K.
*See Figure 3.3.1 for the full decay scheme.

potassium has therefore decreased through the 4.55 Ga course of Earth history. Today ^{40}K makes up only 0.012% of present-day potassium (Figure 10.2). ^{40}K decays in two alternative ways (Table 10.1; Figure 10.1.1), one route leading to the calcium isotope ^{40}Ca, accounting for 89% of decaying ^{40}K nuclei, and the other to ^{40}Ar, an isotope of the **inert gas** argon (the remaining 11% of decaying ^{40}K nuclei). The accumulation of ^{40}Ar in a crystal of a K-bearing mineral from the decay of ^{40}K provides the basis for the K–Ar dating technique.

From the moment that potassium is incorporated into a newly formed mineral (Figure 10.3a), the proportion of ^{40}K present begins to decline (Figure 10.3e,f). Argon – a gas – is not incorporated during the initial crystallization of a mineral (Figure 10.3a), but ^{40}Ar will however form *in situ* in a K-rich crystal (Figure 10.3b,c,e,f) as the product of ^{40}K decay. If none of it escapes, the $^{40}Ar/^{40}K$ ratio (how much ^{40}Ar is present relative to ^{40}K) provides a measure of the time that has elapsed since the mineral crystallized (Equation 10.1):

$$t = \frac{1}{\lambda} \ln \left[1 + \frac{\lambda}{\lambda_{Ar}} \frac{{}^{40}Ar}{{}^{40}K} \right] \tag{10.1}$$

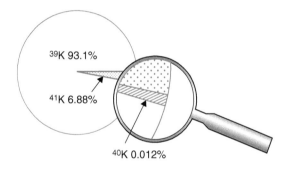

^{39}K 93.1%

^{41}K 6.88%

^{40}K 0.012%

Figure 10.2 Pie chart showing the isotopic composition of potassium.

where t is the time in years since the mineral crystallized (or – more accurately – cooled below the **closure temperature** for Ar diffusion – see Chapter 3); λ_{Ar} is the decay constant for the decay of ^{40}K to ^{40}Ar (y^{-1}); λ is the overall decay constant for both modes of ^{40}K decay (y^{-1}, Table 10.1); and $^{40}Ar/^{40}K$ is the measured present-day daughter/parent abundance ratio: the ^{40}Ar amount in this ratio is determined by mass spectrometry (Box 10.3), whereas the amount of ^{40}K is calculated from the K-content of the mineral sample.

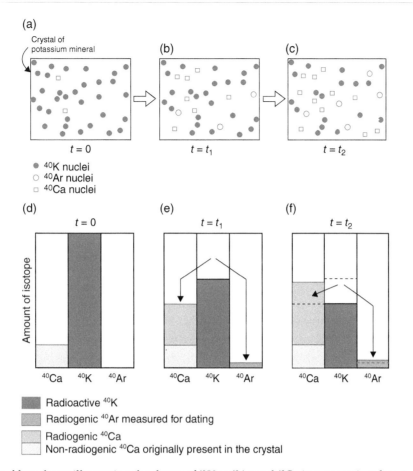

Figure 10.3 Cartoons and bar charts illustrating the decay of ^{40}K to ^{40}Ar and ^{40}Ca in a potassium-bearing crystal.

The complexity of Equation 10.1 is a consequence of the branched decay scheme of ^{40}K.

Various K-rich minerals separated from plutonic or metamorphic rocks may be used for K–Ar dating, including biotite, muscovite and hornblende. For dating volcanic rocks, feldspars (sanidine, anorthoclase, plagioclase) and whole-rock samples are often used as well.

A K-rich mineral is likely to contain a trace of calcium as well, so ^{40}Ca will be present in it at the outset (Figure 10.3a,d).[1] Distinguishing a small contribution of *radiogenic* ^{40}Ca (the product of *in situ* β-decay of ^{40}K) from the non-radiogenic component of ^{40}Ca already present (Figure 10.3a,d) is problematic, and so the ^{40}Ca branch of the ^{40}K decay scheme is not used in geochronology.

K–Ar dating, although straightforward to understand, is prone to **systematic errors** that are hard to

quantify, and so it is rarely used today. It has been superseded by the more reliable ^{40}Ar/^{39}Ar geochronometer (Box 10.2).

Rb–Sr geochronology

The initial absence of ^{40}Ar from the potassium-bearing mineral or rock being dated makes the K–Ar and Ar–Ar dating techniques relatively straightforward, allowing an age to be determined from a single rock sample. For most geochronometers, however, an unknown amount of the daughter isotope occurs naturally in the sample from the outset. In such cases *several **cogenetic** samples* need to be analysed if we are to eliminate this unknown factor and secure an accurate age.

The Rb–Sr isotope system (introduced in Figure 10.1) illustrates the principles. The way in which Sr isotopic composition varies with time is sketched out in Figure 10.4. The vertical axis represents the amount of

[1] ^{40}Ca is the most abundant isotope of calcium, making up 97% of the element.

Box 10.2 ^{40}Ar/^{39}Ar dating: solutions to ^{40}K–^{40}Ar dating problems

Conventional K–Ar geochronology relies on a K-rich mineral having remained sealed with respect to ^{40}K and ^{40}Ar migration ever since the event being dated. Experience shows, however, that ^{40}Ar may leak out of such minerals ('Ar loss') or indeed may diffuse in, for example from a hydrothermal fluid percolating through the sample (leaving 'excess Ar' in the crystal). Such deviations from closed-system behaviour introduce **systematic errors** in measured K–Ar ages that are impossible to detect routinely and which severely limit the method's reliability as a dating tool.

The so-called ^{40}Ar/^{39}Ar (or just 'Ar–Ar') dating technique circumvents these and other shortcomings of K–Ar geochronology in two ingenious ways, as outlined in Table 10.2.1. The same K-bearing minerals (or volcanic whole-rocks) are used as for K–Ar dating.

These systematic errors do not manifest themselves uniformly through a mineral crystal, a fortunate fact that makes them easier to decipher. Argon loss is more marked near the margins of a crystal (or near to cracks). During stepwise heating, early heating steps preferentially extract argon from the ^{40}Ar-depleted (low ^{40}Ar/^{39}Ar) rim of a crystal,

giving apparent ages younger than the geological age, whereas ^{40}Ar in the more retentive interior will only be released in higher-temperature steps. The age spectrum of a sample affected by argon loss therefore exhibits an initial rise (Figure 10.2.1b), and commonly leads to an age plateau from which consistent, accurate age values can be calculated.

'Excess' ^{40}Ar that has diffused into a sample *after* crystallization (most common in metamorphic minerals) tends to reside on subgrain boundaries or in fluid inclusions, from where it escapes at low temperatures giving anomalously old ages from early heating steps (Figures 10.2.1b and 10.2.2). Later steps are more likely to yield accurate geological ages, although sometimes older ages emerge again towards the end of argon release, reflecting ^{40}Ar trapped in melt or mineral inclusions (Kelley, 2002a,b).

Whereas K–Ar dating yields *absolute* ages, Ar–Ar is a *relative* dating method that requires calibration by analysing a standard of known geological age alongside unknown sample(s). Recent re-calibration has brought Ar–Ar dates into closer agreement with other geochronometers (Kerr, 2008).

Table 10.2.1 How ^{40}Ar–^{39}Ar dating overcomes K–Ar problems

Problems encountered with conventional ^{40}K–^{40}Ar dating	*Features of ^{39}Ar–^{40}Ar dating that circumvent these problems*
1. K and ^{40}Ar are determined by *different analytical techniques* on *separate sample **aliquots***, introducing errors and reducing the internal consistency of the ^{40}K/^{40}Ar ratio.	A *single sample aliquot* is irradiated with neutrons in a nuclear reactor to convert ^{39}K (a stable K isotope) to ^{39}Ar.* The ^{39}Ar/^{40}Ar ratio of the irradiated aliquot is readily determined by mass spectrometry (Box 10.3), from which a precise ^{40}K/^{40}Ar parent:daughter ratio for the aliquot can be calculated.
2. The K–Ar technique is prone to *systematic errors* that alter the daughter/parent ratio: • ^{40}Ar loss from the sample, giving *low* age values relative to true age. • 'Excess Ar' (including ^{40}Ar) diffusing into the sample, giving *high* age values.	An irradiated sample aliquot is heated under vacuum in a *series of temperature steps*§ until all the Ar has been extracted. The ^{39}Ar/^{40}Ar ratio is measured by mass spectrometry – and the apparent age calculated – separately *for each heating step* (Figure 10.2.1b). The resulting 'age spectrum' commonly reveals Ar loss or excess Ar in the earlier heating steps. Higher-temperature steps usually define a consistent age 'plateau' from which a reliable mean age for the sample – free of systematic error – can be determined. Data from a sample giving no age plateau can be disregarded.
3. ^{40}Ar is extracted for mass spectrometric analysis by completely melting the sample aliquot under vacuum. This provides *only one* ^{40}Ar determination (and thus *only one* ^{40}K/^{40}Ar ratio) that offers no direct indication of systematic errors (Figure 10.2.1a).	

* ^{39}Ar is radioactive with a half-life of 269 years; being short-lived, it is not shown as a box in Figure 10.1.1. Nonetheless, on a laboratory timescale, it can be treated arithmetically as a stable isotope.

§ In a furnace or using a laser.

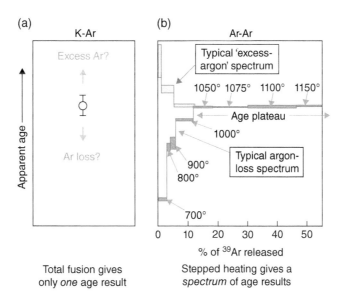

Figure 10.2.1 A cartoon illustrating key differences between K–Ar and Ar–Ar dating techniques. (a) K–Ar geochronology involves total fusion of the sample, which yields only one age result, providing no indication of argon loss or excess argon. (b) Ar–Ar dating entails step-heating of the sample, providing a spectrum of age measurements that can reveal either argon loss during the sample's history (dark shaded spectrum – numbers show illustrative temperatures in °C for each step), or an 'excess argon' spectrum (light shading) indicating exchange of ^{40}Ar with an external reservoir. The vertical thickness of the shaded rectangles in (b) represents **precision** (±1σ). For clarity, only the first half of the heating profile is shown in (b).

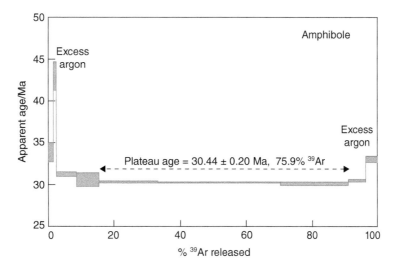

Figure 10.2.2 A typical Ar–Ar age spectrum for a hornblende separate from a volcanic rock. The vertical depth of each shaded rectangle indicates the precision of the age determination for that step (±1σ). (Source: Baker *et al.* (1996). Reproduced with permission of Elsevier.)

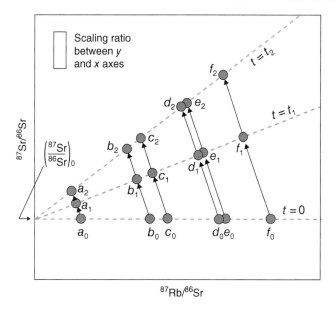

Figure 10.4 How $^{87}Sr/^{86}Sr$ and $^{87}Rb/^{86}Sr$ ratios evolve with time, shown on a graph of $^{87}Sr/^{86}Sr$ versus $^{87}Rb/^{86}Sr$. Points *a*, *b*, *c*, *d*, *e* and *f* represent the whole-rock isotopic compositions of six **cogenetic** igneous rocks (same age, same magma source) from the same intrusive complex. Their compositions differ in composition owing to geochemical differentiation during magma crystallization (over an interval of time short by comparison with the age of the complex). t_1 and t_2 represent different elapsed times after crystallization. The vertical rectangle indicates the relative expansion on the *x* and *y* axes.

the *radiogenic daughter* isotope ^{87}Sr in a sample, ratioed to the amount of a reference, non-radiogenic isotope ^{86}Sr (the reasons for doing so are explained in Box 10.4). The horizontal axis represents the amount of the *radioactive parent* nuclide ^{87}Rb in the sample, also divided by ^{86}Sr.

Imagine we wish to establish the age of an ancient intrusive complex, comprising a variety of igneous rock types related to each other by magma different-iation. Suppose the field evidence is consistent with the rocks being cogenetic, that is, formed at essentially the same time[2] as fractionation products of the same par-ent magma. Let the shaded dots in Figure 10.4 represent samples *a–f* of these diverse rock types. Since magma melting and crystallization fractionate element ratios *but not isotope ratios*, we can assume that all of the crys-tallized igneous rocks *a–f inherited the $^{87}Sr/^{86}Sr$ ratio of the magma source region* and so initially (at *t* = 0) they

formed a horizontal array of points a_0–f_0 in the figure, varying in $^{87}Rb/^{86}Sr$ ratio but not $^{87}Sr/^{86}Sr$.

As time passes, each ^{87}Rb nucleus that decays forms an ^{87}Sr nucleus, and therefore each sample will evolve along a trend of falling $^{87}Rb/^{86}Sr$ and rising $^{87}Sr/^{86}Sr$, repr-esented by the arrows in Figure 10.4. If each axis were plotted at the same scale, the arrows would have grad-ients of −45° since ^{87}Sr increases at the rate at which ^{87}Rb falls. Usually, however, the $^{87}Sr/^{86}Sr$ axis is expanded in such plots (see the scaling ratio shown) so the arrows in Figure 10.4 will appear steeper than −45°. $^{87}Sr/^{86}Sr$ will increase more quickly in samples with a high $^{87}Rb/^{86}Sr$ ratio, so the arrow lengths increase from sample *a* to sample *f*: as time passes each composition will migrate upward and to the left by a distance proportional to $^{87}Rb/^{86}Sr$. After a given time (e.g. at *t* = t_1), the compos-itions will still define a linear trend, but one whose grad-ient depends on the time elapsed since *t* = 0 (an **isochron**).

The intercept of the isochron with the *y*-axis represents the composition of a hypothetical sample having $^{87}Rb/^{86}Sr$ = 0. Because there is no ^{87}Rb in this notional 'sam-ple', there can be no increase in ^{87}Sr in it, which is why the isochron appears to pivot about this point. The intercept preserves the initial value of the Sr isotope ratio shared by all of the cogenetic samples at the outset (a_0, b_0, etc.).

The isochron plot

If we collect and analyse rocks *a–f* today (e.g. at *t* = t_2), we will find that these cogenetic samples define a sloping linear trend (a_2–f_2). Because the trend is determined by the common age of the samples, the best-fit line through a_2–f_2 is called an **isochron** (Greek: 'same age'). The isochron may be represented algebraically (see Box 10.4) by an *isochron equation*:

$$\left(\frac{^{87}Sr}{^{86}Sr}\right)_t = \left(\frac{^{87}Sr}{^{86}Sr}\right)_0 + \left(\frac{^{87}Rb}{^{86}Sr}\right)_t \left(e^{\lambda_{Rb} t} - 1\right) \tag{10.2}$$

Here $\left(\dfrac{^{87}Sr}{^{86}Sr}\right)_t$ represents a sample's current Sr isotope abundance ratio (how much ^{87}Sr there is in relation to ^{86}Sr, in atomic proportions) measured by mass spec-trometry. $\left(\dfrac{^{87}Rb}{^{86}Sr}\right)_t$ is calculated from the ratio of the sample's Rb and Sr element concentrations (determined by routine analytical methods and again given in atomic proportions,), and λ_{Rb} is the ^{87}Rb decay constant

---

[2] We assume the time taken for the parent magma to cool and fractionate is short compared to the age of the complex.

Box 10.3 Mass spectrometry

Because the isotopes of an element have virtually identical *chemical* properties, routine spectrometric methods of analysis that discriminate between elements by means of emitted spectral wavelengths (Box 6.3) cannot distinguish between an element's isotopes. The one property by which isotopes differ significantly is *atomic mass*. A mass spectrometer is an instrument that separates isotopes on this basis and measures their relative abundances.

When ionized, the isotopes of an element share the same ionic charge but differ in ionic mass (reflecting the different *A*-values that distinguish the isotopes) and therefore in mass/charge (m/q) ratio too. Mass spectrometric isotope analysis consists of five stages:

1 *Separation* – before being introduced into the mass spectrometer, the element whose isotopic composition is to be determined needs to be chemically separated from other elements in the sample that might interfere.[3]

2 *Ionization*– the separated element is introduced into the mass spectrometer's ion source chamber as solid, solution or gas, where it is ionized in one of several ways (thermal ionization on a hot filament, inductively coupled plasma, electron impact).

3 Electrostatic *acceleration* of the ions released, through aligned slits to form a narrow ion beam (Figure 10.3.1).

4 *Deflection* and *dispersion* of the ion beam in a strong magnetic field, causing each isotope component to emerge in a slightly different direction, according to m/q ratio (Figure 10.3.1).

---

[3] For example, to prevent measurement of a ^{87}Sr peak being biased by the presence of ^{87}Rb (an example of **isobaric** interference).

5 *Detection* of ion beams by carefully positioned collectors that register the intensity of each ion beam, reflecting the abundance of each isotope.

A mass spectrometer (stages 2–5) needs to be evacuated to high vacuum to avoid the ions in the narrow beam being scattered by collision with air molecules.

Modern mass spectrometers employ a number of adjacent collectors so that all relevant isotope ion beams can be measured simultaneously to deliver high-precision isotope ratios. Automated computer control of accelerating voltage, magnetic field strength and collector position allows a single mass spectrometer to be reconfigured to analyse a range of isotope systems.

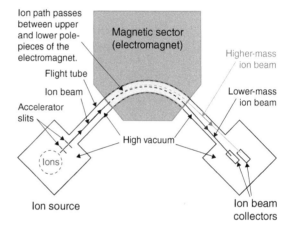

Figure 10.3.1 Sketch of a mass spectrometer viewed from above, showing ion source, flight tube, magnetic sector (with electromagnet pole pieces above and below the page), dispersion of the ion beam according to m/q, and ion collectors. Modern mass spectrometers typically have 5–10 ion collectors whose positions can be varied under computer control.

(Table 10.1). Equation 10.2 contains *two* unknowns that can be measured from the isochron plot: t, the age we are seeking to measure, and the *initial Sr isotope ratio* $({}^{87}\text{Sr}/{}^{86}\text{Sr})_0$.

In the context of Figure 10.4, Equation 10.2 has the same form as the equation for a straight line (see Appendix B):

$$y = c + x.m \qquad (10.3)$$

Where the isochron line intercepts the y-axis (the intercept c) defines the *initial Sr isotope ratio* $({}^{87}\text{Sr}/{}^{86}\text{Sr})_0$, and

the gradient (m) of the line equals $e^{\lambda_{\text{Rb}}t} - 1$. The age of the intrusive complex can be calculated from the measured gradient of the isochron in Figure 10.4 as

$$t = \frac{\ln(\text{gradient} + 1)}{\lambda_{\text{Rb}}} \text{ years}^4 \qquad (10.4)$$

---

[4] The units in which t is expressed here are the inverse of those used for λ_{Rb} (Table 10.1).

Box 10.4 How to derive the isochron equation

The kinetics of the Rb–Sr isotope system was briefly introduced in Box 3.2. The general decay equation may be written (equation 3.2.3):

$$n_P = (n_P)_0 \, e^{-\lambda_P t} \tag{10.4.1}$$

where n_P is the number of parent (radionuclide) nuclei present in the sample at time t, $(n_P)_0$ is the number that were initially present (at $t = 0$), λ_P is the decay constant for the decay of the parent radionuclide, and t is the time elapsed since the event being dated. Writing this in a more reader-friendly style specific to the decay of ^{87}Rb:

$$^{87}Rb_t = {}^{87}Rb_0 \, e^{-\lambda_{Rb} t} \tag{10.4.2}$$

Here $^{87}Rb_t$ represents the number of ^{87}Rb nuclei present at time t, and so on. Equation 10.4.2 represents a curve similar to Figure 3.1a.

Every ^{87}Rb nucleus that decays generates a new ^{87}Sr nucleus in its place. The number of radiogenic ^{87}Sr nuclei accumulating in time t is equal to the number of Rb nuclei that have decayed, which can be equated with the number of ^{87}Rb nuclei initially present *minus* the number remaining at time t:

$$^{87}Sr_t = {}^{87}Rb_0 - {}^{87}Rb_t \tag{10.4.3}$$

Since $^{87}Rb_0$ is generally unknown in a geological dating context, we eliminate it from Equation 10.4.3 by rearranging Equation 10.4.2 as follows:

$$\frac{^{87}Rb_t}{^{87}Rb_0} = e^{-\lambda_{Rb} t} \; \therefore \; \frac{^{87}Rb_0}{^{87}Rb_t} = \frac{1}{e^{-\lambda_{Rb} t}}$$
$$= e^{\lambda_{Rb} t} \; \therefore \; {}^{87}Rb_0 = {}^{87}Rb_t e^{\lambda_{Rb} t} \tag{10.4.4}$$

Substituting $^{87}Rb_t e^{\lambda_{Rb} t}$ in place of $^{87}Rb_0$ in Equation 10.4.3, we get:

$$^{87}Sr_t = {}^{87}Rb_t e^{\lambda_{Rb} t} - {}^{87}Rb_t = {}^{87}Rb_t \left(e^{\lambda_{Rb} t} - 1 \right) \tag{10.4.5}$$

Equation 10.4.5 represents a curve similar to Figure 3.1b. This equation tells us only the amount of *radiogenic* ^{87}Sr that has formed in the sample since $t = 0$. Because ^{87}Sr is a naturally occurring stable nuclide, we need to allow for the ^{87}Sr that was already present in the sample at $t = 0$ (see Figure 10.1b). Calling the amount of ^{87}Sr initially present $^{87}Sr_0$, the *total* ^{87}Sr present in the sample at time t is:

$$^{87}Sr_t = {}^{87}Sr_0 + {}^{87}Rb_t \left(e^{\lambda_{Rb} t} - 1 \right) \tag{10.4.6}$$

In terms of practical laboratory analysis, it is much easier to measure isotope *ratios* in a mass spectrometer than absolute amounts of individual nuclides. Therefore dividing each term in this equation by the amount of a reference stable isotope ^{86}Sr, we arrive at a much more useful equation – called the **isochron** equation – expressed in terms of easily measured, present-day Sr isotope and Rb/Sr ratios:

$$\left(\frac{^{87}Sr}{^{86}Sr} \right)_t = \left(\frac{^{87}Sr}{^{86}Sr} \right)_0 + \left(\frac{^{87}Rb}{^{86}Sr} \right)_t \left(e^{\lambda_{Rb} t} - 1 \right) \tag{10.4.7}$$

Here $\left(\dfrac{^{87}Sr}{^{86}Sr} \right)_t$ is the sample's current Sr isotope ratio measured by mass spectrometry. $\left(\dfrac{^{87}Rb}{^{86}Sr} \right)_t$ is calculated from the ratio of the sample's Rb and Sr element concentrations (determined by routine analytical methods), and λ_{Rb} is the relevant decay constant (Table 10.1).

Figure 10.5 shows a published isochron for a Proterozoic intrusive complex, to illustrate the principles explained here. Note how the various intrusive units defined by field mapping differ in Rb and Sr contents and in $^{87}Rb/^{86}Sr$ ratio. Accurate age determination requires the analysis of samples covering a significant range of Rb/Sr composition.

The samples used in Figure 10.5 were whole-rock samples. Another way to secure cogenetic samples with a range of Rb/Sr ratios for accurate age determination is to analyse mineral separates from a single rock sample, leading to what is called a *mineral isochron* (see Exercise 10.2). Each mineral, upon crystallizing, inherits the magma's $^{87}Sr/^{86}Sr$ value but, owing to element fractionation during crystallization, acquires a different Rb/Sr ratio. Mineral isochrons are more susceptible to resetting by later thermal re-equilibration, and can therefore be used to date episodes of metamorphism.

(a)

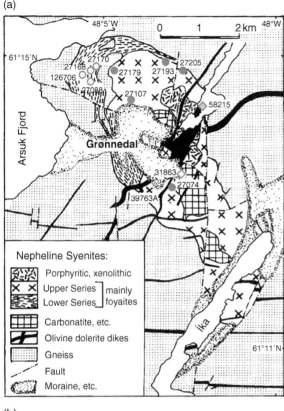

(b)

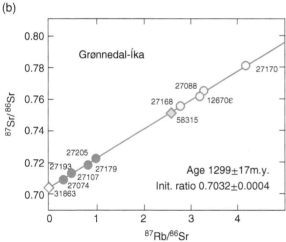

Figure 10.5 (a) Geological map of the Grønnedal-Íka intrusion, South Greenland, showing the distribution of recognized petrological units. (b) An **isochron** plot of Sr isotope data from the Grønnedal-Íka intrusion. The same symbols are used in (a) and (b) to correlate Rb–Sr compositions with map units.[5] Note the expanded scale of the ^{87}Sr/^{86}Sr axis in (b). (Source: Blaxland *et al.* (1978). Reproduced with permission of the Geological Society of America.)

---

[5] The original age of 1327 Ma calculated by these authors was based on a ^{87}Rb decay constant value of 1.39×10^{-11} year^{-1}. This age has been corrected to 1299 Ma to conform to the currently accepted decay constant of 1.42×10^{-11} year^{-1}. One discrepant data point has been omitted for clarity.

The initial Sr isotope ratio – characterizing a magma's origins

A geochemist studying an igneous complex like Grønnedal-Íka (Figure 10.5) is interested not just in its *age*, but also wants information on the *source of the parent magma* from which it formed: was it derived directly from the mantle, from the continental crust, or by some process involving both of these potential sources? Here the initial Sr isotope ratio becomes relevant. To see how, it helps to rewrite the isochron equation in a simplified form. Conveniently it happens that the exponential term $\left(e^{\lambda_{Rb} t} - 1 \right)$ can be quite closely approximated simply by $\lambda_{Rb} t$. To a first approximation, we can also ignore the slight change of ^{87}Rb/^{86}Sr with time, and therefore delete the suffix t. The resulting simplified equation is:

$$\left(\frac{^{87}\text{Sr}}{^{86}\text{Sr}} \right)_t \approx \left(\frac{^{87}\text{Sr}}{^{86}\text{Sr}} \right)_0 + \left[\left(\frac{^{87}\text{Rb}}{^{86}\text{Sr}} \right) \lambda_{Rb} \right] t \qquad (10.5)$$

Treating this as a straight-line equation of the form $y = c + m.x$, we can plot the evolution of ^{87}Sr/^{86}Sr ratio (e.g. of a magma source region) as a function of time, with a gradient equal to $\left(\frac{^{87}\text{Rb}}{^{86}\text{Sr}} \right) \lambda_{Rb}$.

Figure 10.6a illustrates such a 'Sr growth diagram'. Point m represents the ^{87}Sr/^{86}Sr ratio of the primordial Earth's mantle 4.55 Ga ago (the value of 0.69900 has been determined from meteorite studies). The gently rising line m–n represents the 'growth' in radiogenic ^{87}Sr relative to ^{86}Sr in the primitive mantle. Mantle rocks contain much less Rb than Sr (Table 10.2), therefore the limited amount of radiogenic ^{87}Sr formed in a given time will raise mantle ^{87}Sr by a only small amount relative to the ^{87}Sr already present. The low gradient between m and n therefore reflects the low Rb/Sr ratio (=0.03) of primitive mantle peridotite. If extended to the present day (thick dashed line), this Rb/Sr would lead to a mantle ^{87}Sr/^{86}Sr value of around 0.7045, consistent with the average ratio measured in many young basalts.

But suppose that a parcel of mantle underwent *partial melting* around 2.5 Ga ago (n) and, after ascent to crustal depths, the melt solidified to form rocks broadly similar in composition to continental crust. These rocks inherited the ^{87}Sr/^{86}Sr ratio of the mantle source at this time, but because Rb is more **incompatible** in mantle minerals than Sr (Figure 9.1.1),[6] the melt formed by

---

[6] http://earthref.org/KDD/

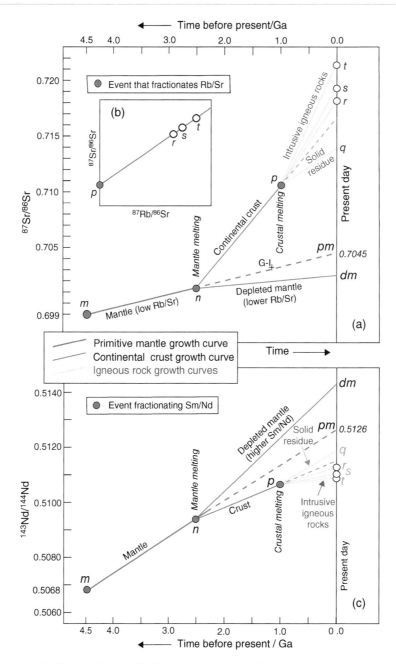

Figure 10.6 (a) Sketch of a 'growth diagram' for the Rb–Sr isotope system, with geological time progressing from left to right. A mantle melting event n leads to the generation of continental crust with higher Rb/Sr (providing steeper $^{87}Sr/^{86}Sr$ growth), leaving a volume of depleted mantle with lower Rb/Sr. This continental crust may undergo subsequent partial melting at p, and fractionation of the melt formed leads to intrusions with a range of Rb/Sr ratios. pm and dm represent the present-day $^{87}Sr/^{86}Sr$ ratios for *primitive mantle* and *depleted mantle* as defined in the text. The cross marked 'G-I' shows for reference the initial ratio and age obtained for the Grønnedal-Íka complex (Figure 10.5b). (b) Cartoon showing the isochron that would be obtained for intrusive rocks r, s and t in Figure 10.5a. (c) The $^{143}Nd/^{144}Nd$ growth curves for the sequence of events envisaged in Figure 10.5a.

partial melting of the mantle would have had a Rb/Sr ratio *higher* than the mantle value. The higher Rb/Sr in continental crust formed by mantle melting causes its $^{87}Sr/^{86}Sr$ ratio to increase with time at a faster rate than in the mantle: segment n–p in Figure 10.6a represents the steeper evolution of $^{87}Sr/^{86}Sr$ in typical continental crust, formed by a hypothetical mantle melting event at 2.5 Ga before the present (BP).

What happens in the mantle during this time? Regions of the mantle unaffected by this (or subsequent) melting

Table 10.2 Rb, Sr, Nd and Sm element concentrations (in **ppm**) in the Earth's mantle and continental crust (as cited in Henderson and Henderson, 2009)

Element	Z	Average primitive mantle	Average continental crust
Rb	37	0.605	49
Sr	38	20.3	320
Rb/Sr		*0.0298*	*0.153*
Nd	60	0.431	20
Sm	62	1.327	3.9
Sm/Nd		*0.325*	*0.195*

event retain their primordial Rb/Sr ratio and continue to evolve along the dashed extension of line *m–n* to *pm* (representing the present-day $^{87}Sr/^{86}Sr$ ratio of 'primitive mantle', i.e. that unaffected by the melting event). What remains of the parcel that *did* undergo partial melting event *n*, on the other hand, is a refractory solid residuum with a Rb/Sr ratio even *lower* than the original mantle (since Rb has been preferentially removed into the partial melt and into the crust formed from it). This 'depleted' parcel of mantle evolves hereafter along a shallower trajectory in Figure 10.6a. Any further melt extracted from this depleted mantle region today will testify to the earlier melting event through its less radiogenic $^{87}Sr/^{86}Sr$ ratio (*dm* in Figure 10.6a).

Now suppose that the crustal rocks resulting from the mantle melting event at *n* themselves undergo partial melting in the crust at around 1.0 Ga BP (*p*), and the magma formed fractionates into a range of high-level intrusive rocks of varying compositions. As before, partial melting in the crust generates melts with the same $^{87}Sr/^{87}Sr$ as the source *p* but with Rb concentrated relative to Sr, leaving the solid residue with a lower Rb/Sr ratio. The high Rb/Sr ratios in the intrusive rocks generate steeper $^{87}Sr/^{86}Sr$ evolution in Figure 10.6a, and exhumed samples of these rocks, collected and analysed today, would give $^{87}Sr/^{86}Sr$ values similar to *r*, *s* and *t*. Plotted in $^{87}Sr/^{86}Sr$ versus $^{87}Rb/^{86}Sr$ space (Figure 10.6b), these samples would define an isochron with an initial ratio ≈ 0.7106, too high to be consistent with direct derivation from the mantle. Two requirements have to be satisfied to generate such a high initial ratio:

- the source region must have *significantly higher Rb/Sr* than mantle rocks, providing steeper growth rate in $^{87}Sr/^{86}Sr$;

- *sufficient time* must elapse between the mantle melting event *n* and the later crustal melting event *p* for significant crustal $^{87}Sr/^{86}Sr$ 'growth' to take place. Had the interval *n–p* been short (e.g. ~100 Ma rather than 1500 Ma), the initial ratio *p* obtained from *r*, *s* and *t* would be barely distinguishable from mantle values, regardless of the higher Rb/Sr ratio.

The solid residue from crustal melting, left behind at depth, would evolve along a *less* steep path, leading to a present-day $^{87}Sr/^{86}Sr$ ratio illustrated by *q*.

It follows that the initial Sr isotope ratio $(^{87}Sr/^{86}Sr)_0$ obtained from an isochron provides a sensitive tracer for the involvement of old continental *crust* in magma genesis. When compared to the mantle growth curve (Figure 10.6a), the initial Sr isotope ratio tells us whether an igneous magma has originated from the Earth's mantle or whether, on the other hand, continental crust with higher $^{87}Sr/^{86}Sr$ has made a significant contribution to magma genesis. Using Figure 10.5b as an example, the parent magma of the Grønnedal-Íka intrusion when the complex formed 1299 Ma ago had an initial ratio of 0.7032. This initial ratio lies very close to the mantle growth curve at that time (see the coordinates labelled 'G-I' in Figure 10.6a), leading Blaxland *et al* (1978) to conclude that the parent magma was essentially mantle-derived, with negligible crustal involvement.

$(^{87}Sr/^{86}Sr)_0$ can also shed light on different *mantle* source regions, as illustrated by the depleted mantle represented by *dm* in Figure 10.6a. As we shall see (Figure 10.8), such depleted source signatures are characteristic of most mid-ocean ridge basalts, suggesting that such basalts tap a depleted 'reservoir' in the Earth's mantle.

Dating Cenozoic sediments using $^{87}Sr/^{86}Sr$

Rb–Sr isochron dating as such does not lend itself to dating the deposition of sedimentary rocks, because the clastic component of sediments significantly predates deposition, and generally consists of minerals too poor in Rb to generate measurable radiogenic ^{87}Sr (a limitation that also applies to limestones). Rb-bearing authigenic minerals such as glauconite may in some cases provide Rb–Sr isochron dates, but such ages may not accurately reflect the age of deposition that is being sought.

The $^{87}Sr/^{86}Sr$ ratio does nonetheless provide a reliable geochronological tool for the dating of Cenozoic marine sediments. Those consisting of carbonate shells inherit

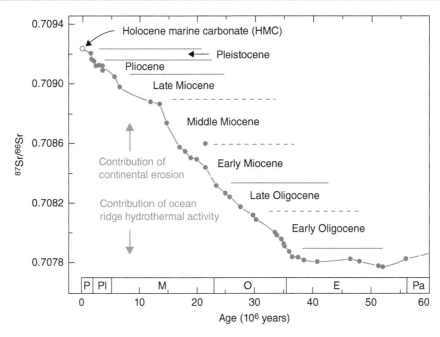

Figure 10.7 The correlation between measured $^{87}Sr/^{86}Sr$ values of Cenozoic marine carbonate sediments and biostratigraphic age. The abbreviations across the bottom refer to geological epochs (Pa = Palaeocene, E = Eocene, O = Oligocene, M = Miocene, Pl = Pliocene, P = Pleistocene). (Source: Adapted from DePaolo & Ingram (1985). Reproduced with permission of American Association for the Advancement of Science.)

the Sr isotope composition of the seawater from which the shells precipitated. The present oceans are known to be well mixed with regard to $^{87}Sr/^{86}Sr$, but oceanic $^{87}Sr/^{86}Sr$ has varied significantly through Phanerozoic time. Figure 10.7 shows how the $^{87}Sr/^{86}Sr$ values of marine carbonates have increased consistently (but not quite linearly) for the last 40 Ma, providing a powerful dating and stratigraphic correlation tool for Cenozoic sedimentary rocks, as illustrated by Exercise 10.1.

How has this remarkably regular evolution of seawater $^{87}Sr/^{86}Sr$ during Cenozoic times arisen? Temporal fluctuations in oceanic $^{87}Sr/^{86}Sr$ reflect a changing global balance between the two dominant Sr inputs into the oceans (Figure 10.7):

- Dissolved Sr from the weathering of continental landmass, delivered to the oceans by rivers with a global average $^{87}Sr/^{86}Sr$ ratio around 0.711; this contribution makes seawater Sr *more* radiogenic.
- Sr leached from ocean-ridge basalts by hydrothermal solutions (averaging 0.7045), tending to make seawater Sr *less* radiogenic.

The past 40 Ma have seen an unusually regular rise in seawater $^{87}Sr/^{86}Sr$. The tectonic uplift of the Himalaya-Tibet region during this period has increased continental weathering and erosion rates considerably, and the resulting dissolved riverine Sr flux from this region – with an average $^{87}Sr/^{86}Sr$ value of about 0.713 – is sufficient to account for most of the steady rise in seawater $^{87}Sr/^{86}Sr$ shown in Figure 10.7 (Richter *et al.*, 1992). Surprisingly, much of this Sr flux (>60%) and its radiogenic signature derives from the weathering of carbonate lithologies rather than silicate rocks (Oliver *et al.*, 2003).

The Sm-Nd radiogenic isotope system

The trace elements samarium (Sm) and neodymium (Nd) are both 'light' rare-earth elements (LREEs, Figure 9.9). Although the α-decay of ^{147}Sm to ^{143}Nd differs physically from the β-decay between ^{87}Rb and ^{87}Sr (Figure 10.1.1), the evolution of the two isotope systems can be represented by the same algebra and the two isotope systems are often used together (Figure 10.9).

The practice developed above for Sr is applicable to Nd too, ratioing the amount of the radiogenic isotope ^{143}Nd to a stable Nd isotope, in this case ^{144}Nd (Figure 10.1.1). By plotting the $^{143}Nd/^{144}Nd$ isotope ratios for a cogenetic suite of rock samples or minerals against the corresponding $^{147}Sm/^{144}Nd$ ratios, we can generate a Sm–Nd isochron from which the age of the rock suite can be determined (see Exercise 10.2 for an example). The equation for this isochron is:

$$\left(\frac{^{143}Nd}{^{144}Nd}\right)_t = \left(\frac{^{143}Nd}{^{144}Nd}\right)_0 + \left(\frac{^{147}Sm}{^{144}Nd}\right)_t \left(e^{\lambda_{Sm}t} - 1\right) \tag{10.6}$$

The value of the decay constant λ_{Sm} is given in Table 10.1. Because REEs like Sm and Nd are less mobile than Rb and Sr under hydrothermal and low-grade metamorphic conditions, Sm–Nd dating is invaluable for dating the eruption ages for moderately altered samples. The longer half-life (lower λ– Table 10.1) of ^{147}Sm makes Sm–Nd most suited to the dating of Precambrian rocks (see Exercise 10.2).

Figure 10.6c shows how $^{143}Nd/^{144}Nd$ evolves in the scenario outlined above for Rb–Sr. By analogy with Equation 10.5, the $^{143}Nd/^{144}Nd$ growth equation may be written:

$$\left(\frac{^{143}Nd}{^{144}Nd}\right)_t \approx \left(\frac{^{143}Nd}{^{144}Nd}\right)_0 + \left[\left(\frac{^{147}Sm}{^{144}Nd}\right)\lambda_{Sm}t\right] \tag{10.7}$$

The first difference to note in Figure 10.6c is the greater expansion of the y-axis compared with Figure 10.6a: in both figures each division represents 1 in the third significant figure. Different scaling is necessitated by the fact that Sm and Nd are LREEs of very similar chemistry and ionic radius, so they have closer incompatibility and are fractionated (relative to each other) to a smaller degree in melting and crystallization than is the case for Rb and Sr. The need to measure smaller changes in $^{143}Nd/^{144}Nd$ ratio also requires very precise mass spectrometric measurement.

Although $^{143}Nd/^{144}Nd$ rises with time in a similar fashion to $^{87}Sr/^{86}Sr$, aspects of Figure 10.6c appear 'upside down' relative to Figure 10.6a. When mantle melting occurs at n, the $^{143}Nd/^{144}Nd$ ratio of the crust formed evolves at a *lower* gradient than the mantle, while the depleted mantle left behind evolves with a *steeper* gradient, contrary to what we see with Rb–Sr. The same differences arise when the crust itself undergoes partial melting at p. The reason is that the parent nuclide ^{147}Sm is *less* incompatible than the daughter element Nd (whereas parent ^{87}Rb is *more* incompatible than daughter ^{87}Sr). The melts formed at n and p

therefore acquire *lower* parent/daughter ($^{147}Sm/^{143}Nd$) ratios than the source rock, and this is the reason why $^{143}Nd/^{144}Nd$ grows more *slowly* in the rocks derived from them.

It follows from Figure 10.6c that continental crust is characterized by low Sm/Nd and therefore by Nd that is *less radiogenic* than primitive mantle, whereas depleted mantle Nd is *more radiogenic* than primitive mantle.

Sr and Nd isotope signatures of young oceanic volcanics – mapping geochemical reservoirs in the mantle

Is the Earth's mantle homogeneous in composition, or has it become segregated over geological time into regions having distinct geochemical fingerprints? Isotopic analysis of young basalts from the ocean basins (where contamination by continental crust cannot confuse the picture) offers a means to probe present-day heterogeneity in the Earth's mantle. Basalts inherit the radiogenic isotope ratios of their source, and if the mantle feeding recent oceanic basalt eruptions[7] were homogeneous in its content of radioactive and radiogenic nuclides like ^{87}Rb and ^{87}Sr, we would expect such basalts to cluster around a single $^{87}Sr/^{86}Sr$–$^{143}Nd/^{144}Nd$ composition, equivalent to *pm* in Figure 10.6. A glance at the global compilation of such data in Figure 10.8 makes clear this is not the case.

Each data point in Figure 10.8 represents a basalt analysis, and collectively they define not a single cluster but an elongated array of basalt compositions, a broad correlation between $^{143}Nd/^{144}Nd$ and $^{87}Sr/^{86}Sr$ sometimes referred to as the 'mantle array'. Because isotope ratios of heavy elements like Sr cannot be altered by partial melting, the array in Figure 10.8 suggests that:

(a) the mantle regions from which oceanic basalt magmas are sourced must vary significantly in their Rb/Sr and Sm/Nd ratios (to account for the observed differences in $^{143}Nd/^{144}Nd$ and $^{87}Sr/^{86}Sr$ as in Figure 10.6); and

(b) this chemical heterogeneity in the mantle must have evolved long ago in Earth history (illustrated by point n in Figures 10.6a and c), to allow sufficient

---

[7] All terrestrial basalts are the product of the partial melting of mantle peridotite.

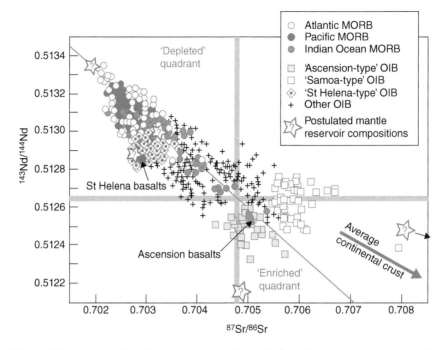

Figure 10.8 A plot of Nd and Sr isotope ratios of recently erupted basalts from the ocean basins, Each data point represents an individual basalt analysis, symbolized differently according to where the basalt was erupted ('MORB' stands for mid-ocean ridge basalt; 'OIB' refers to ocean-island basalt – see text). Shaded horizontal and vertical lines indicate the estimated present-day Nd and Sr isotopic composition of **primitive** mantle. The broad arrow points toward the composition of average continental crust. Star symbols signify the approximate 'end-member' compositions of supposed mantle reservoirs. (Source: Hofmann (1997). Reprinted by permission of Macmillan Publishers Ltd.)

time for the inferred differences in Rb/Sr and Sm/Nd to have generated the $^{87}Sr/^{86}Sr$ and $^{143}Nd/^{144}Nd$ variations observed today.

What other conclusions can we draw from Figure 10.8?

(c) Despite some overlap, basalts from specific oceanic settings form discrete clusters in separate parts of Figure 10.8, which suggest the existence of chemically distinct domains or '*reservoirs*' in the mantle from which these basalts are derived.

(d) Most *mid-ocean ridge basalts* ('MORB', round symbols) form a dense cluster in the top-left corner of the diagram, distinguished by the highest $^{143}Nd/^{144}Nd$ and lowest $^{87}Sr/^{86}Sr$ values. These MORBs evidently originate from a 'depleted' mantle reservoir with low Rb/Sr and high Sm/Nd ratios – analogous to *dm* in Figures 10.6a and c – which is moderately uniform, and extensive enough to supply mid-ocean ridge volcanism in all major ocean basins.

(e) This 'depleted' MORB reservoir differs in composition from '*primitive mantle*', a notional mantle reservoir approximating to the Earth's original mantle 4.55 Ga ago, whose current isotopic composition (analogous to *pm* in Figure 10.6) lies at the intersection of the broad orange lines in Figure 10.8 (lines that helpfully divide the diagram into four quadrants).

(f) *Ocean island basalts* ('OIB', squares and diamonds) – collected from intra-oceanic 'hotspot' islands remote from subduction zones, like Ascension and Hawaii – have less 'depleted' signatures extending across a wider range of isotopic composition towards, and in some cases beyond, the primitive mantle composition shown in Figure 10.8 (into what is sometimes called the 'enriched' quadrant). The range of OIB compositions tells us that the mantle includes not only a depleted reservoir, but a number of 'enriched' domains as well (having Rb/Sr higher and/or Sm/Nd lower than primitive mantle).

(g) Some OIB mantle sources seem to aggregate in distinct *clusters* in Figure 10.8: basalts from the Azores and St Helena in the Atlantic and the Austral and Ballemy Islands in the South Pacific, for example, all plot in the 'St Helena-type' cluster; Pitcairn Island and Tristan da Cunha plot with Ascension Island. Other hotspots plot in a more general continuum in between.

(h) The wide spread of Indian Ocean MORB compositions from the depleted quadrant into the enriched quadrant is the clearest sign in Figure 10.8 of the importance of *mixing* (either between mantle reservoirs themselves or between their partial melting products).

The continuum of basalt compositions in Figure 10.8 can be seen as the product of mixing together, in different circumstances and in varying proportions, a number of mantle reservoirs of contrasting composition: think of each basalt composition as being derived from a heterogeneous mantle 'cake' prepared from a common set of 'ingredients' (reservoirs) that were not fully blended before baking. Five reservoir compositions postulated in the literature are shown by the star symbols in Figure 10.8.

Where are these supposed reservoirs located and how have they formed? Answers to these questions are unavoidably speculative. There is a broad consensus equating the depleted MORB source with the ductile, convecting asthenosphere, which wells up and undergoes decompression melting beneath oceanic spreading centres (Figure 2.5.1b) across the globe. The uniformity of this reservoir reflects mixing associated with convective circulation, and its depleted character is thought to arise from the extraction of the continental crust – much earlier in Earth history – by partial melting of originally homogeneous primitive mantle.

Many OIB centres/hotspots are associated with mantle plumes (Figure 2.5.1b) rising from – and tapping mantle reservoirs located within – the lower mantle. Their diverse compositions are believed to reflect 'contamination' of these parts of the primordial mantle by recycled materials such as subducted terrigenous sediment or altered oceanic crust.

Stable isotope systems

A body suspended from a spring oscillates up and down with a natural frequency that depends upon the body's mass: the heavier the body, the slower the oscillations will be. The same applies to the atoms in a molecule (Figure 9.7.1), where chemical bonds take the place of the spring: introducing a heavier isotope into a molecule of H_2O, for example, slows the thermal vibration (Chapter 1) that stretches the O–H bond, and thereby lowers the molecule's internal energy. The presence of a heavier isotope – whether ^2H in $^1H^2H^{16}O$ or ^{18}O in $^1H_2^{18}O$ – subtly alters the water molecule's thermodynamic and kinetic properties, and thereby affects its distribution between coexisting phases (between water and vapour, for instance), potentially leading to slight fractionation in the isotope ratio between the coexisting phases.

Such *mass-dependent isotope fractionation* is insignificant for heavier elements like Sr and Nd (at least in natural equilibria)[8] because the relative differences in mass number between the isotopes involved are minute (little more than 1% in the case of ^{86}Sr and ^{87}Sr). For lower-A elements like hydrogen, carbon and oxygen, however, the mass difference between isotopes assumes greater significance: the relative mass difference between ^{16}O and ^{18}O, for instance, is $(18-16)/[0.5 \times (18+16)] = 2/17 \approx 12\%$. Using modern high-precision mass spectrometry, measurable variations in the isotopic composition of the low-A elements H, C, N, O and S (Figure 10.9) can be observed between various natural reservoirs, fractionations which provide important insights into the workings of a range of Earth processes. This research field is known as *stable isotope geochemistry*.

Notation

Variations in a stable isotope ratio (e.g. $^{18}O/^{16}O$) seen in Nature mostly amount to only a few parts in a thousand. To optimize analytical precision and minimize inter-laboratory variations, the standard analytical practice is to alternate repeated measurements on an unknown sample with measurements on a universally available standard of known isotopic composition/ratio. The isotope ratio measurement for each sample

---

8   Sr and Nd isotopes are, however, susceptible to mass fractionation in certain artificial circumstances, such as in evaporation from a hot filament during thermal ionization mass spectrometry (for which a correction has to be applied).

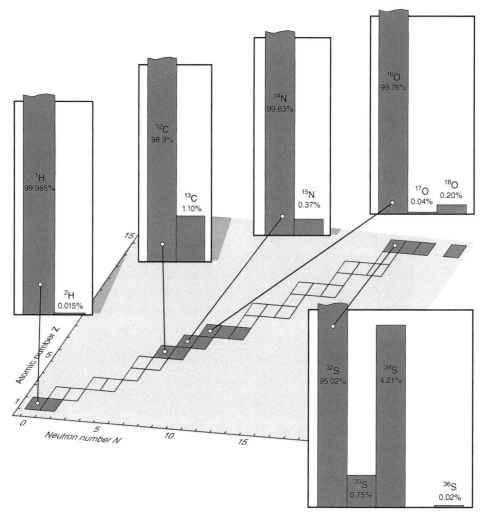

Figure 10.9 The light-element stable isotope systems of interest to geochemists, superimposed on the relevant part of the nuclide chart (cf. Figure 10.1.1); the histograms illustrate the average relative abundances of the minor isotopes.

is then expressed in terms of its 'δ-value'. Using oxygen isotopes as an example:

$$\delta^{18}O = 1000 \times \left[\frac{\left(\dfrac{^{18}O}{^{16}O}\right)_{sample} - \left(\dfrac{^{18}O}{^{16}O}\right)_{standard}}{\left(\dfrac{^{18}O}{^{16}O}\right)_{standard}} \right] \%o \qquad (10.8)$$

Here $\left(\dfrac{^{18}O}{^{16}O}\right)$ represents the measured atomic abundance ratio of the heaviest oxygen isotope to the lightest (and most abundant) one.[9] The symbol ‰ – pronounced

---

9 ^{17}O, the least abundant oxygen isotope, is not usually reported.

'per mil' by analogy with 'per cent' – stands for parts-per-thousand. The δ notation is used for all stable isotope systems. The standards used, on the other hand, vary from one isotope system to another (see Table 10.3).

Hydrogen and oxygen isotopes – keys to past climates

Hydrogen and oxygen are the elements that make up water. Each has more than one stable isotope, and natural variations in the isotopic composition of these two elements provide a means of tracing the origins of natural water samples, and of studying rock–water and mineral–water interactions.

Table 10.3 Summary of low-A stable isotope systems used in Earth and environmental science (see Figure 10.9)

Element	Isotopes	Isotope ratio used	Standard used	Applications*
Hydrogen	^1H, ^2H (=D)	$^2H/^1H = D/H$	VSMOW§	Hydrothermal water–rock interactions, water provenance (Figs. 10.10a,b), palaeoclimates (Figs. 10.11, 10.12), biochemical processes
Carbon	^{12}C, ^{13}C	$^{13}C/^{12}C$	VPDB¶	Composition of Earth's early atmosphere, detection of early life (Fig. 10.13), mantle heterogeneity and origins of diamonds
Nitrogen	^{14}N, ^{15}N	$^{15}N/^{14}N$	Atmospheric N_2 gas	Oceanic nitrate utilization, mixing of fresh and marine waters
Oxygen	^{16}O, ^{17}O, ^{18}O	$^{18}O/^{16}O$	VSMOW§ VPDB¶	Oceanic palaeotemperatures (Figs. 10.11b, 10.12), geothermometry, hydrothermal water–rock interaction, water provenance (Figure. 10.10)
Sulphur	^{32}S, ^{33}S, ^{34}S, ^{36}S	$^{34}S/^{32}S$	Troilite (FeS) from the Canyon Diablo iron meteorite	Origins of sulfide ores, Earth atmosphere evolution (Fig. 10.14)

*After Henderson and Henderson (2009).
§'Vienna Standard Mean Ocean Water' – despite its name, a pure water sample having specific D/H and $^{18}O/^{16}O$ abundance ratios, adopted by the International Atomic Energy Agency (IAEA) in Vienna in 1968.
¶'Vienna Peedee belemnite' is a similar artificial benchmark for $^{13}C/^{12}C$ adopted by the IAEA in 1985, based on belemnite fossil carbonate from the Peedee Formation in South Carolina.

Hydrogen consists of two stable isotopes, ^1H and ^2H (Figure 10.9). It is the only chemical element whose isotopes are distinguished by separate chemical names. ^2H – whose nucleus comprises one proton and one neutron – is known as deuterium (from the Greek *deuteros* meaning 'second'), and the chemical symbol D is sometimes used for it in place of ^2H. Hydrogen also has a third isotope ^3H – known as tritium (1 proton + 2 neutrons) – which is radioactive with a half-life of 12.3 years.

Oxygen consists of three stable isotopes, ^{16}O, ^{17}O and ^{18}O (Figure 10.9).

The terrestrial water cycle
The $H_2^{18}O$ molecule, being 12% heavier than $H_2^{16}O$, is slightly more difficult to evaporate: its vapour pressure at $100\,°C$ is 0.5% lower than that of $H_2^{16}O$, causing its boiling point to be $0.14\,°C$ higher. Vapour in equilibrium with water is therefore slightly deficient in both $H_2^{18}O$ and the other 'heavy' water molecule HDO ($^1H^2H^{16}O$) relative to the coexisting liquid phase. Atmospheric water vapour, being produced by evaporation of seawater, is thus measurably depleted in these heavier molecules. Furthermore, precipitation of

rain depletes the H_2O vapour remaining in the atmosphere still further (Figure 10.10a).

It follows that moist subtropical air masses, depositing rain as they migrate to higher latitudes and lower temperatures, experience progressive depletion in HDO and $H_2^{18}O$ (Figure 10.10a). Accordingly the isotopic composition of rain and snow (and fresh waters derived from them) is found to correlate strongly with latitude (Figure 10.10b). A similar trend may be seen with distance from the ocean towards continental interiors. On the other hand, equatorial freshwater bodies subject to high evaporation rates (such as rivers and lakes in East Africa) may experience *enrichment* in HDO and $H_2^{18}O$ as shown in Figure 10.10b.

H and O isotope ratios enable us to recognize three distinct categories of water that can be involved in geological reactions (Figure 10.10b):

(a) Seawater, with δD and $\delta^{18}O$ close to zero.
(b) Rain-derived (**meteoric**) surface- and groundwater, having variable (but correlated) negative δD and $\delta^{18}O$ values related to latitude of deposition ('GMWL', Figure 10.10b).

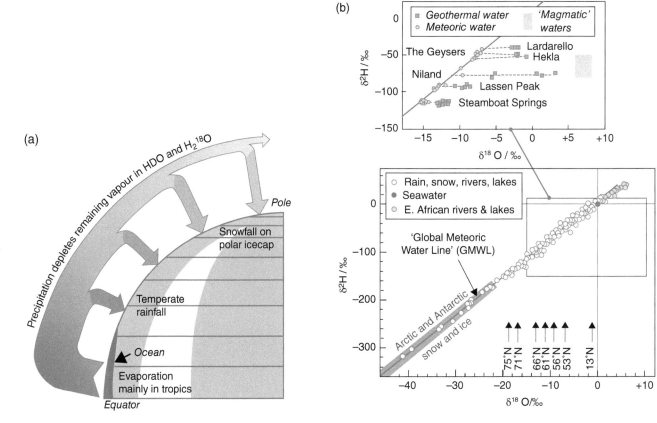

Figure 10.10 (a) Cartoon explaining why rain and snow become progressively more depleted in HDO and $H_2{}^{18}O$ with increasing latitude; downward arrows represent precipitation. (b) δD and $\delta^{18}O$ correlation in rivers, lakes, rain and snow, after Craig (1961). Upward arrows illustrate latitude variation in $\delta^{18}O$ (after Dansgaard, 1964). The enlarged panel (after Craig, 1963) shows how interaction with silicate rocks drives geothermal waters to higher $\delta^{18}O$ values while leaving the δD relatively unchanged, close to that of local meteoric waters from which the geothermal water has evidently been derived. The field of 'magmatic waters' is from Taylor (1974).

(c) Geothermal waters, which are usually meteoric waters that have equilibrated isotopically at high temperatures with silicate rocks. Silicates (in which oxygen is an essential constituent) constitute the dominant oxygen reservoir of the Earth's crust. Waters undergoing high-T isotopic exchange with silicates tend to assimilate their positive $\delta^{18}O$ values. Silicate rocks contain little hydrogen, on the other hand, so geothermal waters commonly retain the latitude-related, negative δD of the local meteoric waters from which they are derived. For this reason, many geothermal waters define subhorizontal arrays of increasing $\delta^{18}O$ lying to the right of the meteoric trend (Figure 10.10b, inset).

When large volumes of water are involved in a geological process, especially at elevated temperatures,

the water leaves its isotopic imprint on the rocks affected by that process. Stable isotope investigation of minerals in continental hydrothermal ore deposits, for example, often reveals negative $\delta^{18}O$ values, pointing to a meteoric origin for the hydrothermal fluids involved.

Stable-isotope palaeothermometry and palaeoclimatology

A calcite crystal growing in equilibrium with seawater at 25 °C is slightly enriched in ^{18}O relative to the water. This can be expressed as a *fractionation factor* α (a kind of **equilibrium constant**):

$$\alpha_{calcite/water} = \frac{\left(^{18}O/^{16}O\right)_{calcite}}{\left(^{18}O/^{16}O\right)_{water}} \qquad (10.9)$$

At 25 °C this fractionation factor has the value 1.0286, but $\alpha_{calcite/water}$ is found to vary significantly with the temperature of equilibration. Because of this useful fact, the $\delta^{18}O$ values of ancient marine carbonates can be used to measure temperatures of deposition in the geological past ('palaeotemperatures'). The calibration equation in terms of $\delta^{18}O$ values can be written:

$$T^\circ C = 16.5 - 4.3 \times \left[\delta^{18}O_{calcite} - \delta^{18}O_{water} \right]$$
$$+ 0.13 \times \left[\delta^{18}O_{calcite} - \delta^{18}O_{water} \right]^2 \quad (10.10)$$

where $\delta^{18}O_{calcite}$ represents the isotopic composition of the CO_2 extracted from the fossil calcite, and $\delta^{18}O_{water}$ represents the composition of the waters from which the fossil shell was deposited (relative to PDB and SMOW standards respectively – Table 10.3).

The very first example of such a study is even now – 60 years on – the most elegant. Urey *et al.* (1951) drilled out tiny samples across a cut section of an inch-wide Jurassic belemnite guard from the Isle of Skye (Figure 10.11a), and measured the $^{18}O/^{16}O$ ratio of each microsample. Using a thermodynamically derived temperature calibration, they were able to show that the creature lived through four summers and four winters (Figure 10.11b), during which time sea temperature varied seasonally between 15° and 20 °C, superimposed upon a cooling **secular** trend.

The preservation of such detailed temperature information within a single belemnite shell for 150 Ma suggests that carbonate sedimentary successions elsewhere might prove to be a valuable repository of palaeoclimate data. This is indeed the case (Zachos *et al.*, 2001), although the need to establish the $^{18}O/^{16}O$ ratio of the waters from which the carbonates were deposited – in order to calculate the temperature in Equation 10.10 – introduces an element of ambiguity. Recognizing this problem, Urey *et al.* (1951) had argued that the Jurassic sea from which their belemnite grew its calcite guard had the same oxygen isotope ratio as today's oceans ($\delta^{18}O = 0.0\permil$). Yet we now know that – as climate fluctuates over time – seawater $^{18}O/^{16}O$ also varies: global warming causes polar ice with $\delta^{18}O$ values as low as $-50\permil$ (Figure 10.10b) to melt and mix with the oceans, thereby lowering mean ocean $\delta^{18}O$, while the locking-up of water in new polar ice during cold periods has the opposite effect on the oceans. This ice-melt effect causes $\delta^{18}O_{seawater}$ to vary by $\sim 1\permil$ between glacial

maxima and interglacial periods. The $^{18}O/^{16}O$ record in dated carbonate sediment cores can in principle be used to monitor both past changes in sea surface temperature *and* variations in the volumes of polar ice through geological time, although the details of how these two factors are disentangled lie beyond the scope of this chapter.

Antarctic and Greenland ice cores provide an alternative isotopic record of late Pleistocene climate change. Both δD and $\delta^{18}O$ in polar ice vary according to the temperature at which the original snow was

(a)

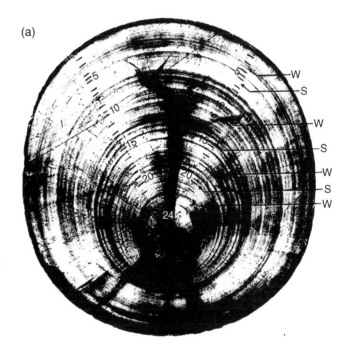

(b)

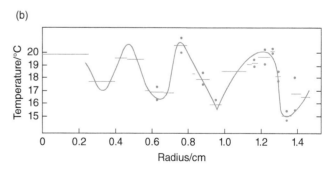

Figure 10.11 (a) Cross-section of a Jurassic belemnite guard showing winter (W) and summer (S) growth rings and microsampling locations. (b) The measured oxygen-isotope temperature profile showing seasonal variations during the life of the belemnite. (Source: Urey *et al.* (1951). Reproduced with permission of the Geological Society of America.)

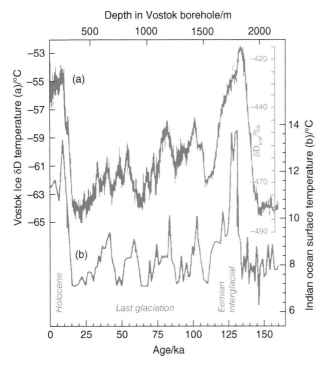

Figure 10.12 (a) Variation in mean annual surface temperature at Vostok Station in East Antarctica over the past 160,000 years (left-hand scale), based on δD measurements on Vostok deep ice cores; the orange scale on the right indicates the corresponding δD values (b) Variation over the same period of summer sea-surface temperature (scale on far right), based on statistical analysis of radiolarian assemblages in borehole RC11-120 in the southern Indian Ocean. (Source: Adapted from Jouzel *et al.*, 1987. Reproduced with permission of Nature Publishing Group; Data from Martinson *et al.*, 1988.)

precipitated, and systematic sampling of ice cores drilled from ice caps allows an isotopic record of Quaternary climate variation to be assembled. The age-calibrated Vostok ice cores from eastern Antarctica have been shown to provide a continuous record of precipitation extending over the past 420,000 years (Petit *et al.*, 1999). Figure 10.12a shows how δD values and calculated mean air temperatures at Vostok have varied over the last 160,000 years. A trace of sea-surface temperatures in the subpolar Indian Ocean (based on diatom populations) over the same period is shown for comparison (Figure 10.12b); although covering very different latitudes and temperature ranges, the two traces show the same glacial–interglacial climate variations. Both are important isotopic proxies that contribute to our knowledge of natural global climate change.

$\delta^{18}O$ can also be used to measure much higher temperatures of equilibration between minerals and igneous melts, although with lower precision.

Carbon stable isotopes – detecting signs of ancient life

Carbon has two stable isotopes, ^{12}C and ^{13}C (Figure 10.9). Owing to the 8% difference in atomic mass between them, geochemical reactions discriminate to a small extent between the two isotopes. Dissolved carbon dioxide and carbonate sediments in the oceans contain a higher $^{13}C/^{12}C$ isotope ratio (by 5–10‰) than atmospheric CO_2 (Figure 10.13).

Organic matter, on the other hand, is strongly depleted in ^{13}C, leading to $\delta^{13}C$ values – in marine phytoplankton, for instance – that are about 20‰ lower than atmospheric carbon dioxide. This reflects a remarkable isotopic fractionation during photosynthesis that can be traced to a key enzyme known as Rubisco,[10] the most abundant protein in green leaves. It is Rubisco that fixes the atmospheric CO_2 absorbed during photosynthesis, and in doing so it exerts a marked preference for ^{12}C. This fact makes $\delta^{13}C$ an invaluable tracer for photosynthesis in the geological record, for instance in detecting the beginnings of photosynthetic life in ancient sedimentary successions. This can be illustrated by reference to Figure 10.13.

Carbon from atmospheric CO_2 is locked up in sedimentary rocks in two forms:

- inorganic carbonate which – although it resides in the shells of living creatures – is broadly in isotopic equilibrium with atmospheric CO_2 with $\delta^{13}C$ values straddling zero (Figure 10.13);
- reduced 'organic' carbon (in forms such as coal and oil) derived from decomposed living soft tissue. Since biological carbon – even in animals – is ultimately derived from the atmosphere by photosynthesis, all organic carbon in sediments carries the negative $\delta^{13}C$ 'Rubisco signature'.

Over much of the geological record, the marine carbonate $\delta^{13}C$ values shown in Figure 10.13 maintain

---

[10] Abbreviation of the enzyme's full name: Ribulose-1,5-bisphosphate carboxylase oxygenase.

(despite their scatter) a level distribution broadly lying within the range of present-day carbonate sediments. There were two intervals of Earth history, however, when marine carbonate isotope compositions underwent wild excursions – both positive and negative – from this steady-state trend. Each episode is believed to represent an explosion of photosynthesis, when a blooming of biota across the globe drew down a significant proportion of the atmosphere's CO_2 and, upon death, these organisms deposited their remains as *reduced* carbon in sediments accumulating on the ocean floor. This 'organic carbon' locked up in sedimentary rocks has the distinctive negative $\delta^{13}C$ values that are the fingerprint of Rubisco's role in photosynthesis and its preference for ^{12}C. The removal of this negative $\delta^{13}C$ carbon left the complementary atmospheric CO_2 reservoir enriched in ^{13}C during these episodes, and by exchange with the oceans this led to the deposition of the high-$\delta^{13}C$ carbonate sediments shown in Figure 10.13.

The older of these two excursions – at the beginning of the Proterozoic eon – marked the first sustained global appearance of O_2 in the Earth's atmosphere, which up to this point had consisted only of gases like H_2O, CO_2, CO and H_2 (see Chapter 11). Oxygen's appearance had profound consequences, and accordingly this first period of $\delta^{13}C$ excursions is known as the *Great Oxidation Event* (Figure 10.13). Although the atmospheric O_2 concentration generated at this time was far lower than today's level (Figure 11.8), the draw-down of CO_2 required to release this photosynthetic oxygen reduced the natural atmospheric 'greenhouse effect' sufficiently to bring about major worldwide glaciations (a phenomenon known loosely as *Snowball Earth*). The wild fluctuations in $\delta^{13}C$ during this episode (Figure 10.13) reflect alternation of (i) periods in which photosynthesis and organic matter deposition predominated (*raising* atmospheric $\delta^{13}C$), with (ii) colder periods in which collapsing biological productivity allowed oxidation of organic matter to exceed the rate of carbon burial, *depressing* atmospheric $\delta^{13}C$ values (Kump *et al.*, 2011).

The second excursion (between 850 and 500 Ma ago) coincided with the time when photosynthesizing life first colonized the land surface towards the end of the Neoproterozoic era, initiating a steeper growth in atmospheric O_2 content towards the breathable range we depend on today (Figure 11.8).

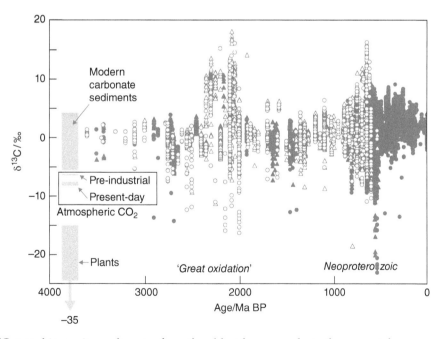

Figure 10.13 The $\delta^{13}C$ record in marine carbonates from the oldest-known rocks to the present day; open symbols signify samples with less precise ages (uncertainty >50 Ma). The shaded bars on the left indicate the isotopic ranges of *present-day* carbon reservoirs; 'pre-industrial' shows the isotope composition of atmospheric CO_2 prior to the industrial revolution. (Source: Adapted from Shields and Veizer, 2002. Reproduced with permission of the American Geophysical Union.)

Mass-independent fractionation of sulfur isotopes

Since oxygen and sulfur both have more than two stable isotopes, we can in principle formulate several δ values ($\delta^{17}O$, $\delta^{18}O$, $\delta^{33}S$, $\delta^{34}S$, $\delta^{36}S$), although normally only those involving the two more abundant isotopes of each element ($\delta^{18}O$, $\delta^{34}S$) are measured. Most natural processes fractionate oxygen and sulfur isotopes in a predictable, *mass-dependent* way. In other words, we would expect the $\delta^{17}O$ value of a sample to be roughly half of its $\delta^{18}O$, since the mass difference relative to ^{16}O is halved ($17 - 16$ compared to $18 - 16$) and the degree of fractionation follows suit. This expectation is borne out by measurements on a range of meteoritic and terrestrial materials:

$$\delta^{17}O \approx 0.52 \times \delta^{18}O \tag{10.11}$$

$$\delta^{33}S \approx 0.515 \times \delta^{34}S \tag{10.12}$$

Do circumstances occur in which this relationship breaks down? One conspicuous example where it has is illustrated in Figure 10.14, which shows – in a time plot similar to Figure 10.13 – how the sulfur isotope composition of sedimentary sulfides and sulfates has varied throughout Earth history. The *y*-axis shows the departure of measured $\delta^{33}S$ from the mass-dependent value expected from Equation 10.12.

Samples younger than 2200 Ma (darkest circles) have $\Delta^{33}S$ values very close to zero, indicating normal mass-dependent fractionation, whereas samples older than 2200 Ma (intermediate circles) deviate to positive values up to 2, and Archaean samples (lightest circles) vary erratically over a much wider range. Prior to 2450 Ma, when atmospheric oxygen levels were negligible, no stratospheric ozone (O_3) layer could form, and therefore – unlike today – the atmosphere and the Earth's surface were exposed to the full force of incoming solar UV radiation. Under such conditions, photochemical reactions fractionate sulfur isotopes in a more chaotic *mass-independent* manner in which Equation 10.12 no longer holds. The transition from a *mass-independent* (kinetic) *fractionation of sulfur* ('MIF-S') signal in Archaean times to mass-dependent (equilibrium) fractionation ($\Delta^{33}S \sim 0.0$) in younger samples provides further evidence for the first emergence of significant atmospheric oxygen (some of which formed ozone) around the close of the Archaean eon (cf. Figure 11.8).

Transition metal stable isotopes

The remarkable capacity of living organisms to fractionate isotopes is not confined to low-mass isotope systems like $^{13}C/^{12}C$. Figure 10.15 shows how $^{66}Zn/^{64}Zn$

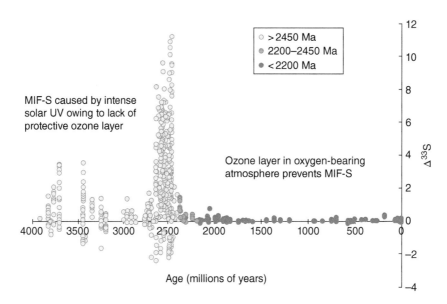

Figure 10.14 Plot of $\Delta^{33}S = \delta^{33}S_{meas} - 0.515 \times \delta^{34}S_{meas}$ (cf. Equation 10.13) in sedimentary sulfides and sulfates as a function of age from 4 Ga to the present. (Source: Reproduced with permission of David Johnston of Harvard University.)

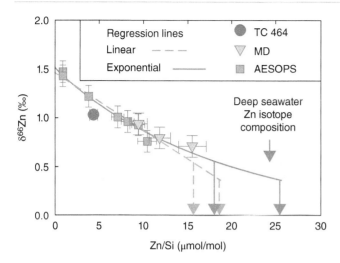

Figure 10.15 Correlation between $\delta^{66}Zn$ and Zn/Si in cleaned siliceous marine diatom frustules (hard parts) in core tops from the Southern Ocean TC464, MD and AESOPS refer to drill cores sampled.

$$\delta^{66}Zn = \left[\frac{\left(^{66}Zn / ^{64}Zn \right)_{sample}}{\left(^{66}Zn / ^{64}Zn \right)_{NIST683}} - 1 \right] \times 1000\,‰$$

where NIST683 is the isotopic standard used. The Zn/Si ratio is given in μmoles Zn per mole Si (= molar ppm). (Source: Andersen *et al.* (2010), reproduced with permission of Elsevier.)

varies in marine diatom skeletal material as a function of Zn availability in the surface waters from which the diatoms grow (represented here by Zn/Si ratios in diatom frustules). Zn is a **micronutrient** which, when in short supply, limits biological productivity. Owing to its take-up by phytoplankton in the photic zone, the element Zn is depleted to varying degrees in surface seawater relative to deep ocean water. Phytoplankton selectively incorporate the lighter isotope ^{64}Zn into their tissues, leaving the remaining dissolved Zn enriched in ^{66}Zn. Increased phytoplankton productivity leads to greater depletion of dissolved Zn (reflected by lower Zn/Si in skeletal material) and depletion in ^{64}Zn relative to ^{66}Zn, resulting in higher $\delta^{66}Zn$ in surface seawater. Diatom skeletal parts grown from Zn-depleted, high $\delta^{66}Zn$ surface waters reflect this correlation (Figure 10.15).

The isotopic compositions of Fe, Cu and Mo also find palaeo-oceanographic applications, providing information on oxygenation and metal cycling in ancient oceans (Anbar and Rouxel, 2007).

Cosmogenic radioisotope systems

Radiocarbon dating

The short-lived radioactive carbon isotope ^{14}C provides an important dating tool in archaeology and Quaternary geology. Terrestrial ^{14}C (half-life 5730 years) is formed by the bombardment of ^{14}N nuclei in the atmosphere by neutrons derived from cosmic rays (^{14}C is the best known example of a **cosmogenic** nuclide). The ^{14}C atoms that are formed rapidly oxidize, becoming part of the atmospheric inventory of CO_2, and are incorporated into plant tissue by photosynthesis. Living organisms (plants and animals) maintain a carbon-isotope composition in equilibrium with atmospheric CO_2 during their lifetime, but this steady state ceases when an organism dies: the ^{14}C trapped in the dead organic matter decays back to ^{14}N with time, providing an isotopic clock.

In **radiocarbon dating**, as this technique is called, the minute amount of ^{14}C remaining in a sample[11] is determined either by measuring its $\hat{\beta}$ activity, or by counting the number of ^{14}C atoms individually using an ultra-sensitive **accelerator mass spectrometer**. The latter technique extends the range of radiocarbon dating back to at least 40,000 years. The age is calculated using Equation 10.13:

$$t = \frac{1}{\lambda_C} \ln \left[\frac{\left(^{14}C / ^{12}C \right)_0}{\left(^{14}C / ^{12}C \right)_{sample}} \right] \qquad (10.13)$$

where t represents the age of the carbon sample in years, λ_C is the decay constant for ^{14}C in year^{-1} (see exercise 10.3), $^{14}C/^{12}C$ represents the atomic ratio of ^{14}C to ^{12}C in the sample (or the ^{14}C count rate per gram), and $(^{14}C/^{12}C)_0$ represents the corresponding ratio or count rate at the time the artefact or rock formed (we can approximate this by using the value for present-day carbon).

From a geological perspective, the radiocarbon method dates younger events (hundreds of years to tens of thousands of years ago) than Ar – Ar and Rb – Sr dating methods can routinely measure. It can only be applied to artefacts or rocks that incorporate carbon of biological origin.

Another cosmogenic isotope of geological interest is beryllium-10 (^{10}Be), which has a half-life of 1.39 Ma. It is

---

11 The proportion of ^{14}C to ^{12}C is typically of the order of 1 part in 10^{12}.

produced in the atmosphere by cosmic ray spallation of N and O nuclei, from where it is washed out by rain into the oceans and becomes incorporated into sediment. Its detection in young island-arc volcanic rocks is important evidence for the subduction of significant amounts of oceanic sediment down subduction zones, since its short half-life rules out any other source within the Earth.

Review

Although some elements are mono-isotopic in nature (see exercise 11.1), most consist of two or more isotopes (Box 10.1). High-precision mass spectrometry (Box 10.3) measurements of isotope abundance ratios of selected elements provide opportunities to quantify a host of geological processes:

- *Radiogenic isotope* systems like K–Ar (including $^{40}Ar/^{39}Ar$ – Box 10.2), Rb–Sr, Sm–Nd and U–Th–Pb give us tools for determining the *ages* of rocks, rock associations and meteorites (Box 10.2, Figures 10.5 and 10.7).
- In addition to geochronology, radiogenic isotope systems – illustrated in the text by Rb–Sr and Sm–Nd, although Lu–Hf, Re–Os and U–Th–Pb are important too (Table 10.1) – shed light on the *origins* of igneous magmas (Figure 10.8), on *mantle heterogeneity* (Figure 10.8), on *crustal evolution*, and on *sediment provenance*.
- Measurements of $\delta^{18}O$ (Equation 10.8) in rocks and minerals allow us to estimate the *temperature of crystallization* of relevant minerals, such as marine carbonates (Figure 10.11). $\delta^{18}O$ and δD together provide important quantitative information on *past climates* (Figure 10.12). $\delta^{18}O$ and δD measurements also provide insights into the terrestrial *water cycle* (Figure 10.10).
- Other *stable isotope* systems provide avenues for quantifying other geological and Earth-surface phenomena (Table 10.3). $\delta^{13}C$ and $\delta^{34}S$ data document the dramatic changes in the composition of the Earth's atmosphere through geological time (Figures 10.13 and 10.14), notably the 'Great Oxidation Event' marking the first appearance of a significant level of oxygen in the global atmosphere at the close of the Archaean, and its rise to higher concentrations with the appearance of land plants in the Neoproterozoic (see Figure 11.8).

- Recent work suggests that the isotopic compositions of *transition metals* like Fe, Cu, Zn (Figure 10.15) and Mo have much to tell us about the history of the Earth's oceans.
- *Cosmogenic radioisotope* systems include ^{14}C, which provides the foundation for radiocarbon dating, and ^{10}Be, which allows estimation of the role of sediment subduction in island arc magma genesis.

Further reading

Allegre, C. (2008) *Isotope Geology*. Cambridge: Cambridge University Press.

Dickin, A.P. (2005) *Radiogenic Isotope Geology*, 2nd edition. Cambridge: Cambridge University Press.

Faure, G. and Mensing, T.M. (2005) *Isotopes –Principles and Applications*, 3rd edition. New York: John Wiley and Sons, Ltd.

Lenton, T. and Watson, A. (2011) *Revolutions that Made the Earth*. Oxford: Oxford University Press.

White, W.M. (2013) *Geochemistry*. Chichester: Wiley-Blackwell.

Exercises

10.1 (a) By examining Figure 10.1.1 and the Periodic Table, list the elements that are monoisotopic (i.e. have only one stable isotope) in Nature.

(b) Which chemical elements have no stable isotopes?

10.2 A Mesozoic limestone from an oceanic drill core is found to have an $^{87}Sr/^{86}Sr$ ratio of 0.707980 ± 0.000022. Determine the age of the sediment, and estimate the precision of the age determination.

10.3 The table below gives Sm-Nd isotopic data for the separated minerals and whole-rock specimen for a gabbro from the Stillwater Intrusion in Montana. Plot a mineral isochron for these samples and, using the constants given in Table 10.1, determine the age and initial Nd isotope ratio of the intrusion.

Sample	$^{147}Sm/^{144}Nd$	$^{143}Nd/^{144}Nd$
STL-100 whole rock	0.20034	0.511814
STL-100 plagioclase	0.09627	0.509965
STL-100 orthopyroxene	0.28428	0.513317
STL-100 clinopyroxene	0.24589	0.512628

Data from DePaolo and Wasserburg (1979).

10.4 The half-life of ^{14}C is 5730 years; calculate its decay constant. Charcoal from charred tree remains preserved in a volcanic pyroclastic flow shows an average β^- count rate of 5.8 disintegrations per minute (dpm) per gram of sample, whereas carbon extracted from living vegetation today produces a ^{14}C count rate of 13.56 dpm per gram. Calculate the age of the volcanic deposit.

10.5 Analysis of a Neogene[12] marine limestone core gives a $\delta^{18}O_{calcite}$ value of -1.3 ± 0.1‰. Calculate (including $\pm$ analytical uncertainty) the temperature of the sea from which it was deposited, assuming seawater isotopic composition at the time of deposition was the same as today. How much additional uncertainty in the temperature value arises from not knowing the actual seawater $\delta^{18}O$ at that time?

10.6 Spot the incorrect statements among the following, and explain how they are wrong:
(a) Every ^{87}Rb nucleus that decays becomes an ^{87}Sr nucleus.
(b) Every ^{87}Sr atom in a rock is the product of the decay of an ^{87}Rb nucleus in that rock.
(c) ^{87}Sr/^{86}Sr grows more rapidly in continental crust than in the mantle.
(d) ^{143}Nd/^{144}Nd grows more rapidly in continental crust than in the mantle.
(e) Radiocarbon provides an important dating technique for all artefacts and rocks.
(f) The fractionation of stable isotopes in Nature is always mass-dependent.

[12] Neogene refers to Miocene and Pliocene epochs lasting from 22 Ma to 1 Ma ago, during which period Antarctic ice sheets had become established.

11 THE ELEMENTS IN THE UNIVERSE

The significance of element abundance

Having considered the behaviour of some important elements in the lithosphere, hydrosphere and atmosphere (Chapters 4, 8 and 9) and their isotopes (Chapter 10), it is natural to ask the question: how were the chemical elements formed in the first place? Have they always existed at their present levels, or has there been a progressive building up of the cosmic inventory through the life of the universe? Can we identify the process(es) by which they were formed?

Current opinion favours the gradual synthesis of heavier elements from lighter ones, by a complex series of nuclear fusion reactions occurring in stars. This process, called *stellar nucleosynthesis*, leaves its fingerprint on all forms of cosmic matter. We can learn how this works by studying the relative abundances of the elements in the universe as a whole, or in some representative part of it.

A second question to be addressed in this chapter is how the Earth has come to have its present composition and structure. In seeking an answer to this problem, we again consider element abundances, noting this time how they differ between the Sun, the Earth and the other planets. Such differences provide clues to the kind of chemical processing that has produced the Solar System and the Earth in their present form.

Measuring cosmic and Solar System abundances

Our knowledge of the overall composition of visible matter[1] in the Universe rests mainly on two kinds of analysis:

---

[1] Invisible (including so-called 'dark') matter will be discussed a little later.

Chemical Fundamentals of Geology and Environmental Geoscience, Third Edition. Robin Gill.
© 2015 John Wiley & Sons, Ltd. Published 2015 by John Wiley & Sons, Ltd.
Companion Website: www.wiley.com/go/gill/chemicalfundamentals

(a) spectral analysis of the light received by telescopes from stars, including the Sun, and from other radiant bodies such as nebulae (gas clouds);

(b) laboratory chemical analysis of meteorites, which represent the solid constituents of the Solar System.

Spectral analysis

Stars are intensely hot nuclear fusion reactors. They derive their high temperatures and radiant output from the energy released when light nuclei (such as ^1H and ^2H) fuse together into more stable, heavier nuclei (such as ^3He). The theory of this **thermonuclear** process is outlined in Box 11.2.

In common with any very hot body – such as a red-hot poker or the filament of an incandescent light bulb – the hot surface of a star radiates light consisting of a continuum of wavelengths (the 'white light' we receive from the Sun). Superimposed on this smooth electromagnetic spectrum are dark lines that constitute the *absorption spectra* (see Chapter 6) of elements present in the cooler outer atmosphere of the star (see Plate 4). Each line is characteristic of a specific transition in a specific chemical element. From the wavelengths and intensities of these absorption lines (called **Fraunhofer lines**),[2] an astronomer can establish the identity and abundance of most elements present in the star. The 'calibration' factors used to translate absorption line intensity into element abundance have to be estimated theoretically, yet astronomers are confident that the abundance data available today, for about 70 elements in a great many stars, are mostly accurate to within a factor of two. As element abundances vary by as much as 10^{12} times (Figure 11.2), such uncertainties are tolerably small. Stars progressively alter their internal element abundances through the nucleosynthesis taking place deep inside them, but the cool outer envelope of a star (where the Fraunhofer lines originate) is thought to remain representative of the material from which the star originally accreted.

In discussing the Solar System, we shall be concerned with abundances in our nearby sun rather than stars in general.

Analysis of meteorites

As anyone knows who has slept in the open on a clear starry night, shooting stars are a common phenomenon. They are the visible manifestation of many tons of Solar-System debris that fall to Earth each day.[3] Smaller infalling bodies may be vaporized completely in the atmosphere by frictional heating, but about 1% of them are large enough to survive and reach the Earth's surface as recoverable **meteorites**.

A simplified classification of meteorites is shown in Box 11.1.

Primitive meteorites

The commonest meteorite type (see Box 11.1) is the *chondrites*, so called because most of them contain *chondrules* (millimetre-sized spheroidal assemblages of crystals and glass – see Plate 5). Chondrules[4] are considered to be solidified droplets of melt formed by impact melting of dust in the early protoplanetary disk. Despite the high temperatures implied by chondrule melting (typically 1500 °C), it is commonly the case that minerals in the silicate–metal-sulfide matrix *surrounding* the chondrules (Plate 5) have highly variable chemical compositions, indicating that they have never achieved chemical equilibrium with each other or with the chondrule minerals. The lack of equilibrium suggests low ambient temperatures during and after accretion of some chondrites. Such 'unequilibrated chondrites' preserve various chemical, mineralogical, and structural signatures inherited from the protoplanetary disk.

Chondrites of one particular class, known as *carbonaceous chondrites,* also contain a complex, tarry organic component and various hydrous silicate minerals. The very limited thermal stability of these components suggests that carbonaceous chondrites have suffered the least thermal and chemical processing of any chondrite group. They are described as being chemically **primitive**; one particular subgroup known as *CI chondrites*[5]

---

[2] After Joseph von Fraunhofer, a German physicist who made the first systematic study of them.

[3] The infall rate of meterorites to the Earth's surface is about 40,000 tons per year.

[4] From the Greek *chondros* for 'grain', a reference to their rounded shape.

[5] An abbreviation of 'carbonaceous Ivuna' named after the locality in Tanzania from which the type example was recovered. Paradoxically, despite being classified as chondrites on chemical grounds, CI chondrites actually contain no chondrules.

Box 11.1 Types of meteorite, shown in proportion to percentage of total falls[1]

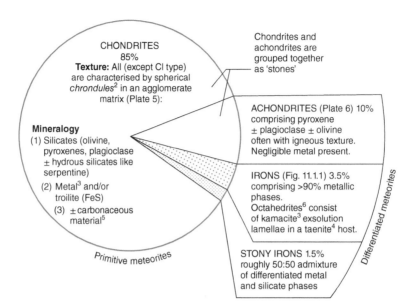

CHONDRITES
85%
Texture: All (except CI type) are characterised by spherical *chrondules*[2] in an agglomerate matrix (Plate 5):

Mineralogy
(1) Silicates (olivine, pyroxenes, plagioclase ± hydrous silicates like serpentine)
(2) Metal[3] and/or troilite (FeS)
(3) ± carbonaceous material[5]

Primitive meteorites

Chondrites and achondrites are grouped together as 'stones'

ACHONDRITES (Plate 6) 10% comprising pyroxene ± plagioclase ± olivine often with igneous texture. Negligible metal present.

IRONS (Fig. 11.1.1) 3.5% comprising >90% metallic phases. Octahedrites[6] consist of kamacite[3] exsolution lamellae in a taenite[4] host.

STONY IRONS 1.5% roughly 50:50 admixture of differentiated metal and silicate phases

Differentiated meteorites

1 'Falls' are meteorite samples that have been seen to fall, as opposed to those just discovered on the ground ('finds'). Relative numbers of 'falls' give the best estimate of the relative abundance of the different meteorite types.

2 Chondrules (see Plate 5) are millimetre-sized spheroidal clusters of crystals and glassy material, thought to have originated as molten droplets in the presence of a gas (Lauretta *et al.*, 2006). Crystals often show evidence of having been rapidly chilled.

3 Kamacite $\sim Fe_{95}Ni_5$.

4 Taenite, typically with a composition of $Fe_{60}Ni_{40}$ (see Figure 3.5.1 in Box 3.5).

5 The carbonaceous component characteristic of the *carbonaceous chondrite* group is a complex tarry mixture of abiogenic organic compounds.

6 So called because, owing to the cubic symmetry of the taenite host, the kamacite plates have four orientations parallel to the faces of a regular octahedron. This distinctive texture, most apparent in a flat surface that has been etched with acid (Figure 11.1.1), is called Widmanstätten structure after the Austrian count who discovered it in 1808.

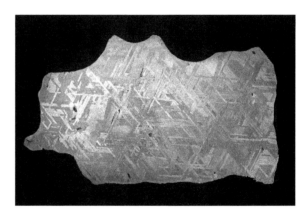

Figure 11.1.1 The acid-etched surface of an octahedrite[6] iron meteorite (Canyon Diablo iron), exposing the internal Widmanstätten texture of kamacite[3] lamellae in a taenite[4] host. The specimen is 15 cm in length. (Source: © The Natural History Museum, London. Reproduced with permission.)

is regarded as relatively pristine relics of the primordial solid matter of the early Solar System, and from their chemistry we can estimate its overall composition (except, of course, for the most volatile elements and compounds, which may have been expelled by high surface temperatures generated during descent through the Earth's atmosphere).

Differentiated meteorites

Meteorites other than chondrites are products of the segregation of metal from silicate (forming 'irons' and achondrites[6] respectively), and are described as **differentiated meteorites** (Box 11.1). It is generally assumed that this segregation is a consequence of incorporation into small planetary bodies, perhaps a few hundred kilometres across (Box 3.5), in which high internal temperatures facilitated gravitational segregation of the two phases, as indeed happened with the separation of the metallic core of our own planet.

Differentiated meteorites – the result of such bodies being broken up by later collisions – are poor representatives of primordial matter, but shed light on the internal development of the rocky planets. A few differentiated meteorites have chemical signatures indicating a Martian (e.g. the Nahkla achondrite illustrated in Plate 6) or a lunar origin.

Dark matter

Current estimates lead us to believe that there is almost 100 times more dark matter (in the universe) than visible matter. (Ferreira, 2006)

So far in this chapter, we have assumed that all matter in the universe consists of atoms of the chemical elements that are familiar to us on Earth, that is to say, consisting ultimately of protons, neutrons and electrons. Protons and neutrons in the nucleus, making up nearly all of the atom's mass, are known collectively as **baryons**,[7] so matter of this familiar atomic kind is commonly referred to as *baryonic matter*. It reveals its existence across the universe by emitting or absorbing recognizable atomic spectra (Chapter 6), and thus its abundance can be estimated from Earth

by astronomical observation. From such measurements, astronomers are able today to estimate, within quite narrow limits, the overall density of visible baryonic matter in our galaxy and in the universe as a whole, at least those parts whose emissions are measurable.

The mass of baryonic matter in our galaxy calculated in this way, however, contrasts starkly with estimates of *overall* galactic mass that we get from gravity.

How can galactic mass be estimated from gravity? Planets in our Solar System orbit the Sun at velocities inversely proportional to the square root of their distance from the Sun: those close to the Sun orbit more quickly (and have shorter years) than those further out. Planets close to the Sun experience stronger gravity, so must have higher orbital velocities (greater 'centrifugal force') to resist this stronger attraction and remain in orbit. This regularity in orbital velocity is a consequence of nearly all of the Solar System's mass being concentrated at its centre, in the Sun itself.

If, on the other hand, we measure the analogous velocities of our Sun and other stars as they orbit around the centre of the Milky Way (our galaxy), we find that:

(a) stellar velocities, unlike orbiting planets, vary little between stars closer to, and further from, the centre of the galaxy; and

(b) stars move much faster than the mass of visible matter in the galaxy would lead us to expect.

From these discrepancies, astrophysicists conclude that the galaxy *must contain much more mass than we can directly see*. To account for (a) above, they envisage a *dispersed halo of invisible mass*, extending to the outer reaches of the Milky Way, whose gravity exerts the dominant control on the motions of the visible stars in the galaxy. Astrophysicists refer to this invisible matter as *dark matter*, and believe that it far outweighs the visible matter in the Universe.

What might this invisible 'dark matter' consist of? There are reasons to believe that perhaps 5–10% of it consists of baryonic matter (i.e. familiar chemical elements) that just happens to remain dark – 'ordinary stuff that doesn't shine' as Ferreira (2006) puts it. It happens that celestial bodies with masses less than about 8% of the solar mass (~80 Jupiter masses) are unable to ignite the nuclear fusion reactions that power

---

[6] Achondrites contain negligible metallic iron.
[7] From the Greek *barus* meaning heavy.

the Sun's radiance, and will therefore stay dark; such substellar bodies, intermediate in mass between Jupiter-like planets and true stars, include the ultra-dense bodies known as 'brown dwarfs'.

The dark matter making up the other 90–95% of the mass of the universe, however, is much more exotic in character and its nature is arguably the biggest unanswered question in cosmology. Many bizarre novel subnuclear particles have been proposed as the constituents of non-baryonic dark matter, but this cosmological conundrum is still a long way from being solved. Ferreira (2006) provides a readable non-specialist summary of our current understanding.

Although the nature of dark matter is a profoundly important question for the cosmologist, it falls outside the scope of this book. Dark matter, whatever its nature, seemingly plays no significant part in the formation or composition of planets such as ours. It is of course important to acknowledge that, when we generalize grandly about the composition of matter in the universe, we are referring solely to the 'visible' baryonic component that may account for as little as 1% of the mass of the whole. Yet it is this 1% that determines the character and composition of the planet we inhabit.

The composite abundance curve

The two sources of information outlined above – solar spectra and analyses of primitive meteorites – allow us to build a composite picture of the relative abundances of the chemical elements in the Solar System as a whole. Gaseous elements – hydrogen, the inert gases, and so on – can of course be determined only from solar measurements; for other elements, like boron, spectral measurements are difficult or impossible and reliance on meteorite data is the only feasible course. Fortunately the abundance of most other elements can be determined by both methods. Since the two approaches involve different assumptions and employ different instrumental techniques, it is reassuring to find a good correlation between them (Figure 11.1).

The composite abundance data so obtained have been plotted against atomic number (Z) in Figure 11.2. Although compiled specifically for the Solar System,

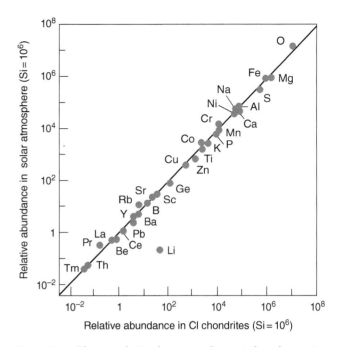

Figure 11.1 The correlation between element abundances in the Sun and in CI carbonaceous chondrites. Abundance is expressed as the number of atoms of each element per 10^6 atoms of silicon (both axes have **logarithm**ic scales).

the key features of this 'abundance curve' are common to practically all stars and luminous nebulae:

(a) Hydrogen and helium are several orders of magnitude more abundant than any other element. In atomic terms, helium has one-tenth of the abundance of hydrogen and together they comprise 98% of the Solar System's mass.

(b) Progressing to higher atomic numbers leads to an overall decrease in abundance, making the heaviest nuclei among the least abundant.

(c) The elements lithium, beryllium and boron are sharply depleted compared with the other light elements. (In the case of Li this depletion is much more marked for the Sun than for CI chondrites – Figure 11.1.)

(d) Elements having even atomic numbers (Z) are on average about ten times more abundant than neighbouring elements having odd atomic numbers. This effect, which is apparent in terrestrial rocks as well, produces a 'sawtooth' profile if adjacent atomic numbers are joined up (see inset showing **REE** abundances in Figure 11.2).

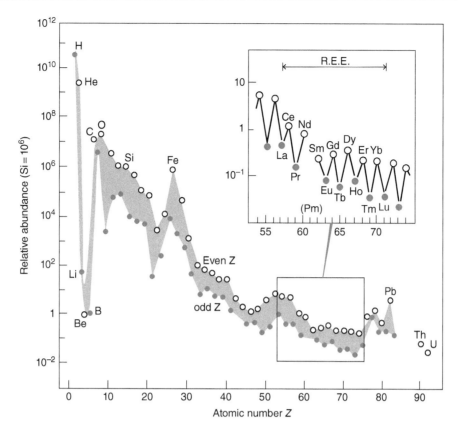

Figure 11.2 The composite Solar-System elemental abundance curve. The vertical axis shows on a **logarithm**ic scale the number of atoms of each element per 10^6 atoms of silicon. The inset shows abundances of the rare earth elements ('REE', La to Lu) and some neighbouring elements. The element promethium (Pm) has no stable isotopes and is not found in meteorites (see Exercise 11.1 at the end of this chapter). ○, elements with Z even; ●, elements with Z odd. The stippled band emphasizes the abundance difference between even-Z and odd-Z nuclides.

(e) The general decline in abundance as Z increases (item (b)) is interrupted by a sizable peak around $Z = 26$, comprising elements in the neighbourhood of iron.

These features provide clues as to how the elements were formed.

Cosmic element production

The Hot Big Bang

Modern cosmology is founded on the standard model of the Hot Big Bang, an event that occurred 13.8 Ga ago whose early moments have been illuminated in astonishing detail by the application of theoretical astrophysics (Weinberg, 1993; Riordan and Schramm,

1993, Ferreira, 2006). Baryonic matter began to take form in the expanding primordial fireball after only the first second of time, at a temperature of about 10 billion degrees (10^{10} K). Furthermore, theory predicts that 75% of the baryons formed were protons and 25% neutrons. After about 15 seconds of cooling, when the temperature of the fireball had fallen below 3×10^9 K, neutrons combined with protons to form a sprinkling of light nuclei such as ^{2}H (deuterium), ^{3}He, ^{4}He and ^{7}Li. But it was far too hot for these nuclei to capture electrons at this stage: matter in the universe had to wait 100,000 years or more before the temperature of the expanding cosmos had fallen low enough (to about 5×10^3 K) for neutral atoms of these light nuclides to form.

Why is temperature such a key factor here? The kinetic energy of any atomic particle increases with temperature. The electrostatic force binding an electron in a hydrogen atom, for example, prevails only as long as the electron's kinetic energy is lower than the atom's ionization energy (Figure 5.6). Raise the temperature high enough (to around 10^3 K), and hydrogen undergoes thermal ionization to form a plasma of free protons and electrons, although to liberate *inner* electrons from heavier atoms (Chapter 6) the temperature must rise further (~5×10^3 K). At ~10^9 K even protons and neutrons bound within a nucleus gain sufficient thermal kinetic energy to surmount the **strong force** and separate from each other. In the cooling, post-Big Bang cosmos these processes operated in reverse, as particles combined for the first time into nuclei and ultimately atoms.

Stars

These light nuclides formed in the Big Bang provided the feedstock for the manufacture of all the heavier chemical elements that make up the Earth and the other baryonic matter in the cosmos. This manufacturing process has been going on throughout the life of the Universe. Evidently it has not been very 'efficient', because ^{1}H and ^{4}He still make up 98% of the mass of the observable baryonic universe. How has this small inventory of heavier elements been formed?

The principal process for generating the elements up to iron is nuclear fusion (Box 11.2). Nuclear fusion can only occur if two conditions are both satisfied:

(i)   a high density of matter to raise the probability of nuclear collisions (which in interstellar space is vanishingly small); and

(ii) a high temperature (at least 10^7 K) to ensure that positively charged nuclei will collide with sufficient kinetic energy to overcome their mutual electrostatic repulsion; nuclei need to approach closer than 10^{-14} m before the **strong force** begins to operate to bind them together into a heavier nucleus.

The interior of a star furnishes both of these requirements, and the abundances of heavier elements that we see today are regarded as the cumulative product of nucleosynthesis that has taken place inside many generations of stars. Despite much subsequent research, the visionary 1957 paper by Burbidge, Burbidge, Fowler and Hoyle ('B^{2}FH') in which this

process was first proposed – though based on earlier insights by Fred Hoyle – remains the foundation of our understanding of element formation today.

We can visualize stellar nucleosynthesis as a long series of consecutive steps, like an industrial assembly line. Not every stellar factory, however, possesses the full assembly line. Fusion reactions in stars take place in a series of stages, hand in hand with the thermal evolution of the star, and how far the process may go depends, as we shall see, on the mass of the star.

Hydrogen is consumed to form helium early in the development of a star (see Table 11.1). When hydrogen in the centre of a star (where these reactions occur most quickly owing to the high density and temperature) is almost used up, the star raises its core temperature (by gravitational contraction) to a level sufficient to allow helium nuclei to combine to form carbon and oxygen. Similarly helium must more or less run out in the core of the star before further contraction and heating can occur, allowing carbon and oxygen to be transformed to heavier elements leading to silicon.

The maximum temperature a star can achieve during normal evolution is related to its mass. A star of the Sun's mass ($M_\odot$) is capable only of stages 1 and 2. A star probably needs to have a mass exceeding $30 M_\odot$ before all fusion reactions leading to iron become possible (Tayler, 1975). Even such massive stars only generate heavy elements like iron at their very centres. Many stars fall below this mass range, and therefore contribute only to the abundance of the lighter elements. The general fall-off in abundance towards heavier nuclides (item (b) on p. 210) reflects the relatively small number of stars capable of generating the heaviest elements.

Fusion reactions can generate most but not all of the stable nuclides between hydrogen and iron. As ^{8}Be is a very unstable nucleus, the main fusion reactions evidently proceed directly from ^{4}He to ^{12}C, largely bypassing the elements Li, Be and B. The small amounts of Be and B shown in Figures 11.2 and 11.3 – item (c) on p. 210 – actually seem to have been produced by the breakdown of heavier nuclei (^{12}C, ^{16}O) under cosmic-ray bombardment, a process called *spallation*.

The balance of nuclear forces gives nuclei in the iron mass range the greatest stability (Box 11.2). Massive stars having sufficiently high temperatures can produce these nuclides relatively efficiently, hence the 'iron

Box 11.2 Nuclear fusion and fission

Nuclei are held together by an immensely powerful, short-range force called the **strong force**. It acts between nucleons only over very short distances similar to the size of the nucleus itself ($\sim 10^{-14}$ m). The more nucleons present in the nucleus, the stronger is the binding force that each one experiences. Counteracting the attractive force exerted by the strong force, however, is the electrostatic repulsion acting between the Z positively charged protons present, which – because the protons are held in such close proximity in the nucleus – is also an extremely powerful force.

The relative stability of nuclei can be expressed in terms of the mean potential energy per nucleon in the nucleus, relative to the potential energy each nucleon would possess as an isolated particle (set by convention at zero). Because every nucleus represents a more stable state than the same number of separate nucleons, the mean potential energy per nucleon is a negative quantity. Its variation with mass number A for the naturally occurring nuclides is sketched in Figure 11.2.1.

The shape of the graph reflects the interplay between the strong force and the electrostatic repulsion between protons. Where the curve drops steeply on the left-hand side the strong force is clearly the dominant force, but the curve flattens out around Fe (a region of maximum nuclear stability) and then rises gently as the proton–proton repulsion exerts a steadily more powerful influence; here the increase in strong force obtained by adding further nucleons to a nucleus is slightly outweighed by the consequent increase in electrostatic repulsion.

Nuclei on the extreme left of the diagram, therefore, can in principle reduce their potential energy by fusing with other light nuclei to form heavier ones. Fusion of these lighter nuclei thus *releases* energy (it is an **exothermic** reaction) and this provides the source for the **thermonuclear** energy output of stars and hydrogen bombs. On the right of the diagram, on the other hand, is a region where fusion, were it to occur, would be energy-consuming (**endothermic**). Nuclei in this A range (>60) cannot be generated by fusion (see main text). On the contrary, the heaviest nuclei, such as thorium and uranium, are radioactive and decay by emitting alpha-particles (Box 10.1; also Box 3.3); this is one mechanism for shedding mass and attaining a

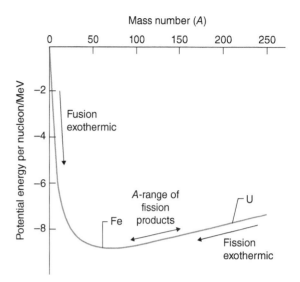

Figure 11.2.1 Sketch of how potential energy per nucleon varies with mass number A for naturally occurring nuclides.

lower energy per nucleon (greater stability). The energy released by the decay of such elements within the Earth constitutes the largest component of terrestrial heat flow.

Certain heavy nuclides (^{235}U being the only naturally occurring example) are also **fissile**: on absorbing a neutron they split into two lower-mass nuclei. These **fission products**, comprising various nuclides in the A range 100–150, have two important properties in common:

(a) They lie on a lower segment of the potential-energy curve than the parent nuclide. Thus fission is an exothermic process: it is the energy source for present nuclear-power reactors and for the original 'atom bomb'.

(b) Although several neutrons are released in the fission process (which by colliding with other ^{235}U nuclei prompt further fission), the fission products still have higher N:Z ratios than stable nuclides in the same A range, which makes them radioactive (Box 10.1). The β-decay of fission products such as ^{90}Sr, ^{131}I and ^{137}Cs (Figure 10.1.1) is the prime cause of the intense initial radioactivity of reactor wastes.[8]

---

[8] Reactor wastes also give rise to longer-term radioactivity which is due to α-emitting isotopes of actinide elements like plutonium, ^{239}Pu.

Table 11.1 Nuclear fusion stages in stellar nucleosynthesis

Stage	Maximum temperature	Range of nuclei produced
1	10^7 K	H $\rightarrow$ He
2	10^8 K	He $\rightarrow$ C, O, etc.
3	5×10^8 K	C, O $\rightarrow$ Si
4	5×10^9 K	Si $\rightarrow$ Fe

peak' in the abundance curve noted above (Figure 11.2). Beyond this point, however, further fusion is impeded because the core temperatures required to overcome electrostatic repulsion between nuclei with such large positive charges exceed those of even the hottest stars. The synthesis of heavier nuclides requires a different process that is not impeded by nuclear charge.

The manufacture of nuclides heavier than iron proceeds instead by the addition of **neutrons**, neutral nuclear particles that experience no repulsion by the target nucleus. Many reactions in stars produce neutrons, particularly in the later stages of stellar evolution. The nuclides to the right of iron in Figure 11.2 are thus believed to be products of cumulative *neutron capture*. Neutrons are absorbed by a nucleus, increasing the N value until an unstable, neutron-rich isotope has been produced, which transmutes by β-decay into an isotope of the next element up (Z increases by one, N falls by one; Box 10.1). The repeated operation of this process can produce all of the heavier nuclides, given a sufficiently high neutron flux (Rauscher and Patkós, 2011). If the rate of neutron capture is *slow* relative to the relevant β-decay rates, the nucleosynthesis pathway lies close to the main band of stable nuclides shown in Figure 10.1.1; this 's-process' (s standing for 'slow') leads as far as bismuth (Bi, $Z = 83$, the nuclide above Pb in Figure 10.1.1). The manufacture of heavier elements like U and Th requires higher neutron fluxes, as discussed under 'Supernovae' below.

The abundance peak around iron in Figure 11.2 (item (e) on p. 211) suggests that neutron capture reactions consume iron-group nuclei more slowly than fusion reactions produce them.

Why are nuclides that have even values of Z – or N, for that matter – more abundant than those with odd values (item (d) on p. 210)? Protons and neutrons reside in **orbitals** inside the nucleus, just as electrons do outside it. According to nuclear wave mechanics, filled orbitals containing two protons or two neutrons

are more stable than half-filled ones. This additional stability manifests itself in the form of more compact orbitals and a smaller nucleus, which reduces the nuclear 'cross-section' (or 'target size') upon which the probability of collision depends, thereby depressing the rate of the fusion or neutron-capture reactions that consume the nuclide, and allowing its abundance to build up. Even-number values of Z and N account for the greatest number of stable nuclides, lending the nuclide chart a 'staircase' appearance in which even values of Z form the treads and even values of N form the steps (Box 10.1). Conversely, nuclei having odd values of N or Z do not enjoy this additional stability, have larger collision cross-sections, and are more susceptible to consumption through fusion or neutron capture reactions or radioactive decay.

Supernovae

When a smaller star ($< M_\odot$.) reaches the end of its life, it can progress into a 'white dwarf' phase and quietly fade away. But theory suggests that massive stars ($> 2 M_\odot$.) follow a different path, leading to a catastrophic collapse of the core, the shock wave from which causes a colossal stellar explosion (Bethe and Brown, 1985). Such *supernovae* are characterized, for a brief period, by energy output of staggering intensity: the luminosity from a single exploding star can rise briefly to levels typical of a whole galaxy ($\sim 10^{11}$ stars), lasting for a few Earth days or weeks. The huge quantities of energy transferred to the zones of the star immediately surrounding the core cause a large proportion of the star's mass to be expelled at high velocity ($\sim 10^7 \, \mathrm{m\,s^{-1}}$). The expanding Crab Nebula is thought to be the remnant of a supernova observed by Chinese astronomers in AD 1054. A supernova was actually observed in the Large Magellanic Cloud on 23 February 1987.

Supernovae contribute to nucleosynthesis in two important ways:

(a) The neutron flux becomes exceedingly high during a supernova, prompting a burst of very rapid neutron-capture ('r-process') reactions leading to U and Th, and even beyond (to heavy unstable nuclides such as plutonium, Pu).

(b) The products of stellar nucleosynthesis, confined up to that moment within a star's interior, are flung

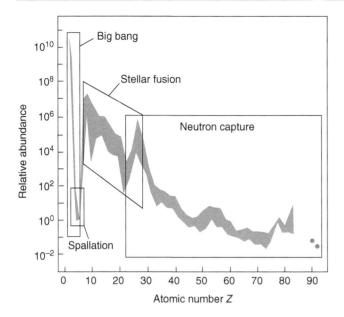

Figure 11.3 The Solar-System abundance curve (Figure 11.2) showing the domains of various nucleosynthetic processes.

out into the interstellar medium, to be eventually incorporated into new generations of stars. Present element abundances (Figure 11.2) reflect recycling of matter through successive generations of stars, each one adding its own contribution to the overall accumulation of heavy elements in the universe.

Figure 11.3 summarizes the contribution of these different processes to the current inventory of chemical elements in the universe. As Hutchison (1983) pointed out, 'we, each one of us, have part of a star inside us'. More detailed accounts of stellar nucleosynthesis can be found in the books by Albarède (2009) and Ferreira (2006), and in a recent review article by Rauscher and Patkós (2011).

Elements in the Solar System

Cosmochemical classification

Differentiated meteorites as a group contain three broad categories of solid material: silicate, metal and sulfide. Analysis of these phases shows that most elements have a greater affinity with one of them than with the others. Magnesium, for example, is overwhelmingly segregated into silicate phases, whereas copper is often concentrated in sulfides. The Norwegian geochemist V.M. Goldschmidt devised the following subdivision:

(a) *lithophile* elements: those concentrated into the silicate phase (from the Greek *lithos*, meaning 'stone');
(b) *siderophile* elements: those preferring the metal phase (from the Greek *sideros*, meaning 'iron');
(c) *chalcophile* elements: those like copper which concentrate in the sulfide phase (from the Greek *chalcos*, meaning 'copper');
(d) *atmophile* elements: gaseous elements (from the Greek *atmos*, meaning 'steam' or 'vapour').

How the elements are divided between these categories is illustrated in Figure 11.4 and Plate 7. Such a compilation involves compromise, and one author's version may differ slightly from another's. Among the metallic elements there is a significant correlation with electronegativity (cf. Box 9.9): the metals that are exclusively lithophile (excluding B and Si) have electronegativities below 1.7, most chalcophile metals have electronegativities between 1.8 and 2.3, and the most siderophile metals are those with electronegativities of 2.2 and above. Goldschmidt's concept is very useful in understanding in what form elements occur in Solar-System matter, in ore deposits, or for that matter in a smelter. For example, the siderophile character of iridium (Ir) means that nearly all of the Earth's Ir inventory is locked away in the metallic core (the same incidentally being true of gold) and its concentration in crustal rocks is extremely low (see Figure 11.2: $Z_{Ir} = 77$). Consequently most of the iridium detected on the Earth's surface, in deep-sea sediments for instance, has been introduced there as a constituent of incoming meteoritic dust; some iron meteorites contain as much as 20 ppm Ir, 20,000 times higher than average levels in crustal rocks. This provides a means of estimating the annual influx of iron meteorites to the Earth's surface. Positive Ir anomalies are also characteristic of clays associated in time with the Cretaceous–Tertiary extinction, one of the factors suggesting that a major impact event occurred around that time.

Some elements exhibit more than one affinity, necessitating areas of overlap in Figure 11.4 and multiple colours in Plate 7. For example, oxygen is considered to be both lithophile – being a major constituent of all silicates, as explained in Chapter 8 – and atmophile (as

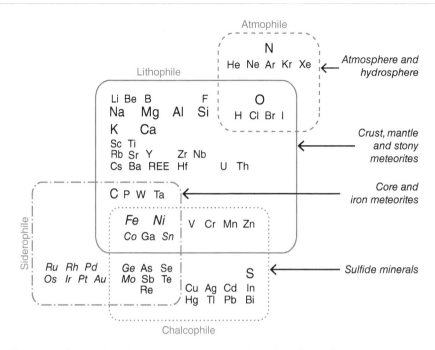

Figure 11.4 Element affinities in the Earth and in meteorites. Areas of overlap show elements common to two or more phases. Larger lettering indicates a major element. Elements found principally in the metal phase are shown italicized. (**REE** – Figure 11.2.)

O_2 and H_2O gases in the Earth's atmosphere). Another prominent example is iron, which exhibits lithophile, siderophile *and* chalcophile tendencies and therefore lies in the area of overlap between all three of these fields in Figure 11.4.

Volatile versus refractory

In considering the development of the Solar System, it is also useful to subdivide chemical elements according to their volatility. **Volatile** elements are those that become gaseous at relatively low temperatures. In cosmochemical terms they include not only the atmophile elements hydrogen, helium (and other inert gases) and nitrogen that are gases at room temperature, but also such elements as cadmium (Cd), lead (Pb), sulfur (S) and most of the alkali metals. **Refractory** elements, on the other hand, are those that remain solid up to very high temperatures. The most refractory elements are the platinum metals (like iridium), and we also include in this category elements like calcium, aluminium and titanium that form highly refractory oxide or silicate compounds (such as the minerals perovskite, $CaTiO_3$ and anorthite, $CaAl_2Si_2O_8$).

Magnesium and silicon, the elements that make up the bulk of the silicate minerals in meteorites and planets, form a 'moderately refractory' category between these two extremes. The major siderophile elements fall within the same range of volatility (Figure 11.5).

The remaining lithophile and chalcophile elements are volatile in varying degrees. We can divide them into moderately volatile (e.g. Na, Mn, Cu, F, S) and very volatile (C, Cl, Pb, Cd, Hg) categories, as shown in Figure 11.5. The atmophile elements can be considered as a third, 'most volatile' category.

Element fractionation in the Solar System

It has long been known that the planets orbiting our Sun vary considerably in composition, evidence that, during the development of the Solar System, the elements have been chemically sorted or **fractionated**.

Because metal, silicate and gas phases themselves differ in density, planetary scientists are able to estimate the proportions of these materials present in the planets (whose mean densities can be determined from astronomical measurements). As one can see in Figure 11.6, the planets differ considerably in their

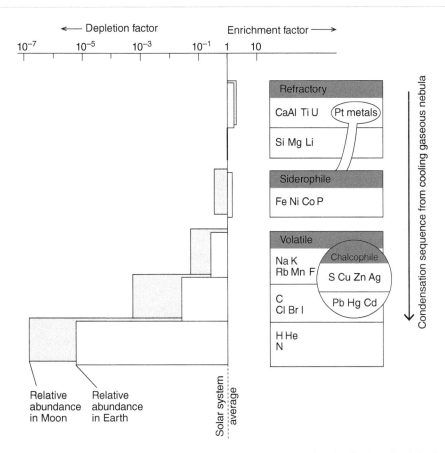

Figure 11.5 Element fractionation in the Earth (white bars) and the Moon (grey bars). The length of the bars indicates the approximate degree of depletion (left) or enrichment (right) of each 'volatility group' of elements relative to Solar-System average abundances – see **logarithm**ic scale at top. The arrow on the right indicates the predicted order of condensation as the solar nebula cooled. Because Si is the reference element for expressing both terrestrial/lunar and Solar-System abundances (Figure 11.2), the Si group of elements registers neither enrichment nor depletion in this diagram.

make-up. The small inner planets Mercury, Venus, Earth and Mars – known as the *terrestrial planets* – have high densities characteristic of mixtures of metal and silicates in varying proportions. Their atmospheres make up only a tiny proportion of the planetary mass.

The remaining planets[9] – the *major planets* – have masses several **orders of magnitude** greater than Earth's. Their low mean densities (0.69 kg dm⁻³ for Saturn to 1.64 kg dm⁻³ for Neptune) indicate compositions closer to the Solar-System average (Figure 10.2), in which atmophile elements predominate: the largest planet, Jupiter, consists almost entirely of hydrogen and helium, although with a small rock and ice core of 10–20 times the Earth's mass.

More is known about the constitution of the Earth and Moon than the other planets, and geochemists have been able to assemble quite detailed models of the overall chemical composition of these two bodies. They are illustrated in Figure 11.5. The Earth and Moon are strongly depleted not just in atmophile (gaseous) elements, but in the other volatile elements too. This depletion is more marked in the Moon than the Earth.

Evolution of the Solar System[10]

The current consensus is that the Sun and the Solar System developed together by gravitational contraction of a large cloud of interstellar gas and dust more than 4.5 billion years ago. The overall composition of this *pre-solar nebula* must have been close to the

[9] Not including Pluto, which is no longer classed as a planet.

[10] The on-line illustrated 'time-line' at http://www.lpi.usra.edu/education/timeline/mural.shtml#right may be a helpful introduction to this section.

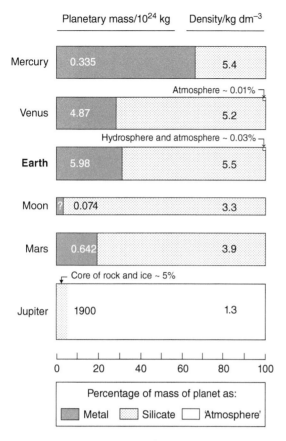

Figure 11.6 Mass proportions of metal, silicate and 'atmosphere' in the terrestrial planets and Jupiter, from astronomical data. Note the rough correlation with the mean density of each body (right-hand figures). It is uncertain in geophysical terms whether the Moon has a metallic core or not (Wieczorek *et al.*, 2006).

present Solar-System average composition shown in Figure 11.2. The heavy elements present in the cloud were the accumulated products of successive cycles of stellar nucleosynthesis in earlier stars, each star's contribution having been recycled into the interstellar medium by the supernova that ended the star's life.

Gravitational collapse of such clouds has two outcomes:

- Mass is accelerated toward the cloud's centre of gravity. The potential energy released by gravitational collapse appears in the form of heat, leading eventually to temperatures at the centre so high that thermonuclear fusion can begin, igniting (in the Solar System case) the infant Sun.
- As the dispersed outer parts of the gently rotating cloud also contract under gravity, conservation of

angular momentum causes them to rotate at increasing velocities, like a pirouetting ballerina. As a result, the shapeless cloud flattens out into a disk (rather like pizza dough being spun by a *pizzaiolo*). It was from this *protoplanetary disk* of gas and dust (examples of which have recently been discovered around many young stars) that the present planets eventually accreted.

Given the heating caused by gravitational collapse, and the large variations observed in the abundances of the volatile elements in the Solar System (Figure 11.5), it is natural to postulate that the inner parts of the solar nebula became very hot and entirely gaseous.[11] The *equilibrium condensation theory* regards the solid constituents of the inner Solar System as **condensates** from a cooling gaseous nebula whose initial temperature may have been as high as 1500 °C. As cooling progressed, elements would **condense** into solids in a predictable 'condensation sequence' which – given one or two assumptions about the density and composition of the gas – can be worked out thermodynamically.

First to condense would be the most refractory elements and compounds (the platinum metals and oxides of Ca, Al, Ti, and so on), appearing as solids at about 1300 °C (~1600 K). Other elements would follow at progressively lower temperatures, broadly in the descending order shown in Figure 11.5.

According to this model, the planets that ultimately formed from these condensates would differ in their content of volatile elements according to (i) the distance from the Sun at which they accreted, and/or (ii) the stage in this condensation sequence at which they accreted to planetary size. Planets accreting early would fail to incorporate moderately volatile elements, not yet available in solid form; whereas accretion at a later stage or in a cooler part of the nebula would lead to assimilation of lower-temperature condensates too, producing bodies of more primitive composition like carbonaceous chondrites. In its infancy, the Sun would also have radiated a much more intense *solar wind* – the outward flux of protons radiating from the Sun – than it does today, and this could have swept uncondensed volatile components

---

[11] It is clear from the low-temperature assemblages preserved in carbonaceous chondrites that the solar nebula could not have been hot throughout.

out of the inner parts of the solar nebula, amplifying their depletion in the inner planets.

Observational evidence for this condensation sequence is provided by *calcium–aluminium-rich inclusions* (CAIs) that are preserved as minute white specks in some carbonaceous chondrite meteorites. They consist of oxides and silicates of Ca, Al and Ti, precisely those refractory substances predicted by thermodynamics to condense at the highest temperatures (Figure 11.5). They also contain tiny refractory metal nuggets rich in platinum metals and other refractory metals like tungsten (W) and molybdenum (Mo). CAIs have long been recognized geochronologically as the earliest solid bodies known in the Solar System: their absolute ages fall in a narrow interval of 4567.30 ± 0.16 Ma (Connelly *et al.*, 2012),[12] and so it seems likely that they were the earliest condensates from a cooling solar nebula (Figure 11.5).

The chondrules characteristic of chondrite meteorites (Plate 5) also had an early, high-temperature origin – dated between 4567.3 and 4564.7 Ma (Connelly *et al.*, 2012) – but their igneous textures (Plate 5) seem to indicate formation by re-melting of dust particles rather than direct condensation from a cooled solar nebula. The processes of re-melting are still poorly understood.

It was only later (at least 1.0–1.5 Ma later) that CAIs and chondrules were incorporated into the meteorites that host them today.

Planet formation

The planets of the Solar System originated as dust and gas in the young Sun's protoplanetary disk. The mechanisms of initial growth towards large bodies are poorly understood but, whether by gravitational instability or simple 'sticking together' of aggregates, the process must have formed a large number of 10-km-sized objects rapidly. … Once bodies reached this critical size, gravitational perturbation became the dominant mechanism for further accretion through collision. (Wood et al., 2006)

How did a protoplanetary disk of dust (including finely dispersed condensates) and gas surrounding the early Sun aggregate into planets, specifically the rocky terrestrial planets familiar to us today? The standard model, accepted since the 1960s, is that the disk became progressively more 'lumpy' as small particles collided and agglomerated into larger ones. Collisions would

have been frequent in the disk, breaking up some bodies but adding to others. As time went by the particle-size distribution evolved through metre-scale and kilometre-scale bodies to 100-km-scale *planetesimals*, similar to bodies known to exist today in the asteroid belt. By sweeping up smaller bodies in their path, such orbiting bodies would generally grow larger and fewer in number (forming what are aptly referred to today as 'oligarchs') and eventually coalesce into bodies resembling the present planets.

This main accretionary stage in the formation of the planets probably lasted only 30–40 Ma (Wood *et al.*, 2006). Yet it is clear from the density of cratering on the Moon – and from radiometric dating of impact melts collected by Apollo missions – that intensive bombardment of the terrestrial planets by smaller planetesimals continued until about 3800 Ma ago,[13] 700 million years after the planets had originally been formed.

Some of these impacts were evidently huge. The anomalously large metallic core of Mercury, for example, has been attributed to a so-called *giant impact* that ejected a large proportion of its original silicate mantle. Indeed, it is widely accepted today that the Earth itself suffered a similar catastrophic impact – perhaps 50–150 Ma after the formation of the Solar System – in which the mantle of the colliding planetary body was ejected, forming an orbiting debris disk from which our present Moon accreted. Such an origin for the Moon would account for its tiny core and low density in relation to other inner-Solar System bodies (Figure 11.6) if, as modelling suggests, the impactor's core was captured by the Earth. An impact origin for the Moon would also explain its extreme depletion in volatiles (Figure 11.5), as the impactor mantle is likely to have been vaporized by the collision. New evidence suggests that the Earth may have been affected by more than one giant impact, only the last of which led to the formation of the present Moon.

Chemical evolution of the Earth

The core

If the Earth was formed by the aggregation of large planetesimals as discussed above, the energy of their collisions would have been sufficient to cause

---

[12] A slightly older age of 4568.2 ± 0.2 Ma was determined by Bouvier and Wadhwa (2010).

[13] Possibly culminating in a so-called 'Late Heavy Bombardment' between 4.0 and 3.8 Ga.

significant melting, giving rise at times to an extensively molten outer layer: a *'magma ocean'*. Isotopic evidence from differentiated meteorites (Wood *et al.*, 2006) suggests that even small precursor planetary bodies segregated metallic cores within 5 Ma of formation, and so impactors accreting to the Earth would already have formed metallic cores. This suggests that core segregation within the Earth took place continuously, hand in hand with planetesimal accretion.

The gravitational accumulation of dense metal into the centre of the molten Earth has two important implications: large amounts of gravitational potential energy would have been released, sustaining a partially molten state of the overlying mantle; and the siderophile elements (Figure 11.4) would have been efficiently scavenged from the mantle into the core.

The giant impact event proposed for creating the Moon, around 4522 Ma ago, would have caused a later phase of extensive (possibly complete) melting of the Earth and an additional contribution (from the impactor) to the Earth's core.

Many physical properties of the present molten outer core are consistent with a major element composition similar to the Fe–Ni alloy found in iron meteorites, although the velocity of compressional seismic waves through the core indicates a lower density than expected for Fe–Ni under the appropriate load pressure. It follows that a significant proportion (~8%) of some less dense element(s) must also be present in the core. Recent research (see Wood *et al.*, 2006) points to a combination of Si, S and O making up this light component of the core.

The mantle

The silicate material surrounding the core – constituting 70% of the Earth's mass (Figure 11.6) – has, over geological time, differentiated into the present-day mantle and crust as a result of igneous activity throughout the Earth's history. When a partial melt (Box 2.4) develops in equilibrium with solid rock, elements are fractionated in two overlapping ways.

(a) The lower-melting major components of the rock (Fe, Al, Na, Si) enter the melt preferentially, leaving the residual solids enriched in refractory (Mg-rich) end-members (Box 2.4).

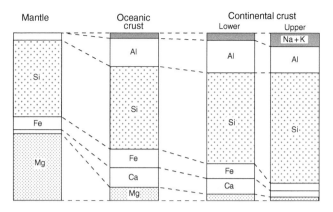

Figure 11.7 Average compositions (oxide percentages) of the Earth's mantle and crust.

(b) Crystals tend to dump into the melt certain trace elements whose ions are difficult to accommodate. The ions of these **incompatible elements** (Box 9.1) are more easily accommodated in the open, disordered structure of a melt than in a crystal lattice.

The extraction of magma from the mantle to form crust has progressively displaced these elements from mantle to crust over geological time. Crustal rocks (e.g. basalt) consist of lower-melting mineral assemblages (Figure 11.7) and are enriched in incompatible elements compared with mantle **peridotite**.

Parts of the mantle have therefore become depleted in these elements. It has been estimated that 20–25% of the mantle's original inventory of highly incompatible elements (K, Rb, U) now resides in the continental crust. Supposing these elements were uniformly dispersed in the primordial mantle, it seems that at least a third of its volume has been tapped for igneous magma during the course of geological time. Whether the mantle was ever homogenous is debatable (although it seems likely if at one time the mantle was completely molten), but its present inhomogeneity is beyond doubt. Figure 10.8 showed how the varying geochemistry of recent volcanic rocks points to a range of chemically distinct source regions ('reservoirs') in the mantle. Most mid-ocean ridge basalts ('MORBs'), for example, come from a global mantle reservoir depleted in the incompatible elements (so having an isotopic composition falling in the 'depleted quadrant' in Figure 10.8) – an inheritance, most geochemists believe, of widespread melt extraction during the course of Earth history.

The crust

The Earth's crust falls into two broad divisions (Figure 11.7). The basaltic crust of the ocean basins, being a product of partial melting of mantle peridotite, has a higher Mg content (although much lower than the mantle itself) and lower Si (less than 50%) than the continents. It has a relatively short lifetime: its passage from ocean ridge to subduction zone, where it is delivered back into the mantle, typically takes less than 200 million years (the age of the oldest known oceanic lithosphere). During this time oceanic crust becomes partially hydrated by chemical interaction with ocean water and it acquires a blanket of sediment. This modified crustal package is returned to the Earth's interior through subduction, and parts of it reappear at the surface as constituents of subduction-related island-arc volcanics and Cordilleran plutonic rocks.

Unlike the ephemeral ocean floor, the **sialic** continental crust has been accumulating throughout known geological time, although probably not at a uniform rate. None of the Earth's earliest crust has survived, having been reworked by the intense planetesimal bombardment that continued until about 3.8 Ga ago. The oldest recognizable remnants of early crust are about 4.0 Ga old (although greater ages have been obtained from detrital zircon grains in younger sedimentary rock) and the present continental crust has been accumulating since that time. It is being extended today by the lateral accretion of island arcs on to continental margins, and by deep-seated igneous intrusions into continental roots. Different mechanisms may have operated in the past, for example during the huge increase in the volume of the continental crust that seems to have occurred between 3.0 and 2.0 billion years ago, at the close of the Archaean era. Average continental crust is equivalent to andesite in composition, having a SiO_2 content of around 57%, significantly higher than oceanic crust (49.5%, Figure 11.7).

Repeated melting and metamorphism within the continental crust have led to its internal differentiation into a lower, more refractory continental crust depleted in incompatible elements; and an upper crustal layer, roughly 10 km thick, which is enriched in Na, K and Si (Figure 11.7). Most of the heat flow we measure in continental areas originates in this top 10 km, into which almost the whole crustal inventory of the radioactive incompatible elements K, U and Th has been concentrated.

It is notable that the other terrestrial planets have dominantly basaltic crusts. Why have none of them developed andesitic or 'granitic' crust resembling the continents on Earth? This unique feature of the Earth probably stems from another: the existence on the surface of liquid water (see reference to the *habitable zone* below). The igneous minerals of the basaltic ocean crust react with seawater to form hydrous secondary minerals. Subduction of altered oceanic crust transports this bound water deep into subduction zones where the hydrous minerals dehydrate (cf. Figure 2.3), thereby releasing water into the overlying mantle wedge. In the presence of water vapour, mantle peridotite undergoes partial melting at lower temperatures and – as experiments show – produces melts that are more SiO_2-rich (andesitic) than when melted 'dry'. This explains the dominance of andesite in many island arcs, and provides the precursor for the formation of granitic upper continental crust.

The early atmosphere

Accreting planets capture a *primary atmosphere* of nebular gases (mainly hydrogen), but this would have been lost from the Earth as soon as the nebula thinned and cleared (Zahnle *et al.*, 2007). Of more interest is the *secondary atmosphere* consisting of volatiles outgassed from within the Earth itself by the high temperatures associated with accretion. Its inferred composition is highly sensitive to the composition of the parental material from which the Earth is presumed to have accreted: an Earth resembling CI carbonaceous chondrites would have outgassed a hot dense atmosphere dominated by H_2O, CO_2, N_2 and H_2 (which of these predominates depends on the temperature), whereas an Earth formed by accretion of other types of chondrite would have generated a more reduced atmosphere dominated by CH_4, H_2, N_2 and CO (Schaefer and Fegley, 2010). Regardless of which starting material is adopted, it is clear that the Earth's early atmosphere contained no free oxygen.

This early atmosphere must have been lost during the Moon-forming giant impact, which melted most of the Earth's mantle and even vaporized part of it, and would have had a dramatic short-term effect on atmosphere composition:

For a thousand years … silicate clouds defined the visible face of the planet. The Earth might have looked something like a small star or a fiery Jupiter wrapped in incandescent clouds. (Zahnle *et al.*, 2007)

The outcome, as the magma ocean beneath cooled, is thought to have been an atmosphere consisting of H_2O, CO_2, CO and H_2 in that order. Again O_2 is notable by its absence.

The subsequent profound transformation of this post-impact[14] anoxic atmosphere into the oxygen-rich atmosphere we depend upon today could not have occurred without the appearance of life around 3.5 Ga ago (possibly even earlier). There is geological evidence for the existence of liquid water on the surface of the Earth since at least 3.8 Ga ago, and some models place its first appearance much earlier (Zahnle *et al.*, 2007). It was in Earth's oceans that life began its complex journey.

Life and oxygenic photosynthesis

Life on Earth relies on a remarkable astrophysical coincidence. The Earth orbits the Sun at a distance lying within the Sun's *habitable zone*,[15] the range of orbital radius within which planets with atmospheres enjoy surface temperatures that allow water to exist in the liquid state. Were the Earth much closer to the Sun, its surface temperature would boil water (as on Venus), whereas if it were further away – like present-day Mars – any surface water could exist only as ice; either departure would make the huge diversity of life found on Earth today unsustainable.

The ultimate origins of life remain shrouded in mystery. The earliest organisms must have thrived in the Earth's early anoxic atmosphere, but they included some – similar to today's *cyanobacteria* ('blue-green' algae) – that produced oxygen in the oceans as a by-product of the photosynthesis of carbohydrate from carbon dioxide and water (Equation 9.3). It is likely that photosynthesis in some form began around 3.5 Ga BP[16] – perhaps even earlier – but it is clear that *oxygenic* photosynthesis was well established in cyanobacteria by 2.7 Ga BP. Marine biogenic oxygen could not accumulate in the atmosphere immediately, however, because the oceans had built up through weathering a legacy of reduced solutes, notably ferrous iron, Fe^{2+}. Until completely oxidized, this 'oxygen sink' would have mopped up free oxygen in the oceans as it was produced, preventing its escape into the atmosphere.

In sediments of the Archaean eon, iron occurs chiefly in *banded iron formation* (BIF), the thin iron-oxide-rich layers of which were precipitated from seawater through the oxidation of dissolved Fe^{2+} to insoluble Fe^{3+} (see Figure 11.8b). Each layer may represent a single short-lived blooming of oxygenic bacteria in an otherwise anoxic ocean, repeated countless times through cyclic fluctuation in bacterial populations or Fe^{2+} supply. Delivery of abundant dissolved Fe^{2+} to the oceans depended, of course, on an atmosphere containing negligible oxygen, an inference that is supported by the MIF-S record in sulfur isotopes shown in Figure 10.14.

In post-Archaean sediments, in contrast, iron occurs mainly as rusty diagenetic coatings on detrital grains in reddish sandstones or shales ('red beds'), suggesting that iron released by weathering was being oxidized subaerially to ferric iron even before it reached the sea. This marked change in the predominant form of iron-bearing sediment, which occurred between 2.4 and 1.8 Ga (Figure 11.8a), is therefore taken to mark the first sustained appearance of free oxygen in the Earth's atmosphere, although its abundance – even after this 'Great Oxidation Event' ('GOE' in Figure 11.8a) – still fell well below the current level of 21 vol.%. The period from 2.4 to 2.3 Ga brought a number of worldwide glaciations, as the surge in oxygenic photosynthesis drew down atmospheric CO_2 and diminished its key contribution to greenhouse warming.

Why does Figure 11.8a show a brief reversion to BIF deposition in mid- to late-Neoproterozoic times? This period, from 750 to 570 Ma, was, like the early Proterozoic, characterized by intense Snowball Earth (global glaciation) episodes and wild fluctuations in

---

[14] The resemblance of the early Earth's surface conditions to a vision of hell led to the name *Hadean* (after *Hades*, the Greek god of the underworld) being coined for this earliest, pre-Archaean eon of Earth's history.

[15] The habitable zone is sometimes dubbed the 'Goldilocks zone', signifying that conditions are – like porridge ready for eating! – 'neither too hot, nor too cold, but just right'.

[16] BP, before present.

(b)

(a)

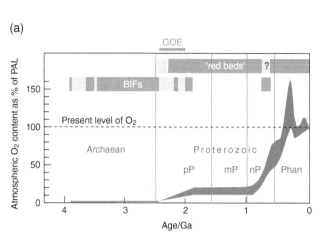

Figure 11.8 (a) Cartoon illustrating, from left to right, the stepwise evolution of atmospheric oxygen content from the Archaean to the present as envisaged by Holland (2006), expressed as volume % of the present atmospheric level (PAL); the thickness of the band reflects estimated uncertainty. The subdivisions of the Proterozoic eon are abbreviated as pP (Palaeoproterozoic), mP (Mesoproterozoic), nP (Neoproterozoic). Phan = Phanerozoic. 'GOE' shows the duration of the Great Oxidation Event referred to in Chapter 10 (Figure 10.13). The horizontal orange bars indicate the periods over which iron sedimentation was dominated respectively by banded iron formation (BIF) and by red beds; the darker BIF bars represent the main episodes of BIF deposition identified by Isley and Abbott (1999). The brief resumption of BIF deposition in the Neoproterozoic correlates with three 'Snowball Earth' episodes. (b) Field picture of early Archaean banded iron formation, Isua, West Greenland. (Sources: Own based (a) on Holland 2007; Reproduced with permission of GEUS).

the $\delta^{13}C$ record (Figure 10.13). Curiously, there is indirect evidence suggesting this was also a period of rapidly *increasing* atmospheric O_2 (Figure 11.8a; Frei *et al.*, 2009), implying conditions quite different to those that promoted BIF deposition in Archaean times. This paradox can be resolved if the Neoproterozoic ice sheets largely isolated the oceans from the atmosphere, so that iron introduced into the oceans from seafloor hydrothermal vents could accumulate as dissolved ferrous iron until glacial retreat allowed oxygen exchange and precipitation of the dissolved iron. In any event, the Neoproterozoic glaciations must again reflect large-scale drawdown of atmospheric CO_2, but whether this had an entirely biological origin (e.g. colonization of continents by photosynthetic biota) or was somehow related to profound tectonic changes then taking place (notably the break-up of the Rodinia supercontinent) remains unclear.

The manner in which life adapted to the prevalence of oxidizing conditions at the Earth's surface, putting oxygen to good use to generate energy, is a fascinating story explored in the book by Lenton and Watson (2011). The earliest life-forms evidently developed on an abundant supply of organic nutrients that could

exist stably in the oxygen-free primordial atmosphere. Such organisms would not be able to develop in the Earth's present atmosphere, where oxidation would rapidly destroy their simple molecular foodstuffs, any more than we could survive in oxygen-free conditions. Life, by introducing free oxygen into the atmosphere and sustaining it there for more than 2 billion years (when the residence time of oxygen in the atmosphere is only a few thousand years), has burned the environmental boat by which it first came into being.

Yet life has also transformed the Earth into the tolerable planet to live on that it currently is. All of the oxygen in the atmosphere has been manufactured by photosynthetic organisms from carbon dioxide, and by lowering atmospheric CO_2 levels such organisms have thereby turned down the heat in the Earth's 'greenhouse' to a much lower level than operates on Venus, whose atmospheric greenhouse maintains the surface temperature at a searing 470 °C. This mechanism for removing CO_2 from the Earth's atmosphere is essentially a reversible equilibrium, balancing oxidized carbon in the air against reduced carbon in the biosphere. However, a small proportion of reduced biosphere carbon has also became fixed in the crust in the form of fossil carbon which – transformed by diagenesis and thermal maturation over many millions of years – has provided the coal, petroleum and natural gas resources that fuel industrial societies today. Equally important from the 'greenhouse' point of view, however, are the countless carbonate-secreting organisms that over the eons of geological time have fixed atmospheric carbon dioxide in the form of limestone (the product of accumulation of calcareous biogenic debris). Together these carbon reservoirs in the crust account for the marked difference in atmospheric CO_2 content between Earth and Venus (0.03% versus 96.5%) and for the temperate climate that we enjoy on Earth.

Future prospects

Life on Earth has survived a range of climates over the course of geological history, encompassing mean global temperatures 6–8 °C higher than at present during Cretaceous times to 10 °C cooler in the depths of the recent ice age. Despite the evident resilience of life in general to such changes, one must recognize that *human* civilization has developed almost entirely during the relatively constant and benign inter-glacial climate of the last 11,700 years (the Holocene epoch), to which our agriculture, settlement patterns and economies are now finely tuned. Through our reliance on fossil fuels, however, we have been returning reduced carbon from the crustal reduced-carbon reservoir into the atmosphere, partially undoing the cumulative work of photosynthesis over the last billion years. Today we are releasing greenhouse gases back into the atmosphere at a faster rate than any in recent geological history, and the mean atmospheric CO_2 content is higher today than at any time for the last half-million years (IPCC, 2013) and probably higher than the last 15 million years (see, for example, Tripati *et al.*, 2009). In the words of one climate expert:

We are (re)creating a prehistoric climate in which human societies will face huge and potentially catastrophic risks. Only by urgently reducing global emissions will we be able to avoid the full consequences of turning back the climate clock by 3 million years.[17]

---

[17] Bob Ward, policy director at the Grantham Research Institute on Climate Change at the London School of Economics. http://bit.ly/1eEJyN2. See also www.epa.gov/climatestudents/scientists/index.html

Human society as we understand it today will survive only if sound climate science prevails *soon* over business-as-usual politics.

Review

- The average chemical composition of baryonic matter making up the Solar System can be determined (a) from spectral analysis of the light reaching us from the Sun and other stars, and (b) from laboratory analyses of primitive meteorite samples (Box 11.1). On the scale of galaxies, however, such 'visible' matter is outweighed many times by 'dark matter' of uncertain composition.
- The Solar System abundance curve (Figure 11.2) shows that H and He are the most abundant elements in the cosmos. Except for Li, Be and B, heavier elements become progressively less abundant with increasing Z, but with peaks around $Z = 26$ and 52. Even-Z nuclides are generally ten times more abundant than neighbouring odd-Z nuclides.
- H, He and Li are predominantly products of the Big Bang, but heavier elements have been formed progressively by fusion ($Z < 26$) and neutron capture reactions ($Z > 26$) in multiple generations of stars (Figure 11.3).
- Element behaviour in the Solar System may be described in terms of four overlapping cosmochemical categories (Figure 11.4): lithophile, siderophile, chalcophile and atmophile. Elements are also subdivided according to whether they – or their oxides – are refractory or volatile (Figure 11.5).
- The four 'terrestrial' planets closest to the Sun have rocky, metal-rich, volatile-depleted compositions (Figure 11.6), modified in the case of Mercury and the Earth–Moon system by post-accretion giant impacts. The larger outer planets have lower-density atmophile-rich compositions closer to the Solar System average.
- Planetesimal accretion, core-segregation and the Moon-forming giant impact (~4522 Ma ago) brought about extensive partial melting in the early Earth, promoting efficient partition of siderophile elements into the core. Heat accumulated from these early exothermic events still contributes to surface heat-flow today. Magmatism through Earth history and extraction of crust have generated mantle

reservoirs with distinct geochemical signatures (Figure 10.8).

- The Earth is the only planet to reside in the Solar System's 'habitable zone' where liquid water exists sustainably at the surface, a key factor in the development of life. Subduction of oceanic crust modified by chemical reaction with seawater probably lies behind another unique feature of the Earth: the chemically evolved continents.

- The Earth's early atmosphere consisted of combinations of H_2O, CO_2, CO, N_2, H_2 and CH_4. Oxygen, the product of photosynthesis by living things, first appeared in the atmosphere at the end of the Archaean eon, remained at a low concentration through most of the Proterozoic eon, and then built up progressively to the present level through the Phanerozoic eon (Figure 11.8a), alongside the evolution of plants.

- The cumulative influence of photosynthesis in drawing down atmospheric levels of CO_2 and other greenhouse gases has been the key to the Earth's present benign climate, and has led to the deposition of a large crustal reservoir of reduced hydrocarbons that we exploit as 'fossil fuels'. Our current rate of releasing fossil carbon back into the atmosphere poses huge risks to the future of mankind.

Further reading

Albarède, F. (2009) *Geochemistry. An Introduction*, 2nd edition. Cambridge: Cambridge University Press.

Bethe, H. and Brown, G. (1985) How a supernova explodes. *Scientific American* **252** (5), 40–48.

Ferreira, P.G. (2006) *The State of the Universe – A Primer in Modern Cosmology*. London: Weidenfeld and Nicolson.

Lenton, T. and Watson, A. (2011) *Revolutions that Made the Earth*. Oxford: Oxford University Press.

Lin, D.N.C. (2008) The genesis of planets. *Scientific American* **298** (5), 50–59.

Riordan, M. and Schramm, D.M. (1993) *The Shadows of Creation – Dark Matter and the Structure of the Universe*, 2nd edition. Oxford: Oxford Paperbacks.

Taylor, S.R. and McLennan, S.M. (1985) *The Continental Crust: Its Composition and Evolution*. Oxford: Blackwell.

Weinberg, S. (1993) *The First Three Minutes*, 2nd edition. New York: Basic Books.

White, W.M. (2013) *Geochemistry*. Chichester: Wiley-Blackwell.

Exercises

11.1 Examine the values of Z and N up to the element lead in Figure 10.1.1 (nuclide chart, Box 10.1) for which there are no naturally occurring nuclides. What do they have in common? Why?

11.2 The decay of the short-lived isotope ^{26}Al (half-life 0.7 million years) is thought to have provided an important source of heating during the early history of the Solar System. Calculate (a) the decay constant of ^{26}Al, and (b) the time required for the rate of ^{26}Al heat production to fall to a hundredth of its initial value.

11.3 Identify the elements represented by the six unlabelled data points in Figure 11.2 (enlarged inset). To which subgroups in the Periodic Table do they belong?

11.4 Why are volatile elements more depleted in the Moon than the Earth (Figure 11.5)? Why does the Moon have a small core compared with other planetary bodies?

11.5 Highlight the mistakes in the following statements:

(a) Nuclides sharing the same number of neutrons are called isotopes.

(b) Chemical elements are the products of fusion reactions in stars.

(c) Dark matter consists of chemical elements that just don't emit light.

(d) The s-process occurs only in supernovae.

(e) Siderophile elements in the Earth occur only in the core.

(f) The beginning of photosynthesis led immediately to atmospheric oxygen.

ANSWERS TO EXERCISES

Chapter 2

2.1 Point X Phases present are calcite + quartz + CO^2 gas.

$\phi = 3. C = 3(CaO, SiO_2, CO_2)$.

$3 + F = 3 + 2 \rightarrow F = 2$.

Point Y, $\phi = $ calcite + quartz + wollastonite + $CO_2 = 4 \rightarrow F = 1$. Temperature and P_{CO_2} can vary independently at point X without changing the equilibrium assemblage.

2.2 The lower density of ice indicates that at 0 °C, $V_{ice} > V_{water}$. For the reaction:

$$\underset{\text{ice}}{H_2O} \rightleftharpoons \underset{\text{water}}{H_2O}$$

$\Delta S = $ +ve (always true for melting).

$\Delta V = $ −ve. Therefore $dP/dt = \Delta S / \Delta V = $ −ve.

The negative slope of the melting curve indicates that the melting temperature falls as pressure is increased.

2.3 $\Delta S = 202.7 + (2 \times 82.0) - 241.4 - 41.5 = 83.8 \, J \, K^{-1} mol^{-1}$

$\Delta V = 32.6 \times 10^{-6} \, m^3 \, mol^{-1}$

$\Delta S / \Delta V = 2.57 \times 10^6 \, J \, K^{-1} m^{-3} = 2.57 \times 10^6 \, Pa \, K^{-1}$

We know one point on the reaction boundary ($10^5 \, Pa$ at 520 °C). At 520 + 300 °C the pressure on the boundary will be $300 \times 2.57 \times 10^6 \, Pa = 7.710 \times 10^8 \, Pa$. As all phases are anhydrous, a straight reaction boundary is expected between these two points. The volume of the grossular + quartz assemblage is less than the anorthite + 2(wollastonite) assemblage, and therefore it will be found on the high-pressure side.

Chemical Fundamentals of Geology and Environmental Geoscience, Third Edition. Robin Gill.
© 2015 John Wiley & Sons, Ltd. Published 2015 by John Wiley & Sons, Ltd.
Companion Website: www.wiley.com/go/gill/chemicalfundamentals

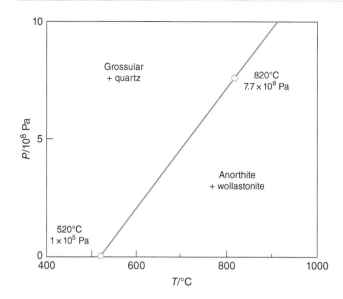

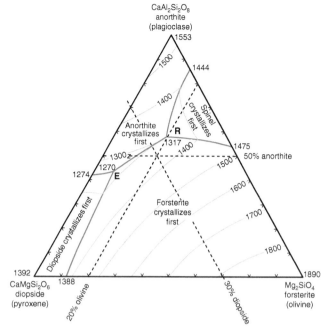

2.4 (a) 1400°: liquidus to vertical dashed line is 4.0 mm.

Solidus to vertical line is 23 mm.

Liquid/crystals = 23/4.0 = 5.8

$$\text{or percent liquid} = \frac{23 \times 100}{(4.0 + 23)} = 85\%$$

Liquid composition = An_{35}. Crystals An_{73}.

(b) Liquid/crystals = 0.55. Percent melt = 36%. Liquid An_{15}. Crystals An_{54}.

(c) Liquid/crystals = 0. Liquid An_7. Crystals An_{40}.

2.5 A rock consisting of plagioclase, diopside, nepheline and olivine can be plotted in a $CaAl_2Si_2O_8$–$CaMgSi_2O_6$–Mg_2SiO_4 ternary diagram, *provided that* the percentages of plagioclase (represented by $CaAl_2Si_2O_8$), diopside (represented by $CaMgSi_2O_6$) and olivine (represented by Mg_2SiO_4) are recalculated to add up to 100% (Box 2.7).

Using the data given, plag + diopside + olivine = only 85%. Therefore, eliminating nepheline (not represented in this diagram), the values to plot are calculated as follows: plag = 42.55% × 100/85 = 50%. Diopside = 25.5% × 100/85 = 30%. Olivine = 17% × 100/85 = 20% (check total = 100%). The composition plots as shown:

The composition plots in the forsterite field, meaning this will be the first mineral to crystallize from a melt of this composition as it cools to the liquidus.

2.6 The required diopside–plagioclase mixture is *d*:

$$\text{Percent diopside in } d = \frac{40}{46 + 40} \times 100 = 47\%.$$

$$\text{Percent plagioclase } c = \frac{46}{46 + 40} \times 100 \times 53\%$$

Solid mixture *a* consists of plagioclase (average composition *f* = An_{31}) and diopside. The lever rule gives 41% *f* and 59% diopside.

a lies on the boundary of the three-phase triangle melt–diopside–plagioclase at 1220 °C.

The equilibrium assemblage is melt *b* 58%, diopside 42% and plagioclase *c* 0%.

2.7 $20 \times 10^8 \, Pa \equiv 67 \, km$ from Equation 2.1.

(a) $Di_{69} \, Fo_{07} \, En_{24}$

(b) $Di_{69} \, Fo_{07} \, En_{24}$

The first melt to form has the composition E in each case.

Chapter 3

3.1 $N_0 = 18{,}032$. Calculate $\ln(N_0/N)$ for each value of t. (As a check on arithmetic, the value for $t = 25$ is 0.09426.) $\ln(N_0/N)$ gives a linear plot against time, indicating that the reaction is first-order.

After one half-life, $N/N_0 = 1/2$, therefore $\ln(N_0/N) = 0.6913$. Reading from the graph, this value is reached at $t = 190$ hours. If $n =$ the number of half-lives required for decay to $1/100$, $(1/2)^n = 1/100$, so $n\log(1/2) = \log(1/100)$, therefore $n = 6.6$ and $t_{1/100} = 1254$ hours.

3.2 Room temperature $= 25\,°C = 298\,K$. The doubling of reaction rate can be written:

$$k_{308} = 2k_{298}$$

The Arrhenius equation in log form gives simultaneous equations:

at $298\,K$: $\ln(k_{298}) = \ln A - E_a/(8.314 \times 298)$
at $308\,K$: $\ln(2k_{298}) = \ln A - E_a/(8.314 \times 308)$
Therefore $\ln A = \ln k_{298} + E_a/2478 =$
$\ln 2k_{298} + E_a/2561$.
Rearranging, $\ln 2k_{298} - \ln k_{298} = \ln 2$
$= E_a(1/2478 - 1/2561)$.
$E_a = 0.6915/0.013 \times 10^{-3} = 52{,}900\,\mathrm{J\,mol^{-1}} = 52.9\,\mathrm{kJ\,mol^{-1}}$

3.3 Calculate $\ln(1/\text{viscosity})$ for each temperature and plot against $1/T$ (e.g. for $T = 1325\,°C = 1598\,K$, $1/T = 0.000626\,\mathrm{K^{-1}}$ and $\ln(1/\eta) = -7.622$). Slope of graph $= -34{,}030\,\mathrm{K} = -E_a/R$.
Thus $E_a = 283\,\mathrm{kJ\,mol^{-1}}$.

3.4 Half-life $= \ln 2/\lambda_{87_{Rb}} = 4.9 \times 10^{10}$ years.
Therefore $\ln(N_0/N) = \lambda t = 1.42 \times 10^{-11} \times 4.6 \times 10^9 = 0.0653$
therefore $N_0/N = 1.068$
$\therefore N/N_0 = 94\%$
Therefore percentage decayed $= 6\%$.

3.5 Equation 3.11 $\left(\text{Fick's First Law of Diffusion:}\right.$
$f_i = -D_i \dfrac{dc_i}{dx}\Big)$ can be rearranged to express the diffusion coefficient D_i in terms of f_i, c_i and x:

$$D_i = \frac{f_i}{(dc_i/dx)}$$

The flux f_i has the units $\mathbf{mol\,m^{-2}\,s^{-1}}$ (see text *above* Equation 3.11). The concentration gradient $\dfrac{dc_i}{dx}$ has the units $(\mathrm{mol\,m^{-3}})\mathrm{m^{-1}} = \mathbf{mol\,m^{-4}}$ (see text *below* Equation 3.11). Therefore the units of D_i are

$\dfrac{\mathrm{mol\,m^{-2}\,s^{-1}}}{\mathrm{mol\,m^{-4}}} = \mathbf{m^2\,s^{-1}}$, since $\mathrm{m^{-2}/m^{-4}}) = \mathrm{m^2}$, and the mol terms cancel out.

Chapter 4

4.1 (a) $\underset{\text{solid}}{BaSO_4} \rightarrow \underset{\text{solution}}{Ba^{2+} + SO_4^{2-}}$ (Equation 4.17)

$$K_{BaSO_4} = a_{Ba^{2+}} \cdot a_{SO_4^{2-}} = \frac{m_{Ba^{2+}}}{m^\theta} \cdot \frac{m_{SO_4^{2-}}}{m^\theta} = 10^{-10}$$

When dissolved in pure water: $a_{Ba^{2+}} \cdot a_{SO_4^{2-}} = 10^{-5}$

Therefore if solution is ideal, $m_{Ba^{2+}} = m_{SO_4^{2-}} = 10^{-5}\,\mathrm{mol\,kg^{-1}}$ $BaSO_4$ in saturated solution.

(b) In $CaSO_4$ solution, $a_{SO_4^{2-}} = a_{Ca^{2+}} = 10^{-3}$
If $x.m\,\mathrm{mol\,kg^{-1}}$ of $BaSO_4$ dissolves:

$$K_{BaSO_4} = a_{Ba^{2+}} \cdot a_{SO_4^{2-}} = 10^{-10}$$
$$= x\left(10^{-3} + x\right) \cong 10^{-3}\,x$$
(since x^2 is very small)

Therefore $x.m = 10^{-7}\,\mathrm{mol\,kg^{-1}}$.

4.2 $\underset{\text{solid}}{CaF_2} \rightarrow \underset{\text{solution}}{Ca^{2+} + 2F^-}$

$$K_{CaF_2} = a_{Ca^{2+}} \cdot \left(a_{F^-}\right)^2$$

In pure water $a_{F^-} = 2a_{Ca^{2+}}$
Therefore $K_{CaF_2} = a_{Ca^{2+}} \cdot \left(2a_{Ca^{2+}}\right)^2 = 4\left(a_{Ca^{2+}}\right)^3 = 10^{-10.4}$

$$\cong 4 \times 10^{11}$$

Therefore $a_{Ca^{2+}} = 0.00022$
$$m_{Ca^{2+}} = 0.00022\,\mathrm{mol\,kg^{-1}}$$

Relative molecular mass of $CaF_2 = 40 + (2 \times 19) = 78$
Therefore $0.00022 \times 78 = 0.017\,\mathrm{g}$ CaF_2 will dissolve in $1\,\mathrm{kg}$ water at $25\,°C$.

4.3 $CO_2 + H_2O \rightleftharpoons H_2CO_3$

$$K = a_{H_2CO_3}/\left(P_{CO_2}\right)^{air} = 0.031$$

$$a_{H_2CO_3} = 0.031 \times 0.00028 = 0.00000868 = 10^{-5.06}$$

According to equation 4.21: $H_2CO_3 \rightleftharpoons H^+ + HCO_3^-$

$$K = \frac{a_{H^+} \cdot a_{HCO_3^-}}{a_{H_2CO_3}} = 10^{-6.4}$$

Therefore $a_{H^+} \cdot a_{HCO_3^-} = a_{H_2CO_3} \times K$

$$= 10^{-5.06} \times 10^{-6.4}$$
$$= 10^{-11.46}$$

$a_{H^+} = a_{HCO_3^-} = 10^{-5.3}$

Therefore pH = 5.73

This is the pH value for pre-industrial rainwater. The values for 1960 and 2012 are 5.71 and 5.66, a pH change of −0.05.

Chapter 5

5.1 2p, 3s, 4f, 5d

5.2 $_6$C $1s^2 2s^2 2p^2$
  $_{11}$Na $1s^2 2s^2 2p^6 3s^1$ or $[Ne]3s^1$
  $_{13}$Al $[Ne]3s^2 3p^1$
  $_{17}$Cl $[Ne]3s^2 3p^5$
  $_{18}$Ar $[Ne]3s^2 3p^6 = [Ar]$
  $_{26}$Fe $[Ar]4s^2 3d^6$

Chapter 6

6.1 The Pauli principle dictates that each orbital can hold no more than two electrons:

Z	Name	Configuration		Block	Group	Valency
		Core	*Valence*			
3	Li	$1s^2$	$2s^1$	s	I	1
5	B	$1s^2$	$2s^2 2p^1$	p	III	3
8	O	$1s^2$	$2s^2 2p^4$	p	VI	−2
9	F	$1s^2$	$2s^2 2p^5$	p	VII	−1
14	Si	$1s^2 2s^2 2p^6$	$3s^2 3p^2$	P	IV	4

6.2 Ti $[Ar] 4s^2 3d^2$: block d
Ni $[Ar] 4s^2 3d^8$: block d
As $[Ar] 4s^2 3d^{10} 4p^3$: block p
U $[Rn] 7s^2 6d^1 5f^3$: block f

6.3 Na valency = 1. Oxygen = −2, therefore there are 2 atoms of sodium per atom of oxygen:

	Na_2O	$x=2$	$y=1$
Similarly	SiO_2	$x=1$	$y=2$
	SiF_4	$x=1$	$y=4$
	$MgCl_2$	$x=1$	$y=2$

Valency of scandium is 3, valency of oxygen = −2. Each Sc combines with 3/2 atoms of oxygen, or 2 Sc atoms combine with 3 oxygen atoms:

	Sc_2O_3	$x=2$	$y=3$
Similarly	P_2O_5	$x=2$	$y=5$
	BN	$x=1$	$y=1$

6.4 (a) Rewrite Moseley's Law in the linear form:

$$(1/\lambda)^{\frac{1}{2}} = k^{\frac{1}{2}}Z - k^{\frac{1}{2}}\sigma$$

i.e. $y = mx + c$

For each point calculate $(1/\lambda)^{\frac{1}{2}}$ and plot against Z. Points fall on a straight line if the equation holds.

$$k^{\frac{1}{2}} = \text{gradient} = \frac{(1/56.1)^{\frac{1}{2}} - (1/83.0)^{\frac{1}{2}}}{47-39}$$
$$= 0.00297 \ \text{pm}^{\frac{1}{2}}$$

$$\sigma = Z - 1/k^{\frac{1}{2}}(1/\lambda)^{\frac{1}{2}}$$

For yttrium (Y), for example, $\sigma = 39 - \dfrac{0.1098}{0.00296}$
$= 2.022$

(b) Read off graph at $Z = 43$:

$$(1/\lambda)^{-\frac{1}{2}} = 0.1216 (\text{pm})^{-\frac{1}{2}}$$

Therefore $\lambda = 67.6 \ \text{pm} = 0.676 \times 10^{-10} \ \text{m}$

$$E_q = h\upsilon = hc/\lambda$$
$$= \frac{(4.135 \times 10^{-15})(2.997 \times 10^8)}{0.676 \times 10^{-10}} \frac{\text{eV s m s}^{-1}}{\text{m}}$$
$$= 18330 \ \text{eV} = 18.33 \ \text{keV}$$

Chapter 7

7.1 Ionic radii in Box 7.2 lead to following radius ratios:
Si$^+$: 0.26. Al^{3+}: 0.36 (tet)/0.46 (oct). Ti^{4+}: 0.52. Fe^{3+}: 0.55. Fe^{2+}: 0.65. Mg^{2+}: 0.61. Ca^{2+}: 0.91. Na$^+$: 0.94 (8-fold). K$^+$: 1.20.
Co-ordination numbers: 4: Si^{4+}, Al^{3+}. 6: Al^{3+}, Ti^{4+}, Fe^{3+}, Mg^{2+}. 8: Ca^{2+}, Na$^+$. 12: K$^+$.

7.2 In the higher oxidation state (Fe^{3+}, Eu^{3+}), the ion has fewer valence electrons and there is less mutual repulsion between them, so ions are smaller than 2+ ions. Ionic potentials: Fe^{2+}: 23. Fe^{3+}: 41. Eu^{2+}: 16. Eu^{3+}: 28. The 3+ ion has a higher polarizing power than the 2+ ion, and therefore forms a more covalent bond.

7.3 He–He: van der Waals interaction; very weak, so He is a monatomic gas at room temperature.
    Ho–Ho: electronegativity of Ho is between 1.1 and 1.3, therefore metallic bond. Crystalline metal at room temperature.
    Ge–Ge: metallic/covalent bond. Electronegativity similar to Si in Figure 7.8b. Semiconductor at room temperature.

7.4 Using Figures 6.3 and 7.8a:

KCl: Electroneg. difference = 2.4. 80% ionic.

TiO_2:   Electroneg. difference = 1.9. 65% ionic.

MoS_2:   Electroneg. difference = 0.4. Mean electroneg. = 2.4. Submetallic bond.

NiAs: Electroneg. difference = 0.3. Mean electroneg. = 2.1. Submetallic bond.

$CaSO_4$:   (a) S–O bond: <20% ionic. Largely covalent.
(b) Ca^{2+}–$(SO_4)^{2-}$: more ionic.

$CaSO_4.2H_2O$: (a) Ca^{2+}–H_2O: ion–dipole interaction (hydration).
(b) S–O: largely covalent as above.
(c) $(Ca^{2+}.2H_2O)$–$(SO_4)^{2-}$: ionic.

Chapter 8

8.1   Edenite Z_8O_{22}, i.e. Z:O = 1:2.75; amphibole (double chain silicate, Table 8.1).

Hedenbergite Si_2O_6, i.e. Z:O = 1:3; single-chain silicate (or perhaps a ring silicate). Hedenbergite is a pyroxene.

Paragonite $Z_4O_{10}(OH)_2$ i.e. Z:O = 1:2.5; sheet silicate (mica).

Leucite $(AlSi_2O_6) \equiv Z_3O_6$, i.e. Z:O = 1:2; framework silicate, i.e. fully polymerized.

Acmite Si_2O_6, i.e. Z:O = 1:3; single-chain (or ring) silicate. Acmite is also a pyroxene.

8.2

	Garnet
Si	17.99
Al	9.57
Ti	0.33
Fe(3)	3.97
Fe(2)	2.92
Mn	0.50
Mg	0.46
Ca	22.58
O	41.27
	99.60

The difference in analysis total represents rounding errors.

8.3

	Garnet		Epidote		Pyroxene		Feldspar	
Si	2.9909	} 2.991	3.0035	} 3.004	1.8589	} 2.000	9.5754	
Al^{IV}	–		–		0.1411		6.3618	} 16.053
Al^{VI}	1.6515		2.8055		0.0217		–	
Ti	0.0321	} 2.014	–	} 3.044	0.0355		–	
Fe(III)	0.3307		0.2389		0.0272		0.1157	
Fe(II)	0.2436		0.0266		0.6730	} 2.058	–	
Mn	0.0419	} 2.997	0.005	} 1.949	0.0379		–	
Mg	0.0876		0.0014		0.4312		–	
Ca	2.6235		1.9205		0.7366		2.3799	
Na	–		–		0.0952		1.4753	} 3.886
K	–		–		–		0.0306	
H	–		0.9548	0.955	–		–	

8.4 Recalculate the molar proportions of the three components to 100%:

	per 6 oxygens	100/1.9059
Ca	0.7366	38.65
Mg	0.4312	22.62
$Fe^{2+} + Fe^{3+} + Mn$	0.7381	38.72
Total	1.9059	100.00

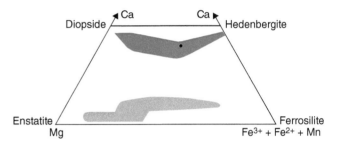

The solid line encloses the *pyroxene quadrilateral*, where most pyroxene analyses would plot. (The shaded areas show the range of pyroxene compositions found in gabbros. The two bands lie on either side of a solvus.)

Chapter 9

9.1 (a) $Fe_2(SO_4)_3 + FeS_2 \rightarrow 3FeSO_4 + 2S$
(b) $CuFeS_2 + 2Fe_2(SO_4)_3 \rightarrow CuSO_4 + 5FeSO_4 + 2S$
(c) $12FeSO_4 + 3O_2 + 6H_2O \rightarrow 4Fe_2(SO_4)_3 + 4Fe(OH)_3$
(d) $MnO_2 + 4H^+ + 2Fe^{2+} \rightarrow Mn^{2+} + 2H_2O + 2Fe^{3+}$
(e) $ZnS + 4Fe_2(SO_4)_3 + 4H_2O \rightarrow ZnSO_4 + 8FeSO_4 + 4H_2SO_4$

Chapter 10

10.1 (a) Be, F, Na, Al, P, Sc, Mn, Co, As, Y, Nb, Rh, Pr, Tb, Ho, Tm, Au and Bi are monoisotopic.
(b) Tc, Pm, Po, At, Rn, Fr, Ra, Ac, Th, Pa and U have no stable isotopes at all.

10.2 A horizontal line from 0.707980 intercepts the curve in Figure 10.7 at 34.0 Ma (see figure).

0.707980 + 0.000022 = 0.708002 which translates into ~34.5 Ma;

0.707980 − 0.000022 = 0.707958 which translates into ~33.5 Ma.

The age of the carbonate sediment is therefore 34.0 ± 0.5 Ma.

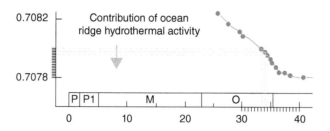

10.3

Exercise 11.4	147/144	143/144
whole rock	0.20034	0.511814
plag	0.09627	0.509965
opx	0.28428	0.513317
cpx	0.24589	0.512628
Max minus min	0.18801	0.00335
Gradient	0.01783	
Gradient +1	1.01783	
ln (grad + 1)	0.01767	
Decay constant	6.54E−12	
Age/y	2.70E+09	
Initial ratio by eye	0.50825	

Draw a line by eye through the data; the extrapolated line intercepts the y-axis at about 0.50826 (=initial ratio). The gradient of the line leads to a mineral age for the Stillwater gabbro of 2.70 Ga.

10.4 From Equation 3.2.5, half-life and decay constant are related by $t_{\frac{1}{2}} = 0.6931/\lambda$ therefore $\lambda = 0.6931/t_{\frac{1}{2}}$, so in this case $\lambda_C = 0.6931/5730 = 1.210 \times 10^{-4}$ year^{-1}.

The count rates are proportional to the concentration of ^{14}C in each carbon sample. Plugging relevant numbers into Equation 10.13:

$$t = \frac{1}{1.210 \times 10^{-4}} \ln\left[\frac{13.56}{5.8}\right] = \frac{0.8493}{1.210 \times 10^{-4}}$$

$$= 7019 \text{ years before present (BP)}.$$

10.5 Using Equation 10.10 for $\delta^{18}O = -1.2, -1.3$ and $-1.4‰$, the temperatures obtained are 21.8, 22.3 and 22.8 °C. The best estimate is therefore 22.3 ± 0.5 °C.

Calcite $\delta^{18}O$	Seawater $\delta^{18}O$	T/°C
−1.4	0	22.8
−1.3	**0**	**22.3**
−1.2	0	21.8

The $\delta^{18}O$ of seawater is believed to vary by about 1‰ (±0.5‰) between glacial maximum and inter-glacial conditions owing to variation in the amount of 'light' (low–$\delta^{18}O$) water locked up in polar ice-sheets. Potentially this factor introduces an uncertainty of ±2.3 °C in the temperature estimate.

Calcite $\delta^{18}O$	Seawater $\delta^{18}O$	T/°C
−1.3	−0.5	20.0
−1.3	**0**	**22.3**
−1.3	0.5	24.7

10.6 (a) Correct.
   (b) *Incorrect:* ^{87}Sr is a naturally occurring isotope; some ^{87}Sr in the rock was already present at the time of its formation (Figure 10.1b).
   (c) Correct.
   (d) *Incorrect:* ^{143}Nd/^{144}Nd grows more *slowly* in continental crust than in the mantle (Figure 10.6b).
   (e) *Partially incorrect:* only artefacts and rocks that (1) incorporate carbon of biological origin (e.g. carbonized tree debris caught up in a volcanic deposit) and (2) are young enough (age < 40 ka) can be dated using ^{14}C.
   (f) *Incorrect:* the mass-independent fractionation of S isotopes in Archaean times illustrates one instance where this statement does not apply (Figure 10.14).

Chapter 11

11.1 There are no naturally occurring nuclides (only short-lived radionuclides) at N values 19, 35, 39, 45, 61, 89 and 123, nor at Z values 43 (technetium) and 61 (promethium). These are all odd numbers. Odd values of Z and N signify half-filled nuclear orbitals, and such nuclei are more prone to trans-mutation into other elements by fusion, neutron capture or radioactive decay than even-valued nuclides.

11.2 (a) The integrated rate equation for radioactive decay (Equation 3.2.4 in Box 3.2) is:

$$\ln\left(n_{26}^0/n_{26}\right) = \lambda_{26}t$$

where n_{26} represents the changing amount of ^{26}Al (e.g. in a planetary body), and n_{26}^0 is the value when $t = 0$, and λ_{26} is the decay constant of ^{26}Al. After one half-life ($t = 0.7$ million years):

$$\ln(1/0.5) = 0.7 \times 10^6 \; \lambda_{26}$$

Therefore $\lambda_{26} = 0.99 \times 10^{-6}$ year^{-1}.
   (b) Heat output is proportional to the rate of decay of ^{26}Al and therefore to its abundance. When ^{26}Al has decayed to 1% of original value:

$$\ln(1/0.01) = 0.99 \times 10^{-6} \; t$$

Therefore $t = \ln(100) \times 1.01 \times 10^6$ years $= 4.6$ million years. For this heat source to have been significant during planet formation, planets must have formed within a few million years of nucleosynthesis of ^{26}Al.

11.3 From their atomic numbers, they are xenon Xe ($Z = 54$, inert gas), caesium Cs ($Z = 55$, subgroup Ia), barium Ba ($Z = 56$, IIa), hafnium Hf ($Z = 72$, IVa), tantalum Ta ($Z = 73$, Va) and tungsten W ($Z = 74$, VIa).

11.4 In the giant impact that is believed to have created the Moon, it is likely that the impactor's mantle completely vaporized before re-condensing to form the Moon. Through taking longer to con-dense, the volatile elements are more likely to have been swept away by the solar wind before incorporation into the solid Moon. Modelling suggests that most of the metallic core of the impactor became incorporated into the Earth's core, rather than being available in orbit for incor-poration in the Moon.

11.5 (a) Isotopes are nuclides that share the same value of *atomic number Z* (i.e. have the same number of *protons* and form part of the same element) but possess *different* numbers of neutrons.
   (b) Only the elements from C to Fe are products of nuclear fusion in stars (Figure 11.3). The ele-ments H, He and Li are believed to be prod-ucts of nucleosynthesis *in the primordial Big Bang*, not in stars (Figure 11.2.1) The elements Be and B are bypassed by stellar fusion reac-tions and are mainly formed by *spallation* of heavier nuclei. Elements heavier than Fe

cannot be formed by fusion (Figure 11.3.1) and are instead products of progressive *neutron capture* in stars and supernovae.

(c) Perhaps 5–10% of dark matter consists of dark **baryonic matter** ('ordinary stuff that doesn't shine', Ferreira, 2006); the remaining 90–95% is believed to consist of *exotic subnuclear particles yet to be identified*.

(d) *The s(=slow)-process neutron capture reactions can occur in stellar interiors*. The heaviest nuclides are however formed by more rapid ('*r-process*') neutron capture that can only operate in the very high neutron fluxes that occur in *supernovae*.

(e) Siderophile elements are indeed concentrated in the Earth's metallic core, but many (notably Fe and Ni) *also have lithophile and/or chalcophile tendencies* (as shown by the overlapping fields in Figure 11.4 and multicoloured cells in Plate 7) and therefore can be found in silicate and sulfide parts of the Earth too.

(f) Oxygenic photosynthesis is believed to have begun in the oceans around 3.5 Ga ago, but the oxygen initially formed there is believed to have been immediately mopped up in oxidizing the accumulated reduced species in seawater like Fe^{2+} (as implied by BIF sediments). Only when this 'oxygen sink' had been completely oxidized could *excess oxygen* escape, around the end of the Archaean, to form an oxygenated atmosphere (Figure 11.8).

APPENDIX A: MATHEMATICS REVISION

SI units of measurement

In 1960, the Système International d'Unités ('SI units') convention was introduced to promote standardization between sciences. The following are the most important features (full details are available at http://physics.nist.gov/cuu/Units/introduction.html):

(a) The basic units are as given in Table A1.

Table A1 Basic SI units

Dimension	Unit	Abbreviation
Length	metre	m
Mass	kilogram	kg
Time	second	s
Electric current	ampere	A
Temperature	kelvin	K
Amount of substance	mole	mol

The kelvin is the unit of temperature measured from absolute zero. $T(K) = T(°C) + 273.15$. Degrees Celsius (°C) continue to be widely used by geoscientists.

Volume is expressed in m^3; in solution chemistry (Chapter 4), the litre (L) has been superseded by dm^3 (cubic decimetres) in SI notation: $1 dm^3 = 10^{-3} m^3 = 1$ litre.

(b) Names are given to various 'derived' SI units (all named after scientific luminaries of the past) which consist of combinations of the basic units above. The derived units are inconvenient to remember at first, but make the system much more concise to use – see Table A2 for examples.

Notice that some derived units have several equivalent forms and may be expressed in alternative ways, either in the basic units or in terms of other derived units. An example is the constant g, the acceleration due to gravity, which may be expressed $m s^{-2}$ or as $N kg^{-1}$, the two forms being exactly equivalent.

Chemical Fundamentals of Geology and Environmental Geoscience, Third Edition. Robin Gill.
© 2015 John Wiley & Sons, Ltd. Published 2015 by John Wiley & Sons, Ltd.
Companion Website: www.wiley.com/go/gill/chemicalfundamentals

Table A2 Derived SI units

Quantity	Unit	Abbreviation	In terms of basic SI units
Force	newton	N	$kg\,m\,s^{-2} = J\,m^{-1}$
Energy, work	joule	J	$kg\,m^2\,s^{-2} = N\,m$
Power	watt	W	$kg\,m^2\,s^{-3} = J\,s^{-1}$
Pressure	pascal	Pa	$kg\,m^{-1}\,s^{-2} = N\,m^{-2} = J\,m^{-3}$
Electric charge	coulomb	C	$A\,s$
Electrical potential difference	volt	V	$kg\,m^2\,s^{-3}\,A^{-1} = W\,A^{-1}$

Table A3 SI prefixes and examples of use

petametre	Pm	$= 10^{15}\,m = 10^{12}\,km$
terabyte	Tb	$= 10^{12}\,bytes$
gigapascal	GPa	$= 10^9\,Pa$
megajoule	MJ	$= 10^6\,J$
kilometre	km	$= 10^3\,m$
millisecond	ms	$= 0.001\,s = 10^{-3}\,s$
micrometre	μm	$= 10^{-6}\,m$
nanogram	ng	$= 10^{-9}\,g = 10^{-12}\,kg$
picometre	pm	$= 10^{-12}\,m$

(c) The units of heat are the same as those of mechanical energy and work, avoiding the need for the 'mechanical equivalent of heat' constant (converting calories to joules) required by earlier systems of units.

(d) The system recognizes prefixes that multiply or divide units by factors of 10^3 (see Table A3).

Note that all prefixes that represent multipliers *greater than* 1 (with the notable historical exception of the k in km) are abbreviated using *capital* letters, whereas prefixes for multipliers *less than* 1 use *lower-case* letters. Popular publications often fail to adhere to this helpful distinction.

Some sciences have been slow to adopt the SI system. For example, units called *bars* (=$10^5\,Pa$) and *kilobars* (kb = $10^8\,Pa$) continued to be used as units of pressure in geological phase diagrams long after formal adoption of the SI system.

Equation of a straight line

Figure A1(a) shows a straight line plotted in the conventional way, against perpendicular x and y axes. For a line with this orientation, increasing the value of x (e.g. from point p to point q) brings about a

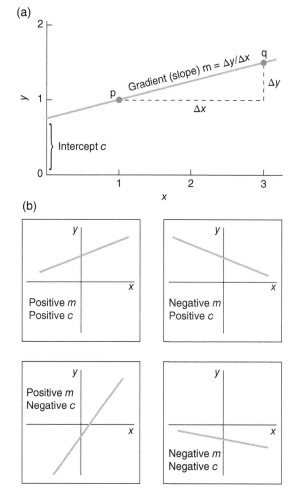

Figure A1 (a) Parameters of a straight line in x–y space. Note that both axes are plotted at the same scale here, although this is not the case for all graphs; this plot also shows the origin (where $x = 0$, $y = 0$), although not all graphs do so. (b) How m and c reflect location and orientation.

corresponding increase in the value of y. The rate at which y rises with increasing x is measured by the *gradient* or *slope* of the line, m:

$$m = \frac{\Delta y}{\Delta x} \tag{A1}$$

Here $\Delta y = 0.5$ and $\Delta x = 2.0$, so $m = 0.5/2.0 = 0.25$. A steep line would have a large value of m, whereas a shallow line would have a low value of m. Note that the value of y at which the line crosses the y axis (where $x = 0$) is called the *intercept*, usually symbolized as c. If x and y had units attached, m and c would need to be given consistent units: for example, if y were measured in km and x were measured in hours, the units of m would be km hour^{-1} and those of c would be km.

By definition, a straight line has a constant value of m throughout its length (variation in the value of m is characteristic of a *curve*). The x and y coordinates of any point on the straight line shown in Figure A1(a) are related by the equation:

$$y = mx + c \tag{A2}$$

The line shown in Figure A1(a) is said to have a *positive* gradient; if it sloped in the other direction, so that an increase in x brought about a *decrease* in y (negative Δy), the gradient in Equation A1 would have a *negative* value. Figure A1(b) shows how positive and negative values of m and c impact on line orientation and location.

Gradient of a curve

Many problems in geochemistry require us to determine the slope or gradient of a curve. The gradient tells us how rapidly one variable (y) is changing in response to variation in the value of another (x) at a particular point on the curve. If we plot the position y of a car travelling along a straight road against time t, for instance (Figure A2), we get a curve whose gradient at each point along the curve tells us the car's speed at the instant concerned: the steeper the graph, the faster the car was travelling. The horizontal portions of the curve, on the other hand, indicate when the car was stationary, for example at a red traffic light.

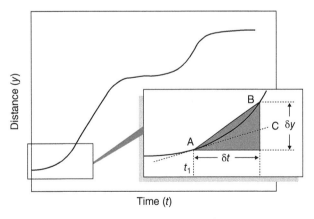

Figure A2 Illustrative plot of a car journey in terms of distance travelled y versus elapsed time.

How do we measure the gradient (the car's speed) at a specific point on this curve? Figure A2 shows a portion of the curve enlarged, and we want to measure the gradient at the point t_1. We can begin by determining the slope of the chord AB, as a first approximation to that of the curve:

$$\text{Gradient of chord AB} = \frac{\delta y}{\delta t} \tag{A3}$$

δt ('little delta t') represents an *increment* (a small increase) in t, and δy is the consequent increment in y. If we imagine making δt (and consequently δy too) smaller and smaller, the slope of the chord grows closer and closer to the gradient of the curve at t_1. If we reduce δt to the point where it is effectively zero (written symbolically $\delta t \to 0$), the chord will coincide with the tangent[1] to the curve, AC, which by definition has the same gradient as the curve itself at A. Expressed in terms of such **infinitesimal** increments, which are written dy and dt, the slope of the curve at point A is:

$$\text{Gradient at A} = \frac{dy}{dt} \tag{A4}$$

Differentiation

This symbolism takes on more meaning when the variation of y with t can be expressed in the form of an equation. Consider a different relationship between y and t, represented by the equation:

$$y = at^2 + c \tag{A5}$$

We can consider a chord AB analogous to that in the enlargement in Figure A2. Since point B lies on the curve, it must also satisfy this equation:

$$\begin{aligned}(y + \delta y) &= a(t + \delta t)^2 + c \\ &= [at^2 + 2at\delta t + a(\delta t)^2 + c]\end{aligned} \tag{A6}$$

Therefore $\delta y = 2at\delta t + a(\delta t)^2$

---

[1] A straight line just grazing a curve, which has the same gradient as the curve has at the point of contact.

So $\qquad \dfrac{\delta y}{\delta t} = 2at + a\delta t$

As $\delta t \to 0$, $\dfrac{dy}{dt} = 2at + 0 = 2at$ \hfill (A7)

This indicates that the gradient at any specific point t_1 on the curve can be calculated simply by evaluating $2at_1$. The expression $2at$ is called the **derivative** of $y = at^2 + c$ with respect to t. The algebraic or arithmetical process by which we calculate the derivative is called **differentiation**, the basic operation in a branch of mathematics called *differential calculus*.

By similar reasoning one can show that the derivative of $y = at^3$ is $3at^2$, and the derivative of $y = at^4$ is $4at^3$ and so on. In general:

if $\qquad y = bx^n$

then $\dfrac{dy}{dx} = nbx^{n-1}$.

For example, if $y = ax^3 + bx^2 + cx + d$
$$= 15.2x^3 + 2.9x^2 - 16.8x - 4.3$$
then $\dfrac{dy}{dx} = 3ax^2 + 2bx + c = 45.6x^2 + 5.8x - 16.8$

To determine the numerical value of the gradient at a specific value of x, one simply introduces that value of x into the dy/dx equation:

For example, $x = 2.0$:

$$\dfrac{dy}{dx} = 182.4 + 11.6 - 16.8 = 177.2.$$

Thus one can evaluate the slope of the curve at any desired point.

For a more extensive introduction to differential calculus, you should refer to the book by Waltham (2000).

Logarithms ('logs') and their inverse

Logarithms provide a handy means of plotting a function that varies over several orders of magnitude, compressing it in a way that allows detail to be seen throughout the range.

Logarithms used in this book are of two kinds:

(1) *'Log to the base 10'*: $\log_{10}(X) = Y$ where $X = 10^Y$. In words, the $\log_{10}$ of a number X is the power Y to which 10 would have to be raised to equal X. Since $100 = 10^2$, the $\log_{10}$ of $100 = 2.00$. 4512 can be expressed as $10^{3.654}$, so $\log_{10}(4512) = 3.654$.

(2) *Natural logarithm* = 'log to the base e' (where $e = 2.71828$, an important mathematical constant), usually symbolized 'ln': $\ln(X) = Z$ where $X = e^Z$.

In words, the natural logarithm of a number X is the power Z to which 2.71828 would have to be raised to equal X. Since 100 can be expressed as $e^{4.605}$, $\ln(100) = 4.605$.

Logarithms also have the useful property of making an exponential plot linear, as discussed below in relation to the Arrhenius plot.

To transform a logarithm ($\log_{10}(X)$ or $\ln(X)$) back into the original number X, use:

$$X = 10^{\log_{10}(X)} \text{ or } X = e^{\ln(X)}.$$

These calculations are easily done on a calculator or in a spreadsheet.

When we want to plot a graph of a quantity that varies over several orders of magnitude (e.g. from 0.02 to 5000), it may be useful to plot the logarithm of the quantity while labelling the axis with the original values, as in Figures 11.1, 11.2, 11.3 and 11.5. A logarithmic axis like this can be recognized by the appearance of *successive powers of 10* at regular intervals.

Inverse square 'law'

The gravitational force between two bodies of mass m_1 and m_2 conforms to the equation:

$$F = G\frac{m_1 m_2}{r^2} \tag{A8}$$

where G is the gravitational constant ($=6.674 \times 10^{-11}$ $m^3 kg^{-1} s^{-2}$ or $N m^2 kg^{-2}$) and r is the distance between the centres of gravity of the two bodies in metres. A similar equation applies to the electrostatic force between two electric charges q_1 and q_2 (e.g. between electron and nucleus in an atom, as in Chapter 5):

$$F = k_e \frac{q_1 q_2}{r^2} \tag{A9}$$

where k_e is known as Coulomb's constant ($=8.99 \times 10^9$ $N m^2 C^{-2}$).

Equations A8 and A9 have the same form, in which the separation distance r appears as $1/r^2$. Both are examples of an algebraic relationship which – for this reason – is called an *inverse-square law.*[2]

Dimensions and units in calculations

In writing the answer to any numerical problem, one should give two items: the numerical answer *and* its units. The number by itself is incomplete (unless it is a **dimensionless number**).

One must pay attention to units at every stage in a calculation, checking that all the quantities are expressed in compatible units. If one variable is entered in millimetres when it should appear in metres, you are immediately introducing an error of 1000 times.

The procedure for carrying units through a calculation can be illustrated by the Clapeyron equation (Chapter 2). Suppose we wish to know the slope of the phase boundary representing the reaction between two (isochemical) minerals A and B:

Table A4 Molar entropy and volume.

Reaction	$A \rightarrow B$		Units in published tables of S and V
Molar entropies:	S_A	S_B	$J K^{-1} mol^{-1}$
Molar volumes:	V_A	V_B	$m^3 mol^{-1}$

$\Delta S = S_B - S_A$ $J K^{-1} mol^{-1}$

$\Delta V = V_B - V_A$ $m^3 mol^{-1}$

In the Clapeyron equation:

$$\frac{dP}{dT} = \frac{\Delta S}{\Delta V} \qquad \frac{J K^{-1} mol^{-1}}{m^3 mol^{-1}}$$

'mol^{-1}' appears on top and bottom, and therefore cancels out, leaving the units for the gradient as $(J K^{-1}) m^{-3} = J m^{-3} K^{-1} = N m^{-2} K^{-1} = Pa K^{-1}$. Using volumes expressed in $cm^3 mol^{-1}$ would make it necessary to introduce a correction factor of 10^6.

Some physical parameters are 'dimensionless' and therefore have no units. They are pure numbers. *Specific gravity* (density of a substance ÷ density of pure water at 4°C) is an example. The numerical values of such numbers are independent of the units being used in their computation. Thus in calculating specific gravity, the units of density cancel out, provided that the two densities are expressed *in the same units.*

Graphs must show the units in which each of the variables is expressed. Current good practice[3] followed in this book is to label each axis in the form 'quantity/units', e.g. $T/°C$.

Experimental verification of a theoretical relationship

When a mathematical equation is proposed (commonly on theoretical grounds) to describe a phenomenon, one often wishes to test it against available experimental observations. Does it describe the experimental results accurately, or would some other form of equation match the experimental results more closely? The simplest way to answer this question is to plot the experimental data and the theoretical equation together in a suitable graph.

It is important that this be done in such a way that the form of the graph is linear. To see why, consider the verification of the Arrhenius equation relating rate constant to absolute temperature:

$$k = A \exp\left(-\frac{E_a}{RT}\right) \quad or \quad k = Ae^{-E_a/RT} \tag{A10}$$

If we were to plot experimental results for k against T, the results would define a curve as in Figure A3. Unless we happened to know the constants A and E_a in advance – and in general we don't – it would be difficult to determine whether the curve defined by the experimental data has the shape predicted by the equation. It would be necessary to use a complicated curve-fitting calculation to establish the agreement between experimental data and the theoretical equation.

---

[2] Note that when we refer to an equation like A8 as a scientific 'law', it should not be taken to mean that this equation actually *governs* what Nature does: it is simply shorthand for saying that the equation provides our best algebraic *approximation* to what is seen to happen in the natural world. Any such 'law' is a human construct.

[3] See www.animations.physics.unsw.edu.au/jw/graphs.htm#Units.

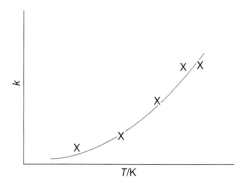

Figure A3 Sketch of variation of rate constant with temperature.

Consider an alternative plot. The Arrhenius equation may be written in a logarithmic form by taking natural logs of both sides:

$$\ln k = \ln A + \ln \left[\exp\left(-\frac{E_a}{RT} \right) \right]$$

$$= \ln A - \frac{E_a}{RT} \qquad\qquad \text{(A11)}$$

$$(y = c + mx)$$

It is clear that when $\ln k$ is plotted against $1/T$, the form of this equation predicts a straight line, with slope $= -\left(\dfrac{E_a}{R} \right)$ and intercept $= \ln A$ (see Figure A4).

Thus there are two reasons for manipulating a theoretical equation into linear form before attempting to verify it against experimental results:

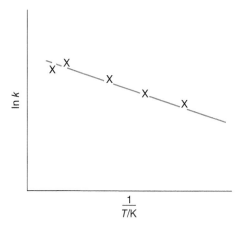

Figure A4 Sketch showing the linear variation of the *logarithm* of rate constant *versus* inverse temperature (in kelvins).

(a) It is easy to see how straight a straight-line graph is. It is much harder to judge the curvature that a curved relationship should have.

(b) It is a simple calculation to determine the values of the constants in the equation (e.g. A, E_a) from a straight-line graph. (A linear relation between y and c may be written $y = mx + c$, where m is the gradient and c is the y-axis intercept.) It is harder to extract the values of these constants from a curved graph.

Significant figures

How many decimal places should be quoted, for example when tabulating a chemical analysis? The answer depends on the **precision** (reproducibility) of the analysis: a precise analysis will justify quoting more digits (e.g. 8.465%) than a less precise analysis (e.g. 8.5%). The number of digits that a measurement's precision will support is called the *number of significant figures*. The value 1.57623 has 6 significant figures, whereas 1.58 has only 3 significant figures.

The accepted practice in tabulating numerical data (e.g. Table 8.4) is to give the number of significant figures that most closely represents the precision of the data. For example, a mass spectrometer may print out a Sr isotope ratio as 0.704249 but, if the measured precision is ±0.0005, the ratio is more truthfully reported as 0.7042.

There is no point in writing down more significant figures than the precision of a measurement will support (although a poorly managed spreadsheet may often do this); thus 1.569821 ± 0.0187 is more objectively written as 1.57 ± 0.02.

When two numbers are combined in arithmetical operations, the answer should be given the same significant figures as the less precise component:

- 3.98595 + 3.2 should be written as 7.2, not 7.18595.
- 4.5 × 6.9877 = 31, not 31.4447.
- 0.79877 ÷ 2.9 = 0.28, not 0.27544.

Further reading

Waltham, D. (2000) *Mathematics: A Simple Tool for Geologists*, 2nd edition. Oxford: Blackwell Science.

APPENDIX B: SIMPLE SOLUTION CHEMISTRY

Acids and bases

The simplest definition of an *acid* is 'a substance capable of contributing hydrogen ions to a solution or reaction' – that is, an H^+ donor. For example, hydrochloric acid (HCl) when dissolved in water undergoes ionization:

$$HCl \rightarrow H^+ + Cl^-$$

In fact this is a simplification. Each ion is actually surrounded in solution by a shell of water molecules attracted electrostatically by the charge of the ion (**hydration** – Box 4.1). One can represent these hydrated ions as H^+_{aq} and Cl^-_{aq}.

An acid like HCl, which contributes only one hydrogen ion (H^+ = proton) per acid molecule to the solution, is said to be *monoprotic*. Phosphoric acid (H_3PO_4), on the other hand, is a *polyprotic* acid,[4] because each molecule has more than one H^+ to donate.

A *base* is a substance that accepts or absorbs H^+ ions, thereby *depleting* the host solution in free H^+ and making it *less* acidic. The reaction between KOH (a base) and H_2SO_4 (sulfuric acid) may be written:

$$\left(K^+ + OH^-\right) + \left(H^+ + HSO_4^-\right) \rightarrow \left(K^+ + HSO_4^-\right) + H_2O$$
$$\quad\quad\text{base}\quad\quad\quad\quad\text{acid}$$

showing that KOH is behaving as a base whose net effect is to convert H^+ into H_2O.

Other reactions between acids and bases show the same pattern:

$$\underset{\text{base}}{NaOH} + \underset{\text{acid}}{HCl} \rightarrow \underset{\text{salt}}{NaCl} + \underset{\text{water}}{H_2O}$$

---

[4] Sometimes such acids are alternatively described as *monobasic* and *polybasic* respectively.

Chemical Fundamentals of Geology and Environmental Geoscience, Third Edition. Robin Gill.
© 2015 John Wiley & Sons, Ltd. Published 2015 by John Wiley & Sons, Ltd.
Companion Website: www.wiley.com/go/gill/chemicalfundamentals

When acids and bases react together, they tend to *neutralize* each other, forming **salts** (plus water).

Acids, bases and salts form **electrolyte** solutions, in which the solute is partly or completely ionized, resulting in electrical conductivity through migration of charged ions. Compounds like HCl, which are more or less completely ionized in solution, are called '*strong* electrolytes' (in this case a **strong acid**). *Weak* electrolytes (like carbonic acid, H_2CO_3, a **weak acid**) are those that exhibit only slight ionization in aqueous solution.

Salts are almost always strong electrolytes, but acids and bases may be strong or weak, depending on the bond holding the compound together (Chapter 7).

Ionization of water: pH

Pure water at room temperature undergoes partial self-ionization:

$$H_2O \rightarrow H^+ + OH^+ \quad K_{H_2O} = \frac{a_{H^+} \cdot a_{OH^-}}{a_{H_2O}} = a_{H^+} \cdot a_{OH^-} \quad \text{(B1)}$$

since the activity a_{H_2O} of pure water = 1.00. Water is a weak electrolyte: the equilibrium constant (Chapter 4) for this reaction, K_{H_2O}, has a value of 10^{-14} at room temperature. It follows that the activities (concentrations) of free H^+ and OH^- ions in pure water are both about 10^{-7} mol kg^{-1}. This can be expressed most concisely by saying that the **pH** of pure water is 7.0, where

$$pH = -\log m_{H^+} \quad \text{(B2)}$$

So if a solution has a pH of 2 it means that the concentration of free hydrogen ions (m_{H^+}) is 10^{-2} mol kg^{-1}. The pH notation may be used to describe the acidity of a solution. Values of pH less than 7.0 denote higher concentrations of the H^+ ion than are found in pure water (*acidic* behaviour), whereas values above 7.0 indicate lower H^+ concentrations than pure water (*basic* behaviour). Other equilibrium constants such as K_1 in Equation 4.21 can be expressed in an analogous way:

$$K_1 = \frac{a_{H^+} \cdot a_{HCO_3^-}}{a_{H_2CO_3}} = 10^{-6.4} \quad \text{therefore } pK_1 = 6.4.$$

The pH of a solution can be measured in two alternative ways:

(a) using a paper treated with a pH-sensitive dye whose colour indicates the pH of the solution (litmus paper is the traditional acid/base indicator, but more specific pH papers are available, whose colours relate to a range of pH values);

(b) using a special electrical meter called a pH meter, which, when a sensing electrode is dipped into a solution, gives a direct digital reading of its pH.

Figure B1 shows the pHs of some familiar solutions.

To summarize, an acid solution is one whose H^+ concentration is greater than that found in pure water (10^{-7} mol kg^{-1}). An acid solute is one that raises the H^+ concentration of a solution, and a base is one that depresses it. An equivalent definition of a base is a solute that increases the concentration of hydroxyl (OH^-) ions in solution: because additional OH^- will associate with

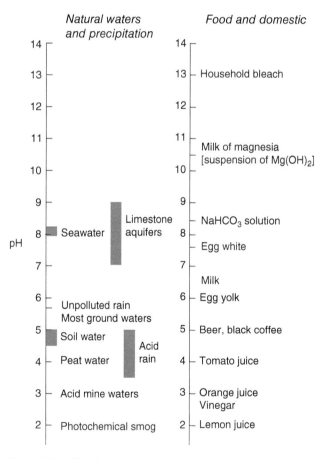

Figure B1 pH values of geological environments and everyday solutions.

H^+ ions to form water, the addition of OH^- is equivalent to a reduction of m_{H^+}.

The term *alkali* describes a water-soluble *base*, in particular the hydroxides of sodium (Na) and potassium (K). For this reason Na and K are known as *alkali metals*; in geological usage this term is often abbreviated to 'alkalis'. Sodium and potassium form basic oxides (Box 8.1).

Further reading

Barrett, J. (2003) *Inorganic Chemistry in Aqueous Solution*. Cambridge: Royal Society of Chemistry.

http://pmel.noaa.gov/co2/story/A+primer+on+pH

APPENDIX C: ALPHABETICAL LIST OF CHEMICAL ABBREVIATIONS AND ELEMENT NAMES, WITH ATOMIC NUMBER AND RELATIVE ATOMIC MASS

Symbol	Name	Z	Relative atomic mass[†]
Ac	actinium	89*	227.03
Ag	silver	47	107.87
Al	aluminium	13	26.98
Ar	argon	18	39.95[†]
As	arsenic	33	74.92
At	astatine	85*	209.99
Au	gold	79	196.97
B	boron	5	10.81
Ba	barium	56	137.34
Be	beryllium	4	9.01
Bi	bismuth	83	208.98
Br	bromine	35	79.91
C	carbon	6	12.01
Ca	calcium	20	40.08
Cd	cadmium	48	112.40
Ce	cerium	58	140.12
Cl	chlorine	17	35.45
Co	cobalt	27	58.93
Cr	chromium	24	52.01
Cs	caesium	55	132.91
Cu	copper	29	63.54
Dy	dysprosium	66	162.50

Symbol	Name	Z	Relative atomic mass[†]
Er	erbium	68	167.26
Eu	europium	63	151.96
F	fluorine	9	19.00
Fe	iron	26	55.85
Fr	francium	87[H]	223.02
Ga	gallium	31	69.72
Gd	gadolinium	64	157.25
Ge	germanium	32	72.59
H	hydrogen	1	1.008
He	helium	2	4.00[†]
Hf	hafnium	72	178.49
Hg	mercury	80	200.59
Ho	holmium	67	164.93
I	iodine	53	126.90
In	indium	49	114.82
Ir	iridium	77	192.2
K	potassium	19	39.10
Kr	krypton	36	83.80
La	lanthanum	57	138.91
Li	lithium	3	6.94
Lu	lutetium	71	174.97

Chemical Fundamentals of Geology and Environmental Geoscience, Third Edition. Robin Gill.
© 2015 John Wiley & Sons, Ltd. Published 2015 by John Wiley & Sons, Ltd.
Companion Website: www.wiley.com/go/gill/chemicalfundamentals

Symbol	Name	Z	Relative atomic mass[†]
Mg	magnesium	12	24.31
Mn	manganese	25	54.94
Mo	molybdenum	42	95.94
N	nitrogen	7	14.01
Na	sodium	11	22.99
Nb	niobium	41	92.91
Nd	neodymium	60	144.24
Ne	neon	10	20.18
Ni	nickel	28	58.71
Np	neptunium	93*	237.05
O	oxygen	8	16.00
Os	osmium	76	190.2
P	phosphorus	15	30.97
Pa	protactinium	91*	231.04
Pb	lead	82	207.19[†]
Pd	palladium	46	106.4
Pm	promethium	61*	146.92
Po	polonium	84*	208.98
Pr	praseodymium	59	140.91
Pt	platinum	78	195.09
Pu	plutonium	94*	239.05
Ra	radium	88*	226.03
Rb	rubidium	37	85.47
Re	rhenium	75	186.20
Rh	rhodium	45	102.91
Rn	radon	86*	222.02
Ru	ruthenium	44	101.07

Symbol	Name	Z	Relative atomic mass[†]
S	sulfur	16	32.06
Sb	antimony	51	121.75
Sc	scandium	21	44.96
Se	selenium	34	78.96
Si	silicon	14	28.09
Sm	samarium	62	150.35
Sn	tin	50	118.69
Sr	strontium	38	87.62[†]
Ta	tantalum	73	180.95
Tb	terbium	65	158.92
Tc	technetium	43*	98.91
Te	tellurium	52	127.60
Th	thorium	90*	232.04
Ti	titanium	22	47.90
Tl	thallium	81	204.37
Tm	thulium	69	168.93
U	uranium	92*	238.03
V	vanadium	23	50.94
W	tungsten	74	183.85
Xe	xenon	54	131.30
Y	yttrium	39	88.91
Yb	ytterbium	70	173.04
Zn	zinc	30	65.37
Zr	zirconium	40	91.22

* Signifies an element with no stable isotope.

[†] Relative to $^{12}C = 12.000$.

[‡] RAM may vary slightly with proportion of radiogenic isotopes.

APPENDIX D: SYMBOLS, UNITS, CONSTANTS AND ABBREVIATIONS USED IN THIS BOOK

The units given below are SI units (see below and http://physics.nist.gov/cuu/Units/). Some older publications may give the same quantities in traditional units (e.g. pressure in kbar); for on-line conversion to SI, visit http://www.megaconverter.com/Mega2/index.html. Greek and other specialized symbols are listed separately lower down.

Items in *italics* are symbols used as names of variables.

A　atomic **mass number** of an isotope $= Z + N$

A_n　amplitude of $(n-1)$th harmonic (Equation 5.1)

An_{65}　symbol representing a plagioclase solid solution consisting of 65 mol.% of the anorthite ($CaAl_2Si_2O_8$) end-member and 35 mol.% of albite ($NaAlSi_3O_8$)

BP　before present (when stating age)

c　speed of light ($=2.998 \times 10^8\,m\,s^{-1}$ *in vacuo*)

c_i　concentration of **component** i

C　number of **components** in a system at equilibrium (**Phase Rule**)*

dm^3　cubic decimetre, an **SI** measure of volume equal to one litre.

D_i　diffusion coefficient for component i

E_a　activation energy of a chemical reaction

E_k　kinetic energy

E_p　potential energy

Eh　redox/oxidation potential

F　force

f　number of degrees of freedom (variance) associated with an equilibrium (**Phase Rule**)*

f_i　flux of component i through unit area per unit time ($mol\,m^{-2}\,s^{-1}$)

Fo_{90}　symbol representing an olivine solid solution consisting of 90 mol.% of forsterite (Mg_2SiO_4) and 10 mol.% of fayalite (Fe_2SiO_4)

g　acceleration due to gravity ($=9.81\,m\,s^{-2}$)

Chemical Fundamentals of Geology and Environmental Geoscience, Third Edition. Robin Gill.
© 2015 John Wiley & Sons, Ltd. Published 2015 by John Wiley & Sons, Ltd.
Companion Website: www.wiley.com/go/gill/chemicalfundamentals

G	molar free energy	p_i	**partial pressure** of component i in a gas mixture
$\boldsymbol{G}$	universal gravitational constant (when printed bold in this book).	P	load pressure, in pascals (Pa), megapascals (MPa = 10^6 Pa) or gigapascals (GPa = 10^9 Pa)
Ga	'giga-annum' = 1 billion years = 10^9 years	pH	= $-\log_{10} a_{H^+}$ (measure of acidity/alkalinity)
h	height in m	p_{H_2O}	**partial pressure** of H_2O in a gas mixture (e.g. air), in same units as P
H	molar enthalpy		
k	rate constant	q	ionic charge, measured as a multiple of the charge on an electron
K	equilibrium constant		
K_i^A	**partition coefficient** describing the distribution of element i between melt and mineral A. See Box 9.1*	r	(a) radial co-ordinate; (b) ionic radius (Box 2.6), in pm (picometres = 10^{-12} m)
L	litre (see dm³)	S	molar entropy
L	length of guitar string (Equation 5.1)	t	time, in seconds, minutes, years (a), thousands of years (ka) or millions of years (Ma)
ln	natural logarithm = $\log_e$		
m	mass	$t_{1/2}$	**half-life**
Ma	'mega-annum' = 1 million years = 10^6 years	T	temperature, in °C or kelvins (K)
meq	**milliequivalents**	v	velocity
$n+$	nominal cation charge in atomic charge units*	x	spatial co-ordinate
n_D	number of daughter nuclei present in a sample*	x^a	the mass proportion ($0 < x^a < 1.00$) of component (e.g. mineral) a in a mixture
n_P	number of parent (radionuclide) nuclei present in a sample*	X	(a) composition in general terms; (b) composition of a mineral expressed in molar proportions of its end-members*
N	neutron number (Box 10.1)*		
N_c	co-ordination number (Table 7.1)*	y	years
p	number of non-bridging oxygens per silicon atom in a silicate crystal structure – a measure of polymerization*	y	spatial co-ordinate, transverse displacement of guitar string (Equation 5.1)
		z	spatial co-ordinate
		Z	atomic number*

*Dimensionless quantity.

Greek and other symbols (in Greek alphabet order)

α	alpha	notional stirring rate in heterogeneous kinetics experiment (Equation 3.4)
α^{2+}	alpha-particle	^{4}He nucleus, consisting of 2 protons and 2 neutrons
β^-	beta	beta-particle = high-energy electron emitted from a nucleus
β^+		positron (**antimatter** particle equivalent to electron)
$\delta^{18}O$	delta^{18}O	$^{18}O/^{16}O$ ratio in a mineral or rock, expressed as the positive or negative deviation in parts per thousand (‰ = 'per mil') from the $^{18}O^{16}O$ value in a reference material (see Glossary)
ΔG	delta†-G	free energy change accompanying a reaction
ΔH	delta†-H	ethalpy change accompanying a reaction
ΔS	delta†-S	entropy change accompanying a reaction
ΔT	delta†-T	temperature change, degree of supercooling (=T – $T_{solidus}$) in °C
η	eta	viscosity (Equations 3.14 and 3.15)
λ	lamda	(a) wavelength; (b) radioactive decay constant
v	nu	neutrino
$\bar{v}$	nu-bar	anti-neutrino (antimatter particle)

† These are the capital forms of the relevant Greek letter.

π	pi	signifies a covalent bond formed by sideways overlap between orbitals
σ	sigma	signifies a covalent bond formed by end-on overlap between orbitals. σ also represents the shear stress applied to a viscous fluid (Equation 3.11)
Σ	sigma[†]	represents the summation of a series of similar terms in an algebraic equation. For example, in the equation for **ionic strength**, $\sum_i m_i z_i^2$ (Equation 4.28) symbolizes adding together $m_i z_i^2$ terms for all of the major solute ions present ($=m_1 z_1^2 + m_2 z_2^2 + m_3 z_3^2 + \ldots$ where $i = 1$ might represent Na^+, $i = 2$ might represent Cl^-, $i = 3$ might represent Ca^{2+} and $i = 4$ might represent HCO_3^-, for example)
θ	theta	the diffraction angle in X-ray diffraction measurements (Equation 5.3.1)
ϕ	phi	number of phases participating in an equilibrium (**Phase Rule,** Equation 2.2)
ω	omega	frequency of oscillation
‰	per mil	= parts per thousand. See $\delta^{18}O$ above

GLOSSARY

α-emitter   A **radioactive** nuclide that decays by emitting α-particles. See Box 10.1.

α-particle   (alpha particle, α) See Box 10.1.

Absolute   Measured from a fundamental zero-point, e.g. 'absolute temperature'.

Absorb, absorption   Incorporation of a chemical species into the *interior volume* of a solid substance (cf. **adsorption**).

Accelerator mass spectrometry (AMS)   An ultrasensitive **mass spectrometer** incorporating a particle accelerator, used to measure minute traces of **cosmogenic nuclides** such as ^{14}C. See http://bit.ly/1iWWzSt.

Acid, acidic   See Appendix B and Box 8.1.

Accurate, accuracy   Describes a measurement whose **systematic error** is small.

Adiabatic   Describes a process in which the system of interest exchanges no heat with its surroundings.

Adsorb, adsorption   Attachment of a chemical species to the *surface layer* of a mineral or other solid (cf. **absorption**). See Box 8.2.

Aerosol   A colloidal suspension of fine droplets or particles dispersed in a gas (usually air). See Box 4.4.

Aliquot   A representative fraction of a sample used for chemical analysis.

Alkali, alkaline   See Appendix B.

Allochromatic   Colour of a crystal that is caused by an impurity element (Gr. *allos* = other), e.g. the red colour of ruby is due to the presence of Cr^{3+}.

Aluminosilicate   Chemical term for a class of silicate mineral (e.g. albite, $NaAlSi_3O_8$ = sodium aluminosilicate) in which aluminium partially substitutes for silicon in the *tetrahedral* (**network-forming**) sites in the crystal structure (Figure 8.2 and Table 8.1).

Aluminium silicate   A class of silicate minerals consisting solely of aluminium, silicon and oxygen, including

Chemical Fundamentals of Geology and Environmental Geoscience, Third Edition. Robin Gill.
© 2015 John Wiley & Sons, Ltd. Published 2015 by John Wiley & Sons, Ltd.
Companion Website: www.wiley.com/go/gill/chemicalfundamentals

the Al_2SiO_5 polymorphs kyanite, andalusite and sillimanite (Figure 2.1). Some or all of the aluminium occurs in *octahedral* co-ordination (Figure 8.2 and Table 8.1).

Amorphous (of a solid) Lacking crystalline structure and morphology.

Amphoteric Describes an **oxide** that can behave as an acidic oxide in reaction with a strong base, or as a basic oxide in reaction with a strong acid – see Box 8.1.

Anaerobic In the absence of oxygen (= anoxic).

Anhydrous Incorporating no water or OH^-. See **dehydration.**

Anion A negatively charged ion, produced when a neutral atom accepts one or more additional **electrons** (*an-* comes from the Greek meaning 'up', referring to the increased number of electrons).

Anisotropy, –ic (of a material, e.g. a crystal) Having physical properties that vary according to the direction of measurement.

Anoxic Describes an environment devoid of oxygen.

Antimatter Matter consisting of **antiparticles**.

Antiparticle Any subatomic particle of **antimatter**, having the same mass as the equivalent conventional particle but the opposite charge; e.g. a **positron** (a contraction of 'positive electron') has the same mass as an electron but a positive charge.

Atomic number (Z) The number of protons in the nucleus of an atom. Also the total number of electrons in the neutral atom.

Atomic weight See **relative atomic mass.**

β-particle (beta-particle, β^-) A high-energy electron released by a nuclear reaction.

Bar A unit of pressure widely used in petrological literature. It approximates to one atmosphere. It has been superseded in the SI system by the pascal (Pa) $= N\,m^{-2}$. 1 bar $= 10^5\,Pa$.

Baryon A term (from the Greek for 'heavy') embracing both **protons** and **neutrons**. **Baryonic matter** is matter consisting of atoms and molecules (and ultimately of protons, neutrons and electrons) as distinct from more exotic forms of dark matter (Chapter 11).

Base, basic See Appendix B.

Biolimiting Describes essential nutrient elements (N, P, Si) whose low concentration may limit biological productivity.

Birefringence (a) A property of a crystal whereby the refractive index varies according to the vibration direction of the incident polarized light. (b) The numerical difference between the maximum and minimum refractive indices of the crystal.

Blocking temperature See **closure temperature.**

Buffer, -ing A chemical equilibrium that has the capacity to resist externally induced change in key properties of a system (e.g. pH, Equation 4.24; f_{O_2} Figure 9.5).

Catalyst A substance that accelerates a chemical reaction without being consumed by it.

Cation A positively charged ion, produced by the loss of one or more **electrons** from a neutral atom (*cat-* comes from the Greek meaning 'down', referring to the reduced number of electrons compared with the neutral atom).

CFC Abbreviation for chlorofluorocarbon: class of inert derivatives of simple hydrocarbons formerly used as refrigerants, foam expanders and aerosol propellants. See Chapter 9.

Chalcophile Cosmochemical category of elements with an affinity for sulfide minerals – see Figure 11.4.

Closure temperature A temperature (specific to a particular reaction or diffusion process) below which diffusion rate becomes slower than cooling rate, and the reaction effectively ceases.

Cogenetic Describes a suite of rocks of the same age that share a common origin (e.g. crystallized from the same parent magma).

Complex, -ing (Formation of) a cluster of **ligands** around a central metal atom or ion, held together by co-ordinate bonds (Chapter 7) that involve the overlap of ligand lone pairs with empty orbitals of the central atom.

Component The chemical components of a **system** comprise the minimum number of chemical (atomic, isotopic or molecular) species required to specify completely the compositions of all the phases present.

Compound A substance in which different elements are combined in specific proportions.

Concentration A parameter indicating how much of a particular chemical species (component) is present in a unit amount of the medium (phase) in which it resides (which may be a gas, a liquid or a solid). See Chapter 4.

Condensate   Liquid or solid phase(s) that **condense** from a **saturated** gas or vapour.

Condense, -ate   To form a liquid or solid from a cooling gas; product of condensation.

Condensed phases   Liquid and solid phases, in which the atoms or molecules are in mutual contact. See Box 1.3.

Connate   (of waters) Seawater trapped in the interstices of a sedimentary rock at the time of deposition.

Co-ordination polyhedron   The hypothetical three-dimensional shape obtained by joining the centres of the anions immediately surrounding a cation (or vice versa).

Core electron   An electron in an inner shell, too tightly bound to participate in bonding.

Corona   A rim of different minerals surrounding a crystal, resulting from incomplete metamorphic reaction between the crystal and surrounding minerals (see Box 3.1).

Cosmogenic nuclide   A nuclide/isotope formed by cosmic-ray bombardment of certain elements in meteorites, the solid Earth and the Earth's atmosphere.

Cotectic   One of several liquidus phase boundaries (thermal valleys) in a ternary phase diagram.

Critical temperature   The temperature above which the liquid and gas states of a given substance merge into one homogeneous phase, i.e. a **supercritical** fluid (Figure 2.2.2).

Cryptocrystalline   Consisting of crystals too small to be readily seen under the optical microscope.

Dehydration   The removal of water from a substance. In mineralogy, a dehydration reaction is one in which a hydrous (water-bearing) mineral such as mica breaks down at high temperature to form an assemblage of anhydrous (water-free) minerals plus water vapour. (See Figure 2.3.)

Derivative   An expression giving the rate of change of one variable in relation to another upon which it depends. A body's speed, the rate at which its position x changes with respect to time t, can be regarded as the derivative of x with respect to t $\left(\text{written } \dfrac{\mathrm{d}x}{\mathrm{d}t}\right)$. See **differentiation** below, Appendix A and Waltham (2000).

Diagenesis   A term embracing post-depositional chemical, mineralogical and sometimes microbiological changes that take place during burial of a sediment (Box 4.7).

Diatomic   (of a molecule of gas) Consisting of two atoms.

DIC   Dissolved inorganic carbon.

Differential equation   An equation in which one or more terms is a **derivative**.

Differentiation   *Mathematics (calculus):* the process of calculating the gradient or **derivative** of a dependent variable with respect to another variable (Appendix A).

   Geochemistry: the physical or chemical segregation of one type of material from another. A homogeneous or intimately mixed body becomes segregated into two bodies or phases of differing composition, hence *differentiated.*

Diffract, -ion   Optical interference phenomenon observed when electromagnetic rays interact with a regularly spaced pattern, e.g. of atoms in a crystal. See Box 5.3.

Diffusion   The dispersal of one substance in another by atom-by-atom (or molecule-by-molecule) migration.

Dilute   Describes a solution having a low solute concentration.

Dimensionless number   A number expressing the magnitude of a quantity that has no associated dimensions or units. An example is π, the ratio of the circumference of a circle to its diameter: the circumference and diameter both have dimensions of length, expressed in units of metres, which therefore cancel each other out when ratioed. The value of a dimensionless number is independent of the system of units adopted.

Dimer   A **polymer** consisting of only two basic units.

Dipole   A system of two electrostatic charges of equal but opposite magnitude, held a specific distance apart. Molecules may be permanent dipoles (owing to internal differences in electronegativity between atoms, as in water) or induced dipoles, caused by the polarizing effect of an electrostatic field.

Dissociation   A reaction in which a compound decomposes into two or more simpler species. See Chapter 4.

Divalent   (of an atom or element) Having a **valency** of 2.

dm^{-3}   'Per cubic decimetre' = 'per litre'. See Appendix A.

Electrolysis   The extraction of an element (e.g. Cl) from a solution containing its ions (Cl$^-$) by passage of an electric current.

Electrolyte   A compound which (as solid, melt or solution) conducts electricity. See Appendix B.

Electron (e$^-$ or β^-)   A negatively charged atomic particle responsible for bonding between atoms. See Table 5.1.

Electron core   The tightly held inner electron shells not participating in bonding.

Electron-volt (eV)   The unit of energy used to measure quantum energies and energy differences within atoms. The **kinetic energy** possessed by an electron that has been accelerated from rest by an electrostatic field of one volt: $1\,eV = 1.6021 \times 10^{-19}\,J$.

Element   A substance consisting of a single type of atom (all atoms having the same atomic number Z).

Empirical   Determined by experiment or observation, not by theoretical reasoning or calculation.

End-member   One of two or more chemical **components** or formulae in terms of which the composition of a solid solution may be expressed.

Endothermic   (of a reaction) Absorbing heat (ΔH positive).

Energy   See Box 1.1.

Energy level   One of a small number of permitted (= 'quantized') energy values that an electron in an atom may possess according to wave mechanics.

Enthalpy (H)   Thermodynamic variable representing the total **kinetic** energy of the atoms or molecules in a substance.

Entropy (S)   Thermodynamic variable representing the internal disorder of a substance (Box 1.4).

Equilibrium constant   A parameter of a particular chemical equilibrium, expressing the relative proportions of reactants and products at which the forward and reverse reaction rates are equalized under specific conditions – see Equation 4.8.

Equivalent, – mass   The mass of a substance that will neutralize one mole of H$^+$ in an acid–base reaction. For a cation, a mass equal to molar mass divided by the charge on ion. See also **milliequivalent**.

Eutectic   An invariant point in a phase diagram where the melt field projects to the lowest temperature (e.g. E in Figure 2.4; E in Fig. 2.8). A eutectic marks the final destination of a melt undergoing fractional crystallization, or the initial melt composition during partial melting (Box 2.4).

Evaporite   A sedimentary rock consisting of mineral salts deposited from supersaturated brine as a result of evaporation.

Exothermic   (of a reaction) Giving out heat (ΔH negative).

Exsolve, exsolution (-lamellae)   The unmixing of a homogeneous phase into two immiscible phases, commonly as a lamellar intergrowth (e.g. perthite).

Extensive   (of variables) Depending in magnitude on the size of the system being considered (e.g. mass).

Ferromagnesian   (of silicate minerals) Containing essential Mg and Fe: olivine, pyroxene, amphibole and dark mica.

Fissile   Describes a nuclide capable of fission – splitting into two lighter nuclei when irradiated with neutrons (Box 11.2).

Fission product   A radioactive nuclide formed by nuclear fission. Most fission products are neutron-rich (Box 11.2) and prone to β-decay (Box 10.1).

Fluid   Either liquid or gas. The term is applied to anything that can flow. In geology, it often connotes a **supercritical** aqueous fluid (see Box 2.2).

Fluorescence   The excitation of light emission by absorption of shorter-wavelength light.

Foliation   Rock texture in which platy crystals tend to lie parallel to each other.

Fractionated, -ation   The progressive partial separation of one element (or isotope) from another owing to chemical differences between them. (Hence 'fractionated', the antonym of **primitive**.)

Fraunhofer lines   Dark lines on the optical spectrum of white light from a star (Plate 4), caused by absorption of light by electronic transitions in atoms of elements present in the cool outer envelope of the star.

Free energy (G)   See Chapter 1.

Free radical   A temporarily unattached atom (e.g. Cl•) or group of atoms (e.g. CH$_3$•) having an unbonded **valence electron** (symbolized by '•') that makes it highly reactive.

Functional group A peripheral group of atoms attached to the basic –C–C–C– skeleton of an organic molecule, that bestows a particular chemical functionality.

Gas constant R A fundamental physical constant. $R = 8.314\,\mathrm{J\,mol^{-1}K^{-1}}$.

Geothermometry Science of estimation of a rock's temperature of formation (the temperature at which its mineral assemblage equilibrated) from the compositions of the equilibrated minerals.

Glass A state of matter having the disordered structure of a melt but the mechanical properties of a solid.

Half-life $t_{1/2}$ The characteristic time it takes for a radioisotope to decay to *half* its original concentration or radioactivity.

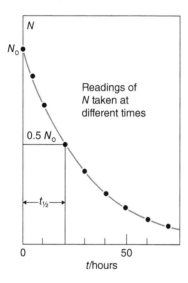

Hard water, hardness Describes fresh water with a relatively high natural content of Ca^{2+} and Mg^{2+} leached from aquifer minerals (usually carbonates) by the reverse of reaction 4.2.1 (Box 4.2). Hardness is the molar concentration of $Mg^{2+} + Ca^{2+}$ in mol l^{-1}.

Heterogeneous (of a reaction or equilibrium) Involving the participation of more than one chemical **phase** (e.g. reaction 4.12), cf. **homogeneous**.

Homogeneous (of a reaction or equilibrium) Taking place within a single chemical **phase** (e.g. reaction 3.1), cf. **heterogeneous**.

Hybrid, -ization (of electron orbitals in atoms) The merging of the wave functions of several valence electrons in an atom to generate a multi-electron waveform with a different shape and energy structure, e.g. the tetrahedral sp^3 hybrid characteristic of many carbon compounds (Figures 7.5a and d) by combining one 2s and three 2p orbitals.

Hydration (a) Incorporation of water or OH$^-$, cf. **dehydration**. (b) The stabilization of an ion in aqueous solution by the gathering of polar water molecules around it. See Box 4.1.

Hydrothermal Refers to hot saline aqueous fluids circulating in the crust, or to veins and ore deposits crystallized from such fluids.

Hydroxyl The OH group or OH$^-$ ion.

Ideal solution A solution sufficiently **dilute** for ion–ion interactions to be insignificant.

Idiochromatic Colour of a crystal that is derived from one of its major elements (Gr. *idios* = own).

Incompatible element See Box 9.1.

Incongruent melting Melting of a mineral where a new mineral is formed in addition to melt. See Box 2.6.

Inert gas The family of gaseous elements helium (He), neon (Ne), argon (Ar), krypton (Kr), xenon (Xe) and radon (Rn) on the extreme right of the Periodic Table.

Infinitesimal Describes something that is extremely small.

Integral The graph shows y varying as a function of x. The **integral** of y is an algebraic function expressing the area beneath the y curve between any specified limits x_1 and x_2. It is written $\int_{x_1}^{x_2} y\, dx$.

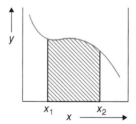

Integrate, –ation The mathematical process of calculating the **integral** of, or the area under, a curve. Integration can be considered the reverse of differentiation: thus the integral of velocity as a function of time indicates the distance travelled.

Intensive (of a variable) Having a magnitude independent of the size of the system considered (e.g. temperature).

Interfere, -ence Term in optics describing interactions between superposed waves (e.g. light rays) of the same wavelength leading to reinforcement (if 'in-phase') or attenuation (if 'out of phase'). See Figure 5.3.1(b) and (c).

Interstitial   Situated in cavities between larger atoms or ions in a regular crystal structure.

Invariant   Describes a point in a phase diagram that has zero variance F (see discussion of Figure 2.1).

Inversion   Recrystallization of a crystalline phase to a more stable polymorph.

Ion   Atom with a net electrical charge, due to the acquisition or loss of **electron**(s). See **anion** and **cation.**

Ion pair   Temporary electrostatic association in solution of two ions of opposite charge.

Ionic potential   Ratio of an ion's nominal charge over its ionic radius.

Ionic strength $I = \frac{1}{2}\sum_i m_i z_i^2$, a measure of how far an ionic solution departs from **ideal solution** behaviour. I = ionic strength, m_i = molality of ion species i, z_i = charge on ion species i. See Appendix D for explanation of Σ.

Ionization energy   The energy (in eV or kJ mol^{-1}) required to remove the most easily detached **electron** from an atom to a state of rest at infinity (Figure 5.6).

Ionization potential   Obsolete synonym for **ionization energy**.

Iso-   Prefix meaning 'having the same …' .

Isobar, isobaric   (a) A pressure 'contour', a hypothetical line or surface in P–T–X space at which pressure is everywhere uniform. 'Isobaric' signifies a process that occurs at constant pressure. (b) **Nuclide** having the same mass number A (Box 6.1) as another.

Isochron   A line on an isotope-ratio plot drawn through data points having the same age.

Isotherm, isothermal   A temperature contour in a phase diagram. A line or surface on which temperature is constant. 'Isothermal' signifies a process operating (or a phase diagram constructed) at constant temperature.

Isotone   See Box 10.1.

Isotope   An isotope of an element consists of nuclei that share the same values of **atomic number** Z and **neutron number** N. Other isotopes of the element have the same value of Z but different value(s) of N. See Box 10.1.

Isotopic tracer   An isotopic ratio differing between one type of source material and another, that indicates the derivation of a rock or solution.

Isotropic   Describes a substance whose physical properties have values that are independent of the direction in which they are measured.

Kelvin, K   Unit of temperature expressed on the absolute temperature scale. $T/K = T/°C + 273.15$. The kelvin (K) and degree Celsius (°C) have the same magnitude but differ in their zero points (absolute zero and the melting point of ice respectively).

Kilobar (kbar, kb) $1 kb = 10^8 Pa$. See **bar**.

Kiloelectron-volt (keV)   1000 electron-volts.

Kinetic energy (E_k) The energy a body possesses by virtue of its motion. See Chapter 1.

Lanthanide   One of the 14 elements following lanthanum in the Periodic Table, characterized by the entry of **electrons** into the 4f orbitals.

Latent heat of fusion   The enthalpy required to transform 1 kg of a solid completely into a liquid at constant temperature (J kg^{-1})

Latent heat of vaporization   The enthalpy required to transform 1 kg of a liquid into a gas at constant temperature (J kg^{-1})

Ligand   (in a co-ordination **complex** or crystal) The ions or molecules surrounding the central ion or atom. For example in the dissolved complex $Cu(HS)_2^-$, the ligands are the two HS^- anions attached to the central Cu^+ ion.

Linear   A relationship between y and x is said to be linear if plotting y against x yields a straight line.

Liquidus   The temperature at which the first crystals begin to form in a cooling melt. A liquidus curve (or surface) shows the variation of liquidus temperature with composition or with pressure.

Logarithm (log, ln)   The magnitude of a number expressed as a power of 10 (log_{10}) or of e (natural logarithm = ln), as explained in Appendix A.

Lone pair   Paired valence electrons in a filled lobe of (usually) a hybrid orbital (e.g. Figure 7.5b, c). Lone pairs influence hybrid shape, and are involved as electron donors in the formation of co-ordination **complexes**.

Macroscopic   Visible to the naked eye or measurable with normal laboratory apparatus, as opposed to **microscopic.**

Magma   Any sort of igneous melt, including suspended matter such as crystals and/or vapour bubbles.

Mass spectrometer   See Box 10.3.

Megascopic   Synonym of **macroscopic.**

meq   Abbreviation for **milliequivalent.**

Metalloid   A metal having both metallic and non-metallic properties.

Metastable   Describes a mineral preserved outside its *P-T* field of stability owing to the slowness of the reaction converting it to a more stable mineral (e.g. aragonite at surface conditions – Fig. 1.4).

Meteoric   Describes water ultimately derived from natural precipitation.

Micronutrient   A trace element required by an organism for its metabolism to function.

Microscopic   Visible only under the microscope. Too small to see or measure with normal apparatus.

Milliequivalent (meq)   A mass of substance equal to one-thousandth of its **equivalent mass.**

Miscible   Two compounds are said to be miscible if they can be combined in any desired proportions to form a single, stable, homogeneous phase.

Mole (abbr. 'mol'), molar   Refers to an amount of a compound (or element) whose mass, expressed in grams, is numerically equal to its **relative molecular (or atomic) mass.** E.g. the relative molecular mass of $H_2O = 2 + 16 = 18$; a mole of water is therefore defined as $18\,g$ water.   An amount of water weighing $54\,g$ therefore constitutes $54/18 = 3$ moles.

Mole fraction   The fraction of the total number of molecules in a phase that is represented by a specific component. Mole fraction is a **dimensionless number.**

Molecular weight   See **relative molecular mass.**

Monatomic   Gases consisting of separate atoms, not combined in molecules.

Monovalent   (of an atom or element) Having a **valency** of 1.

Native   Describes an element (e.g. gold) occurring naturally in the uncombined elemental or metallic state (not as a compound).

Network-forming element   An element whose relatively covalent bond with oxygen allows it to form part of the O–Si–O framework (e.g. chains, sheets) of a silicate crystal structure.

Network-modifying element   An element or group whose ions (e.g. Na^+, OH^-) pack in between the framework components of silicate crystal structure.

Neutron (n)   Uncharged massive nuclear particle, marginally heavier than the **proton.**

Neutron number (*N*)   The number of neutrons in a nucleus.

Noble gas   The gaseous elements in the extreme right-hand column of the Periodic Table are collectively known as the noble (or **inert**) gases on account of their unreactivity.

Nucleon   A term embracing both protons and neutrons in a nucleus.

Nuclide   A substance consisting of atoms with particular values of *Z* and *N*, i.e. a specific isotope of a specific element.

Opaque, opacity   Property of a material through which light cannot pass

Orbital   Term used in wave mechanics to identify the spatial characteristics of the waveform adopted by an electron in an atom (analogous to the classical orbit of a planet migrating around the Sun). See Chapter 5.

Order of magnitude   Estimate of a quantity to the nearest power of 10. Two measurements are said to 'differ by an order of magnitude' when one is more than ten times the size of the other. 10,000 is two orders of magnitude (i.e. 10^2) greater than 100.

Organic chemistry   Branch of chemistry dealing with compounds in which carbon is combined with hydrogen and other elements.

Organo-metallic   Organic compound incorporating metal atoms.

Oxidation   *Original meaning:* a chemical reaction resulting in the addition of oxygen to an element or compound. *Current meaning:* a chemical reaction involving the removal of **electrons** from an atom or ion (Box 4.7), or an increase in the number shared with other atoms. Cf. **reduction.**

Oxidation potential *Eh*   See Box 4.1.

Oxidation state   (of an atom in a molecule/compound) The hypothetical charge the atom would possess if the compound were held together by purely ionic bonds.

Oxide   Compound in which an element is chemically bound to oxygen. See Figure 9.3.

Oxy-anion   A polyatomic anion incorporating oxygen (and/or hydroxide) alongside another element. e.g. nitrate NO_3^-. Alternatively known as oxo-anions.

Partial (vapour) pressure (p_i) The pressure in Pa exerted individually by one volatile component i in a gas or other phase.

Partition coefficient K_i^A See Box 9.1.

Parts per million (ppm)   Unit of (low) concentration of an element (not oxide). 1 ppm = 1 g per 10^6 g = 1 μg per g = 1 mg per kg. 10^3 ppm = 0.1%. In gas mixtures, 'ppm' usually refers to parts per million by volume (ppmv).

Peridotite   Rock type comprising more than 40% Mg-rich olivine (together with other Mg-rich minerals such as pyroxene) making up the bulk of the Earth's mantle.

Perthite   The name for a potassium feldspar crystal containing **exsolution lamellae** of sodium feldspar – see Figures 2.6b and 2.7.

Petroleum   Term embracing all naturally occurring liquid and gaseous hydrocarbon deposits.

pH   See Appendix B.

Phase   (chemical usage) A part of a **system** occupying a specific volume and having uniform[1] physical and chemical characteristics which distinguish it from all other parts of the system. For example, olivine crystals, melt and gas bubbles in a crystallizing magma constitute three distinct phases.

Phase boundary   Line or surface on a phase diagram marking the limit of P–T–X stability of a particular phase or assemblage.

Photodissociation   Break-up of a molecule caused by a high-energy **photon** (e.g. Equation 3.8).

Photon   Quantum of light. Behaves in some respects like a particle.

Polarizability   Susceptibility of a substance to **polarization** on the atomic or molecular scale: a measure of how easily **electron** density can be deformed in an electric field.

Polarization   (a) Attraction of negative charge toward one side of an atom, ion or bond, leaving the other side with a net positive charge. (b) A light beam whose electric vector vibrates only in one plane is said to be plane-polarized.

Polybasic   Synonym of **polyprotic.**

Polymer   Material whose molecules are built up from a series of identical smaller units. Thus polyethylene consists of chains of ≤80 ethylene (ethene, C_2H_4) molecules bonded together.

Polyprotic   Describes an acid whose formula contains more than one hydrogen atom, and whose dissociation therefore produces more than one hydrogen ion (Appendix B).

Polymorph   One of several alternative crystal structures that a given substance can adopt.

Positron (β^+) Anti-matter counterpart of an **electron.**

Potential energy   The energy a body possesses by virtue of its position in a force field such as gravity.

ppm   See **parts per million.**

Precipitate, precipitation   Formation of insoluble solid from solution, or formation of liquid droplets from vapour. Requires solution or vapour to be **supersaturated** (Chapter 4).

Precise   Describes a measurement that is subject to only a small **random error.**

Precision   The magnitude of the **random error** associated with a measurement. (Confusingly, 'high-precision' describes a measurement that has a low random error.)

Primitive   A material that has undergone little or no **differentiation** or **fractionation.**

Product   A chemical species formed by a chemical reaction (appearing on the right-hand side of the written reaction).

Prograde   (of metamorphic reactions) Proceeding to higher grade, or higher temperature.

Proton (p^+) Positively charged massive nuclear particle.

Pseudo-binary, pseudo-ternary   System which for practical purposes approximates to a two-component (binary) system, but which in certain details exhibits more complex behaviour, e.g. crystallizing phases whose compositions lie outside the system. Likewise 'pseudo-ternary'.

Quantum number   A number – usually an integer – whose multiple values (0, 1, 2, etc.) identify alternative electron orbitals/waveforms/stationary states in an atom (Table 5.3) or the harmonics of any standing wave (see Figure 5.2).

Radioactive   Undergoing spontaneous nuclear decay.

---

[1]   Or continuously varying, as in a **zoned** mineral crystal.

Radiogenic   A daughter **isotope** that results from the decay of a **radioactive** parent isotope, and whose abundance increases with time. The proportion of such an isotope that is the product of the decay of a radioactive parent.

Radionuclide   A **nuclide** consisting of unstable nuclei, undergoing **radioactive** decay.

RAM   Abbreviation of **relative atomic mass**.

Random error   an error that varies randomly (in magnitude and sign) about a mean value when a measurement is repeated under identical conditions. The random error of a measurement can be reduced by repeating it several times and taking the mean.

Rare earth element   Alternative name for **lanthanide** elements, but embracing lanthanum (La) too.

Rare gas   See **inert gas.**

Reactant   A chemical species participating in a chemical reaction, appearing on the left-hand side of the written reaction.

Reaction rim   An overgrowth of new minerals rimming an early-formed crystal owing to its reaction with later-stage melt. See Plate 1.

Redox   A term embracing both _red_uction and _ox_idation reactions.

Redox potential _Eh_   See Box 4.7.

Reduction   _Original meaning:_ a chemical reaction resulting in the removal of oxygen from a compound. _Current meaning:_ any chemical reaction resulting in the addition of one or more **electrons** to an atom or ion (Box 4.7), cf. **oxidation**.

REE   Abbreviation for **rare earth elements**.

Refractory   (of elements or their compounds) Having particularly high melting and vaporization temperatures, cf. **volatile**.

Relative atomic mass (A_r) Mass of an atom, expressed on a scale in which $^{12}C = 12.0000$.

Relative molecular mass, relative molar mass (M_r) Mass of a molecule expressed on scale in which $^{12}C = 12.0000$.

RMM   Abbreviation of **relative molecular mass**.

Salinity   'Saltiness' of a water sample, expressed most simply as total grams of dissolved salts per kg solution.[2] Seawater salinity averages $35\,g\,kg^{-1}$.

Salt   A class of compounds that result from reactions between acids and bases, in which one or more hydrogen ions of the acid have been replaced by metal ions from the base – see Appendix B.

Saturated   Describes (1) a solution containing the maximum solute content, that coexists stably with excess solute in solid or gas form; (2) an **organic** compound containing only single C–C bonds.

Screened, –ing   Lessening of the electrostatic interaction between two charges owing to another charge interposed between them. Outer **electrons** in an atom are screened from the nucleus by inner electron shells, making them easier to detach.

Secular   (of variations in the value of a physical quantity) Non-cyclic, changing in a consistent direction over a period of time.

SI   Abbreviation for the Système International d'Unités (International System of Units), the currently accepted metric system used in this book. See Appendix A.

Sialic   Describes rock compositions rich in Si and Al (e.g. continental crust).

Silica   Silicon dioxide (SiO_2), in any crystalline or **amorphous** form.

Silicate   Compound in which one (or more than one) **metal** is combined with silicon and oxygen. See Chapter 8.

Solidus   Temperature at which the last fraction of melt crystallizes, or the temperature at which a substance begins to melt. (Cf. **liquidus.**)

Solute   The dissolved species in a solution.

Solvation   The formation of a sheath of polar solvent molecules around a dissolved ion, inhibiting reaction with other ions and stabilizing the ion in solution. See **hydration** and Box 4.1.

Solvent   The dominant component of a solution, the medium in which dissolved species are dispersed.

Solvus   A line or surface in a phase diagram indicating the compositions of immiscible phases in mutual equilibrium at each temperature.

Speciation   The identification of the different chemical forms in which an element or compound occurs naturally (e.g. mercury metal as distinct from methylmercury).

Specific heat   The enthalpy required to raise the temperature of $1\,kg$ of a substance by one kelvin $(J\,kg^{-1}\,K^{-1})$.

---

[2]   Salinity defined in this simple way is difficult to measure in practice, and other definitions are used today.

Spectrum The various wavelength components present in a beam of light or other electromagnetic radiation (such as an X-ray beam), displayed in order of their wavelength or photon energy. The emission spectrum of an element consists of a series of lines or peaks, representing the specific wavelengths (or photon energies) that are characteristic of that element (e.g. Figure 6.5).

Stoichiometry, stoichiometric The proportions, determined by valency, in which elements combine in a compound.

Strong acid (or base) An acid (or base) – such as HCl (or NaOH) – that is completely ionized in aqueous solution (see Appendix B).

Strong force (or strong nuclear interaction) The powerful short-range force that binds nucleons together in atomic nuclei. See Box 11.2.

Sublimate A solid crystallized directly from the vapour, e.g. frost. Sublimation is the vaporization of a solid without the intermediate formation of a melt.

Supercritical Describes a fluid whose temperature exceeds its **critical point**. See Box 2.2.

Supersaturated (of a solution or solute) Having a molality (or activity) product exceeding the solubility product of that solute at that temperature. Metastable with respect to **precipitation**, tending to precipitate solute until **saturation** is attained.

System Any part of the world to which we wish to confine attention. May refer to a specific domain of composition space, e.g. 'the system $NaAlSi_3O_8$–$CaAl_2Si_2O_8$'.

Systematic error A deviation between a measured value and the true value of a quantity that is not reduced by repeated measurement (cf. **random error**), but introduces a consistent bias.

Texture Describes the geometrical relationships between the constituent mineral grains in a rock.

Thermal energy Total **kinetic energy** possessed by a substance by virtue of individual molecular motions.

Thermonuclear Describes energy released by nuclear fusion reactions (in stars and the hydrogen bomb, etc.).

Tie-line Isothermal line in a phase diagram linking two phases of different chemical compositions that are in mutual chemical equilibrium at the temperature concerned. See Box 2.3.

Tracer A geochemical variable (e.g. $^{18}O/^{16}O$) whose value is altered by a specific geochemical process, and therefore records the operation of that process.

Triple point An **invariant** point where three phase boundaries (and the three phase-fields they separate) meet in a phase diagram.

Trivalent (of atom, element) Having a **valency** of 3.

Twin, twinned Describes a crystal comprising two or more domains with distinct crystal lattice orientations (illustrated by the dark and light grey fields in Fig. 2.7) related by a symmetry operation (e.g. a mirror plane).

Unsaturated Describes an organic compound containing one or more double C=C bonds or triple C≡C carbon–carbon bonds.

Valence electron An electron occupying an orbital in the valence (highest energy) shell, and available for bonding.

Valency The number of chemical bonds that an atom (element) can make in forming a molecule (compound). Some elements have more than one valency.

Vapour (vapor in N. America) Refers to any gas phase present during a geochemical reaction.[3]

Volatile Describes an element or compound readily converted to the gaseous state (at relatively low temperature), cf. **refractory**.

Weak acid An acid (e.g. carbonic acid, H_2CO_3) that exhibits only slight ionization in aqueous solution (see Appendix B).

WHO World Health Organization.

Zoned, zoning Continuous or abrupt (or sometimes oscillatory) changes in composition between the core and rim of a single crystal (see rear cover image) or between neighbouring sectors.

Zwitterion (from German *zwitter* = hybrid) An ion carrying both a positive and negative charge (a property of amino acids, Chapter 9).

---

[3] Chemists reserve the term 'vapour' for a gas that is below its **critical temperature** (Box 2.2) and which can therefore be liquified by compression, but this restriction in meaning is rarely adhered to in geochemical usage.

REFERENCES

Albarède, F. (2009) *Geochemistry – An Introduction*. Cambridge: Cambridge University Press.

Allegre, C. J. (2008) *Isotope Geology*. Cambridge: Cambridge University Press.

Anbar, A.D. and Rouxel, O. (2007) Metal stable isotopes in paleoceanography. *Annual Review of Earth and Planetary Science* **35**, 717–48.

Andersen, M.B., Vance, D., Archer, C., *et al.* (2010) The Zn abundance and isotopic composition of diatom frustules, a proxy for Zn availability in ocean surface seawater. *Earth and Planetary Science Letters* **301**, 137–45.

Archer, D. (2011) *Global Warming – Understanding the Forecast*, 2nd edition. Chichester: John Wiley and Sons, Ltd.

Archer, D., Eby, M., Brovkin, V., *et al.* (2009) Atmospheric lifetime of fossil fuel carbon dioxide. *Annual Review of Earth and Planetary Sciences* **37**, 117–34.

Baas Becking, L.G.M., Kaplan, I.R. and Moore, D. (1960) Limits of the natural environment in terms of pH and oxidation-reduction potentials. *Journal of Geology* **68**, 243–84.

Baker, J., Snee, L. and Menzies, M. (1996) A brief Oligocene period of flood volcanism in Yemen: implications for the duration and rate of continental flood volcanism at the Afro-Arabian triple junction. *Earth and Planetary Science Letters* **138**, 39–55.

Barker, A.J. (1998) *An Introduction to Metamorphic Textures and Microstructures*. Abingdon: Routledge.

Barnes, H.L. (ed.) (1979) *Geochemistry of Hydrothermal Ore Deposits*, 2nd. edition New York: John Wiley & Sons, Ltd.

Berner, R.A. (1971) *Principles of Chemical Sedimentology*. New York: McGraw-Hill.

Best, M.G. (2002) *Igneous and Metamorphic Petrology*. Oxford: Wiley-Blackwell.

Bethe, H. and Brown, G. (1985) How a supernova explodes. *Scientific American* **252**, 40–48.

Blaxland, A.B., van Breeman, O., Emeleus, C.H. and Anderson, J.G. (1978) Age and origin of the major syenite centers in the Gardar province of south Greenland: Rb–Sr studies. *Geological Society of America Bulletin* **89**, 231–44.

Bloss, F.D. (1971) *Crystallography and Crystal Chemistry: An Introduction*. New York: Holt Rinehart and Winston.

Borders, R.A. and Birks, J.W. (1982) High-precision measurements of activation energies over small temperature intervals: curvature in the Arrhenius plot for the reaction

Chemical Fundamentals of Geology and Environmental Geoscience, Third Edition. Robin Gill.
© 2015 John Wiley & Sons, Ltd. Published 2015 by John Wiley & Sons, Ltd.
Companion Website: www.wiley.com/go/gill/chemicalfundamentals

$NO + O_3 \rightarrow NO_2 + O_2$. *Journal of Physical Chemistry* **86**, 3295–3302.

Bouvier, A. and Wadhwa, M. (2010) The age of the Solar System redefined by the oldest Pb–Pb age of a meteoritic inclusion. *Nature Geoscience* **3**, 637–41, 10.1038/NGEO941.

Burbidge, E.M., Burbidge, G.R., Fowler, W.A. & Hoyle, F. (1957) Synthesis of the elements in stars. *Reviews of Modern Physics* **29**, 547–650.

Carlson, W.D. and Johnson, C.D. (1991) Coronal reaction textures in garnet amphibolites of the Llano Uplift. *American Mineralogist* **76**, 756–72.

Connelly, J.N., Bizzarro, M., Krot, A.N., *et al.* (2012) The absolute chronology and thermal processing of solids in the Solar protoplanetary disk. *Science* **338**, 651–5, 10.1126/science.1226919.

Craig, H. (1961) Isotopic variations in meteoric waters. *Science* **133**, 1702–3.

Craig, H. (1963) The isotopic geochemistry of water and carbon in geothermal areas. In *Conference on Isotopes in Geothermal Waters*. Spoleto: Laboratorio di Geologia Nucleare, 53–70.

Dansgaard, W. (1964) Stable isotopes in precipitation. *Tellus* **16**, 436–68.

DePaolo, D.J. and Ingram, B.L. (1985) High-resolution stratigraphy with strontium isotopes. *Science* **227**, 938–41, 10.1126/science.227.4689.938.

DePaolo, D.J. and Wasserburg, G.J. (1979) Sm–Nd age of the Stillwater Complex and the mantle evolution curve for neodymium. *Geochimica et Cosmochimica Acta* **43**, 999–1008, 10.1016/0016-7037(79)90089-9.

Dickin, A.P. (2005) *Radiogenic Isotope Geology*, 2nd edition. Cambridge: Cambridge University Press.

Edmond, J.M. and Von Damm, K. (1983) Hot springs on the ocean floor. *Scientific American* **248** (April), 78–93, 10.1038/scientificamerican0483-78.

Faure, G. and Mensing, T.M. (2005) *Isotopes – Principles and Applications*. New York: John Wiley & Sons, Ltd.

Ferreira, P.G. (2006) *The State of the Universe – A Primer in Modern Cosmology*. London: Orion Books.

Frei, R., Gaucher, C., Poulton, S.W. and Canfield, D.E. (2009) Fluctuations in Precambrian atmospheric oxygenation recorded by chromium isotopes. *Nature* **461**, 250–54.

Fyfe, W.S. (1964) *Geochemistry of Solids – An Introduction*. New York: McGraw-Hill.

Garrels, R.M. and Christ, C.L. (1965) *Solutions, Minerals and Equilibria*. New York: Harper.

Geim, A.K. and Kim, P. (2008) Carbon wonderland. *Scientific American*, April **2008**, 90–97.

Geim, A.K. and Lovoselov, K.S. (2007) The rise of graphene. *Nature Materials* **6**, 183–91.

Geological Survey of Japan (2005) Atlas of Eh–pH diagrams – intercomparison of thermodynamic databases. Available from www.gsj.jp/data/openfile/no0419/openfile419e.pdf. *Open File Report No 419*, 285 pp.

Gill, R. (ed.) (1996). *Modern Analytical Geochemistry*. Harlow: Longman

Gill, R. (2010) *Igneous Rocks and Processes – A Practical Guide*. Chichester: Wiley-Blackwell.

Goldstein, J.I. and Short, J.M. (1967) Cooling rates of 27 iron and stony-iron meteorites. *Geochimica et Cosmochemica Acta* **31**, 1001–23.

Helmy, H.M., Yoshikawa, M., Shibata, T., *et al.* (2008) Corona structure from arc mafic-ultramafic cumulates: the role and chemical characteristics of late-magmatic hydrous liquids. *Journal of Mineralogical and Petrological Sciences* **103**, 333–44.

Henderson, P. (1982) *Inorganic Geochemistry*. Oxford: Pergamon.

Henderson, P. and Henderson, G.M. (2009) *The Cambridge Handbook of Earth Science Data*. Cambridge: Cambridge University Press.

Hofmann, A.W. (1997) Mantle geochemistry: the message from oceanic volcanism. *Nature* **385**, 219–29.

Holland, H.D. (2006) The oxygenation of the atmosphere and the oceans. *Philosophical Transactions of the Royal Society B* **361**, 903–15, 10.1098/rstb.2006.1838.

Holland, T.J.B. and Powell, R. (1998) An internally consistent thermodynamic data set for minerals of petrological interest. *Journal of Metamorphic Geology* **16**, 309–43.

Hutchison, R. (1983) *The Search for our Beginning*. Oxford: Clarendon Press.

IPCC (2013) *Climate Change 2013: The Physical Science Basis*. Working Group I Contribution to the IPCC Fifth Assessment Report (AR5). Geneva: Intergovernmental Panel on Climate Change.

Isley, A.E. and Abbott, D.H. (1999) Plume-related mafic volcanism and the deposition of banded iron formation. *Journal of Geophysical Research* **104**, 15461–77.

Jouzel, J., Lorius, C., Petit, J.R., *et al.* (1987) Vostok ice core: a continuous isotope temperature record over the last climatic cycle (160,000 years). *Nature* **329**, 403–8.

Kelley, S. (2002a) Excess argon in K–Ar and Ar–Ar geochronology. *Chemical Geology* **188**, 1–22.

Kelley, S. (2002b) K–Ar and Ar–Ar dating. *Reviews in Mineralogy and Geochemistry* **47**, 785–818, 10.2138/rmg.2002.47.17.

Kerr, R.A. (2008) Two geologic clocks finally keeping the same time. *Science* **320**, 434–5.

Killops, S.D. and Killops, V.J. (2005) *An Introduction to Organic Geochemistry*, 2nd edition. Chichester: John Wiley & Sons, Ltd.

Koschinsky, A., Garbe-Schönberg, D., Sander, S., *et al.* (2008) Hydrothermal venting at pressure–temperature conditions above the critical point of seawater, 5°S on the Mid-Atlantic Ridge. *Geology* **36**, 615–18.

Krauskopf, K.B. and Bird, D.K. (1995) *Introduction to Geochemistry*. New York: McGraw-Hill.

Kump, L.R., Junium, C., Arthur, M.A., *et al.* (2011) Isotopic evidence for massive oxidation of organic matter following the Great Oxidation Event. *Science* **334**, 1694–6, 10.1126/science.1213999.

Lauretta, D.S., Nagahara, H. and Alexander, C.M.O.D. (2006) Petrology and origin of ferromagnesian silicate chondrules. In D.S. Lauretta and H.Y.J. McSween (eds), *Meteorites and the Early Solar System II*. Tucson, Arizona: University of Arizona Press, 431–59.

Lenton, T. and Watson, A. (2011) *Revolutions that Made the Earth*. Oxford: Oxford University Press.

Libes, S.M. (2009) *An Introduction to Marine Biogeochemistry*. San Diego: Academic Press.

Lowry, R.K., Henderson, P. and Nolan, J. (1982) Tracer diffusion of some alkali, alkaline-earth and transition element ions in a basaltic and an andesitic melt, and the implications concerning melt structure. *Contributions to Mineralogy and Petrology* **80**, 254–61.

Martinson, D.G., Pisias, N.G., Hays, J.D., *et al.* (1987) Age dating and the orbital theory of the Ice Ages: development of a high-resolution 0 to 300,000-year chronostratigraphy. *Quaternary Research* **27**, 1–29.

McKie, D. and McKie, C. (1974) *Crystalline Solids*. London/New York: Nelson/Wiley.

Morse, S.A. (1980) *Basalts and Phase Diagrams*. New York: Springer-Verlag.

Nickson, R.T., McArthur, J.M., Ravenscroft, P., *et al.* (2000) Mechanism of arsenic release to groundwater, Bangladesh and West Bengal. *Applied Geochemistry* **15**, 403–13.

Oliver, L., Harris, N., Bickle, M., *et al.* (2003) Silicate weathering rates decoupled from the Sr-87/Sr-86 ratio of the dissolved load during Himalayan erosion. *Chemical Geology* **201**, 119–39, 10.1016/s0009-2541(03)00236-5.

O'Neill, P. (1998) *Environmental Chemistry*, 3rd edition. London: Blackie (Thomson Science).

Osborne, E.F. and Tait, D.B. (1952) The system diopside–forsterite–anorthite. *American Journal of Science* **Bowen Volume**, 413–33.

Petit, J. R., *et al.* (1999). Climate and atmospheric history of the past 420,000 years from the Vostok ice core, Antarctica. *Nature Geoscience* **399**, 429–36.

Qian, Q. and Hermann, J. (2010) Formation of high-Mg diorites through assimilation of peridotite by monzodiorite magma at crustal depths. *Journal of Petrology* **51**, 1381–1416, :10.1093/petrology/egq023.

Rankin, A. (2005) Fluid inclusions. In R.C. Selley, L.R.M. Cocks and I.R. Plimer (eds), *Encyclopedia of Geology*. Amsterdam: Elsevier, 253–60.

Rauscher, T. and Patkós, A. (2011) Origin of the chemical elements. In *Handbook of Nuclear Chemistry (http://arxiv.org/pdf/1011.5627v2)*. New York: Springer, **2**, 611–65.

Read, W.T.J. (1953) *Dislocations in Crystals*. New York: McGraw-Hill.

Richter, F.M., Rowley, D.B. & Depaolo, D.J. (1992) Sr isotope evolution of seawater – the role of tectonics. *Earth and Planetary Science Letters* **109**, 11–23, 10.1016/0012-821x(92)90070-c.

Riordan, M. and Schramm, D.M. (1993) *The Shadows of Creation – Dark Matter and the Structure of the Universe*, 2nd edition. Oxford: Oxford Paperbacks.

Rose, A.W. (1976) The effect of cuprous chloride complexes in the origin of red-bed copper and related deposits. *Economic Geology* **71**, 1036–48.

Rose, A.W. (1989) Mobility of copper and other heavy metals in sedimentary environments. In R.S. Boyle, *et al.* (eds), *Sediment-hosted Stratiform Copper Deposits*: Geological Association of Canada Special Paper, **36**, 97–110.

Samson, I., Anderson, A. and Marshall, D. (2003) *Fluid Inclusions: Analysis and Interpretation*. Québec: Mineralogical Association of Canada.

Scarfe, C.M. (1977) Viscosity of a pantellerite melt at one atmosphere. *Canadian Mineralogist* **15**, 185–9.

Schaefer, L. and Fegley, B. (2010) Chemistry of atmospheres formed during accretion of the Earth and other terrestrial planets. *Icarus* **208**, 438–48.

Schoonen, M., Elsetinow, A., Borda, M. and Strongin, D. (2000) Effect of temperature and illumination on pyrite oxidation between pH 2 and 6. *Geochemical Transactions* **1**, 4, 10.1186/1467-4866-1-23.

Shields, G. and Veizer, J. (2002) Precambrian marine carbonate isotope database: Version 1.1. *Geochemistry Geophysics Geosystems* **3**, DOI: 10.1029/2001GC000266, 10.1029/2001GC000266.

Smedley, P.L. and Kinniburgh, D.G. (2002) A review of the source, behaviour and distribution of arsenic in natural waters. *Applied Geochemistry* **17**, 517–68.

Tayler, R.J. (1975) *The Origin of the Chemical Elements*. London: Wykeham Publications

Taylor, H.P. (1974) The application of oxygen and hydrogen isotope studies to problems of hydrothermal alteration and ore deposition. *Economic Geology* **69**, 843–83.

Taylor, S.R. (1982) *Planetary Science: A Lunar Perspective*. Houston: Lunar and Planetary Science Institute (available at http://bit.ly/1gC9KVY).

Taylor, S.R. and McLennan, S.M. (1985) *The Continental Crust: Its Composition and Evolution*. Oxford: Blackwell.

Todd, D.K. and Mays, L.W. (2006) *Groundwater Hydrology*, 3rd edition. New York: John Wiley & Sons, Ltd.

Tripati, A.K., Roberts, C.D. and Eagle, R.A. (2009) Coupling of CO_2 and ice sheet stability over major climate transitions of the last 20 million years. *Science* **326**, 1394–97.

Urey, H.C., Lowenstam, H.A., Epstein, S. and McKinney, C.R. (1951) Measurement of paleotemperatures and temperatures of the Upper Cretaceous of England, Denmark and the southeastern United States. *Geological Society of America Bulletin* **61**, 399–416.

Vink, B.W. (1986) Stability relations of malachite and azurite. *Mineralogical Magazine* **50**, 41–7.

Waltham, D. (2000) *Mathematics: A Simple Tool for Geologists*, 2nd edition. Chichester: Wiley-Blackwell.

White, W.M. (2013) *Geochemistry*. Chichester: Wiley-Blackwell.

WHO (2011) *Guidelines for Drinking-Water Quality*. Geneva: World Health Organization.

Wieczorek, M.A., *et al.* (2006) The constitution and structure of the lunar interior. *Mineralogy and Geochemistry* **60**, 221–364, 10.2138/rmg.2006.60.3.

Wood, B.J., Walter, M.J. and Wade, J. (2006) Accretion of the Earth and segregation of its core. *Nature* **441**, 825–33, 10.1038/nature04763.

Wood, J.A. (1964) The cooling rates and parent planets of several iron meteorites. *Icarus* **3**, 429–59.

Zachos, J., Pagani, M., Sloan, L., *et al.* (2001) Trends, rhythms, and aberrations in global climate 65 Ma to present. *Science* **292**, 686–93.

Zahnle, K., Arndt, N.C.C., Halliday, A., *et al.* (2007) Emergence of a habitable planet. *Space Science Reviews* **129**, 35–78, 10.1007/s11214-007-9225-z.

Zhang, Y. (2008) *Geochemical Kinetics*. Princeton, NJ: Princeton University Press.

INDEX

Page references in *italics* refer to Figures; those in **bold** refer to Tables and Boxes

Chemical Fundamentals of Geology and Environmental Geoscience, Third Edition. Robin Gill.
© 2015 John Wiley & Sons, Ltd. Published 2015 by John Wiley & Sons, Ltd.
Companion Website: www.wiley.com/go/gill/chemicalfundamentals